Electrical Machines

Fundamentals of Electromechanical Energy Conversion

Electrical Machines

Fundamentals of Electromechanical Energy Conversion

Jacek F. Gieras

CRC Press
Taylor & Francis Group
Boca Raton London New York

CRC Press is an imprint of the
Taylor & Francis Group, an **informa** business

CRC Press
Taylor & Francis Group
6000 Broken Sound Parkway NW, Suite 300
Boca Raton, FL 33487-2742

First issued in paperback 2020

© 2017 by Taylor & Francis Group, LLC
CRC Press is an imprint of Taylor & Francis Group, an Informa business

No claim to original U.S. Government works

ISBN-13: 978-1-4987-0883-8 (hbk)
ISBN-13: 978-0-367-73694-1 (pbk)

Library of Congress Cataloging-in-Publication Data

Names: Gieras, Jacek F., author.
Title: Electrical machines : fundamentals of electromechanical energy conversion / Jacek F. Gieras.
Description: Boca Raton : CRC Press, 2017. | Includes bibliographical references and index.
Identifiers: LCCN 2016015589 | ISBN 9781498708838 (alk. paper)
Subjects: LCSH: Electromechanical devices--Textbooks.
Classification: LCC TK146 .G457 2017 | DDC 621.31/042--dc23
LC record available at https://lccn.loc.gov/2016015589

Visit the Taylor & Francis Web site at
http://www.taylorandfrancis.com

and the CRC Press Web site at
http://www.crcpress.com

Contents

Preface

The book is intended to serve as a textbook for basic courses on electrical machines covering the fundamentals of the electromechanical energy conversion, transformers, classical electrical machines, i.e., DC brush machines, induction machines (IM), wound-field rotor synchronous machines and modern electrical machines, i.e., switched-reluctance machines (SRM) and permanent magnet (PM) brushless machines. The author breaks the stereotype that a basic electrical machine course is limited only to transformers, DC brush machines, AC induction machines and wound-field synchronous machines.

In addition to academic research and teaching, the author has been working for over 18 years in the United States high-technology corporative business, being directly involved in solution to real problems as design, simulation, manufacturing and laboratory testing of a large variety of electrical machines for energy generation, conversion and utilization. Supervising young engineers the author is fully aware what the 21st century industry requires from them and how they should be trained to meet the demands of employers. The author wants to leverage his industrial experience to the classroom teaching. The structure of this book is as follows. Chapter 1 gives a brief overview of electromechanical energy conversion which is needed to understand the next chapters and have a broad vision of operation of electromechanical energy devices in power trains. Chapter 2 discusses single phase and three-phase transformers used in power electrical engineering. Chapter 3 considers SRMs as the simplest electrical machines. It also shows that such machines must operate in a machine-solid state converter environment. Chapter 4 briefly discusses DC brush (commutator) machines. Nowadays, these machines are gradually replaced by vector-controlled induction and PM brushless motors, so the presented material is limited to fundamentals of standard DC brushless motors while emphasis is given to modern brush PM slotless motors. Chapter 5 deals with armature windings of AC electrical machines with focus on coils construction, connection diagrams, induced EMF, distribution of the magnetomotive force (MMF), and effect of magnetic saturation. The reader can familiarize with such terms as winding factors, saturation of magnetic circuit, effect of slotting and effect of higher space harmonics. This material is crucial to understand AC induction, synchronous and PM brushless machines. Chapter 6 on induction machines is the largest chapter in this book because induction motors are the most common machines in industrial, traction, energy systems and public life applications. This chapter also introduces fundamentals of inverter-fed induction motors. Chapter 7 is devoted to wound-field rotor synchronous machines with focus on turboalternators. This chapter ends with salient-pole rotor synchronous motors, which is a bridge to Chapter 8 on PM brushless motors. PM brushless motors are categorized into sine-wave (synchronous) motors and square wave brushless motors. In this chapter the reader can also find fundamentals of inverter-fed PM brushless motors.

Compared to other textbooks on the subject, the presented book has a number of exclusive features:

(a) It includes SRMs and PMBMs, which are nowadays more popular than DC brush machines;
(b) It deals exhaustively with electrical machines without going deep into mathematical analysis;
(c) It shows examples of application of electrical machines to modern electromechanical drive systems;
(d) It shows technical problems and their possible solutions.

Most of the material of this textbook has been classroom-tested with successful results.

What are the key benefits of the textbook for the readers? Why should they purchase this book?

- The material is presented in as simple a way as possible.
- In addition to standard electrical machines, students can familiarize themselves also with modern electrical machines;
- The book is written by the author with both academic and industrial experience;
- Each chapter ends with a summary containing the most important knowledge presented in the given chapter and numerical problems for solution by students, while similar numerical examples are solved in the relevant chapter;
- The book contains about 100 carefully selected numerical examples attractive for students and encouraging them to study;
- The book highlights the elements of power electronics used in systems with electrical machines.

Students using this textbook should have taken courses in circuit theory and electromagnetic field theory as prerequisites. For a one-semester course and three lectures per week, the author recommends all eight chapters. For two lectures per week, the author recommends Chapters 2 and 4 to 8. A solutions manual for lecturers can be obtained from Taylor & Francis CRC Press.

The author has produced this textbook without any support from funding agencies and/or industry both in European Union countries and the United States.

Any suggestions for improvement, constructive criticism and corrections both from students and professors are most welcome.

Prof. Jacek F. Gieras, PhD, DSc, IEEE Life Fellow
Glastonbury, CT, E-mail: jgieras@ieee.org

1

INTRODUCTION TO ELECTROMECHANICAL ENERGY CONVERSION

1.1 What is electromechanical energy conversion?

Electromechanical energy conversion is a conversion of mechanical energy into electrical energy (generator) or vice versa (motor) with the aid of rotary motion (rotary machines) or translatory (linear) motion (linear machines and actuators).

Electrical machines, solenoid actuators and electromagnets are generally called *electromechanical energy conversion devices* (Fig. 1.1).

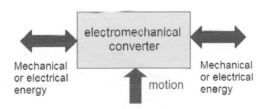

Fig. 1.1. Electromechanical energy conversion.

Transformers and *solid-state converters* do not belong to the group of electromechanical energy conversion devices because they only convert one kind of electrical energy into another kind of electrical energy with different parameters (change in voltage, current, frequency, number of phases, conversion of DC into AC current, etc.) without any motion.

1.1.1 Block diagrams of electromechanical energy conversion devices

Fig. 1.2a shows a block diagram of a motor, while Fig. 1.2a shows a block diagram of a generator. An example of application of an electric motor is shown in Fig. 1.3. An example of application of an electric generator is shown in Fig. 1.4.

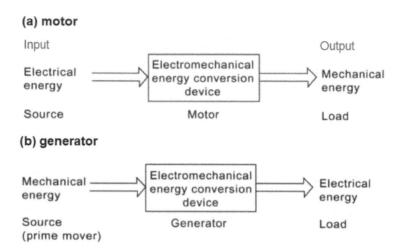

Fig. 1.2. Block diagrams of electromechanical energy conversion devices: (a) motor; (b) generator.

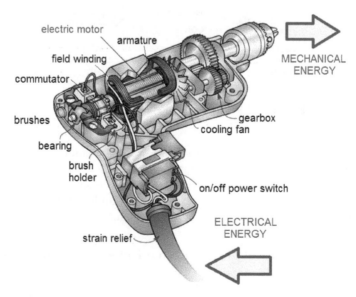

Fig. 1.3. Power tool: an example of conversion of electrical energy into mechanical energy.

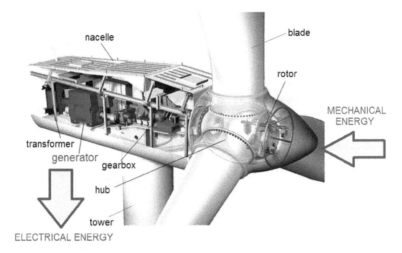

Fig. 1.4. Wind turbine generator: an example of conversion of mechanical energy into electrical energy.

1.1.2 Left-hand and right-hand rule

The left-hand rule (Fig. 1.5a) indicates the direction of the phasor of the electrodynamic force (EDF), i.e.,

$$d\mathbf{F} = I d\mathbf{l} \times \mathbf{B} \tag{1.1}$$

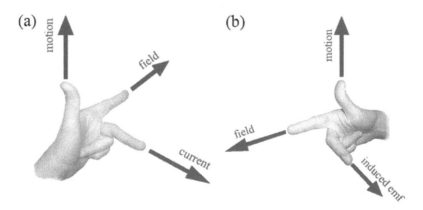

Fig. 1.5. Left-hand and right-hand rules: (a) left-hand rule shows the direction of electrodynamic force (EDF); (b) right-hand rule shows the direction of electromotive force (EMF).

or, in scalar form

$$F = BIl \tag{1.2}$$

The right-hand rule (Fig. 1.5b) indicates the direction of the phasor of the electromotive force (EMF), i.e.,

$$d\mathbf{E} = \mathbf{v} \times \mathbf{B}\ d\mathbf{l} \tag{1.3}$$

or, in scalar form

$$E = Blv \tag{1.4}$$

1.1.3 Energy flow in an electromechanical energy conversion device

Fig. 1.6 illustrates the conversion of electrical energy into mechanical energy, according to the following equation:

$$dW_e = dW_f + dW_{loss} + dW_{mech} \tag{1.5}$$

where dW_e is the electrical energy (input energy) dW_f is the energy stored in the magnetic field (coil), dW_{loss} are all the power losses and dW_{mech} is the mechanical energy (output energy).

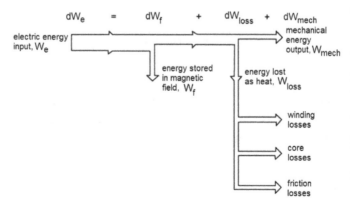

Fig. 1.6. Energy flow in electromechanical energy conversion device.

1.2 Analogies between electric and magnetic circuits

Table 1.1 contains analogies in electric and magnetic circuits, while Table 1.2 compares fundamental equations and laws for electric and magnetic circuits.

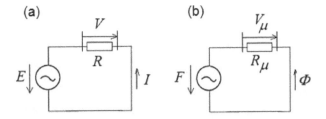

Fig. 1.7. Simple circuits: (a) electric circuit; (b) equivalent magnetic circuit.

Table 1.1. Analogies in electric and magnetic circuits.

Quantity	Electric circuit	Magnetic circuit
Voltage	Electric voltage, V [V]	Magnetic voltage drop, V_μ [A]
Source voltage	Electromotive force (EMF), E [V]	Magnetomotive force (MMF), F [A]
Current/flux	Electric current, I [A]	Magnetic flux, Φ [Wb]
Resistance /reluctance	Resistance R [$\Omega = [1/S]$]	Reluctance R_μ [1/H]
Constant	Electric conductivity, σ [S/m]	Magnetic permeability, μ [H/m]

Table 1.2. Comparison of fundamental laws for electric and magnetic circuits.

Law	Electric circuit	Magnetic circuit
Ohm's law	Resistance $R = \frac{V}{I}$	Reluctance $R_\mu = \frac{V_\mu}{\Phi}$
	Conductance $G = \frac{I}{V}$	Permeance $\Lambda_\mu = \frac{\Phi}{V_\mu}$
2nd Ohm's law	Resistance $R = \frac{l}{\sigma S}$	Reluctance $R_\mu = \frac{l}{\mu S}$
	Conductance $G = \frac{S}{\rho l}$	Permeance $\Lambda_\mu = \frac{\mu S}{l}$
Kirchhoff's current law	Sum of currents $\sum I = 0$	Sum of magnetic fluxes $\sum \Phi = 0$
Kirchhoff's voltage law	Sum of voltage drops $\sum V - \sum RI = 0$	Sum of magnetic voltage drops $\sum V_\mu - \sum R_\mu \Phi = 0$
Faraday's law/Ampere's law	EMF $E = \pi\sqrt{2}fN\Phi$	MMF $F = NI$

1.3 Losses in ferromagnetic cores

The *magnetization curve* $B(H)$ of isotropic silicon cold-rolled steel Armco-DI-Max M19, which is the most popular ferromagnetic material in construction of electrical machines, is plotted in Fig. 1.8. The *specific losses curve* $\Delta p(B)$ of the same silicon steel at 50 Hz is plotted in Fig. 1.9. An addition of silicon improves the magnetic properties of steel, i.e., increases the saturation magnetic flux density and reduces the losses.

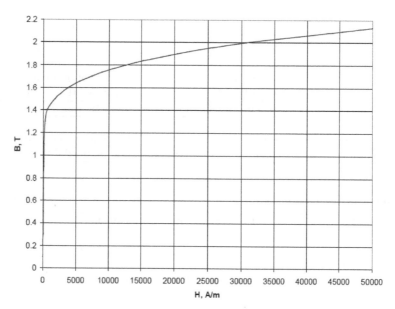

Fig. 1.8. Magnetization curve $B(H)$ of isotropic silicon cold-rolled steel Armco-DI-MAX M19.

As the alternating magnetic flux magnetizes the core, the energy is lost in the core due to the hysteresis effect. The energy loss, called the hysteresis loss, is proportional to the area of the hysteresis loop (Fig. 1.10). The hysteresis loss depends on the ferromagnetic material of the core. The first empirical formula for hysteresis losses was proposed by C.P. Steinmetz and published in the 1892 [37], i.e.,

$$\Delta P_h = k_h B_m^n \tag{1.6}$$

where k_h and n are curve fitted coefficients of actual experimental data and B_m is the peak value of the magnetic flux density. A more accurate empirical formula for the hysteresis loss contains the frequency f of the magnetic flux density, i.e.,

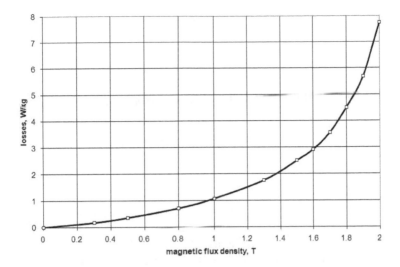

Fig. 1.9. Specific core loss curve $\Delta p(B)$ of isotropic silicon cold-rolled steel Armco-DI-MAX M19 at 50 Hz.

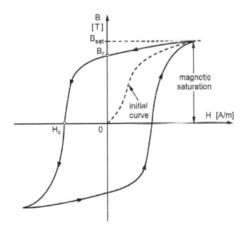

Fig. 1.10. Hysteresis loop. B_r – remanent magnetic flux density, B_{sat} – saturation magnetic flux density, H_c – coercivity.

$$\Delta P_h = c_h f B_m^n \qquad (1.7)$$

where B_m is the peak value of the magnetic flux density, f is the frequency and the hysteresis constants c_h and n vary with the core material. The constant n is often assumed to be $1.6 \ldots 2.0$.

Another source of the power loss in the ferromagnetic core is the eddy currents induced by the alternating magnetic flux. If the magnetic flux is perpendicular, directed toward the plane of this page and increasing with the time, it induces voltages in conductive material of the core (Fig. 1.11). Under

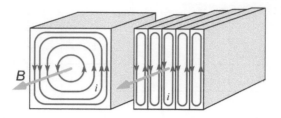

Fig. 1.11. Eddy currents in: (a) solid ferromagnetic core; (b) laminated ferromagnetic core.

action of these voltages, eddy-currents flow in closed loops (paths) producing power losses i^2R, which are converted into heat. The eddy-current losses can be reduced by decreasing the current i or increasing the resistance R. This can be done by replacing a solid ferromagnetic core with laminated ferromagnetic core. The eddy current losses are proportional to the frequency f square and the peak magnetic flux density B_m square, i.e.,

$$\Delta P_e = c_e f^2 B_m^2 \tag{1.8}$$

The eddy-current constant c_e depends on the electric conductivity of the material of the core and the thickness square of laminations. An addition of silicon reduces the electric conductivity of steel, i.e., reduces the eddy-current losses and increases the saturation magnetic flux density.

There are also the so-called *excess eddy current losses*, which can be estimated as [2]

$$\Delta P_{ex} = 8\sqrt{\sigma_{Fe} G S_{Fe} V_0} B^{1.5} f^{1.5} \tag{1.9}$$

where σ_{Fe} is the electric conductivity of steel sheet, $G = 0.1356$ is a unitless constant, V_0 is the curve fitted coefficient and S_{Fe} is a cross-sectional area of the core.

The total power losses, neglecting the excess eddy current loss, are:

$$\Delta P_{Fe} = \Delta P_h + \Delta P_e = c_{Fe} f^{1.3} B_m^2 \tag{1.10}$$

In practical calculations of AC magnetic circuits the core losses ΔP_{Fe} can be estimated on the basis of the specific core losses $\Delta p_{1/50}$ and masses, say, of legs and yokes of a transformer, i.e.,

$$\Delta P_{Fe} = \Delta p_{1/50} \left(\frac{f}{50}\right)^{4/3} \left(k_{adl} B_{ml}^2 m_l + k_{ady} B_{my}^2 m_y\right) \tag{1.11}$$

where $k_{adl} > 1$ and $k_{ady} > 1$ are the factors accounting for the increase in losses due to metallurgical and manufacturing processes, $\Delta p_{1/50}$ is the specific core loss in W/kg at 1 T and 50 Hz, B_l is the magnetic flux density in the

Table 1.3. Magnetization and specific core loss characteristics of three types of cold-rolled, nonoriented electrotechnical silicon steel sheets, i.e., Dk66, thickness 0.5 mm, $k_i = 0.96$ (Sweden); H-9, thickness 0.35 mm, $k_i = 0.96$ (Japan); and DI-MAX EST20, thickness 0.2 mm, $k_i = 0.94$, Italy).

B	H, A/m			Specific core losses Δp, W/kg				
T	Dk66	H9	DI MAX EST20	Dk66 50 Hz	H9 50 Hz	H9 60 Hz	DI-MAX EST 20 50 Hz	400 Hz
0.1	55	13	19	0.15	0.02	0.02	0.08	0.30
0.2	65	20	28	0.24	0.06	0.10	0.15	0.70
0.4	85	30	37	0.50	0.15	0.20	0.25	2.40
0.6	110	40	48	0.90	0.35	0.45	0.42	6.00
0.8	135	55	62	1.55	0.60	0.75	0.63	
1.0	165	80	86	2.40	0.90	1.10	0.85	
1.2	220	160	152	3.30	1.30	1.65	1.25	
1.4	400	500	450	4.25	1.95	2.45	1.70	
1.5	700	1500	900	4.90	2.30	2.85	1.95	
1.6	1300	4000	2400		2.65	3.35	2.20	
1.7	4000	6500	6500					
1.8	8000	10,000	17,000					
1.9	15,000	16,000						
2.0	22,500	24,000						
2.1	35,000							

leg, B_y is the magnetic flux density in the core (yoke), m_t is the mass of legs, and m_y is the mass of yokes.

Fig. 1.12. Stacking factor.

To reduce eddy-current losses in sheet steels, all electrical steels are coated double-sided with a thin layer of insulation, usually oxide insulation. The *stacking factor* is the ratio of thickness of bare sheet to the thickness of sheet with insulation (Fig. 1.12), i.e.,

$$k_i = \frac{d}{d + 2\Delta_i} \tag{1.12}$$

where d is the thickness of bare sheet and Δ_i is the thickness of one-sided insulation.

1.4 Inductor

1.4.1 Ideal inductor

An ideal inductor (Fig. 1.13) has:

- The resistance of the coil $R = 0$;
- Infinitely large magnetic permeability of its core;
- No core losses;
- No leakage magnetic flux, which means that the whole magnetic flux is within the ferromagnetic core and coupled wit the coil.

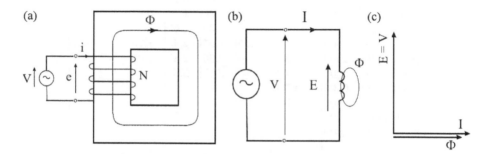

Fig. 1.13. Ideal inductor: (a) coil with zero resistance wound on ferromagnetic core without losses; (b) equivalent circuit; (c) phasor diagram.

Assuming a linear relationship between the current and magnetic flux, the sinusoidal current, i.e.,

$$i = I_m \sin(\omega t) \tag{1.13}$$

produces a sinusoidal magnetic flux

$$\Phi = \Phi_m \sin(\omega t) \tag{1.14}$$

The voltage induced in N-turn coil is

$$e = N\frac{d\Phi}{dt} = N\omega\Phi_m \cos(\omega t) = E_m \cos(\omega t) \tag{1.15}$$

where $E_m = N\omega\Phi_m$. The *rms* voltage is

$$E = \frac{E_m}{\sqrt{2}} = \frac{N\omega\Phi_m}{\sqrt{2}} = \frac{2\pi}{\sqrt{2}}NF\Phi_m = \pi\sqrt{2}Nf\Phi_m \tag{1.16}$$

In eqn (1.16) $\pi\sqrt{2} \approx 4.44 = 4 \times 1.11$ where 1.11 is the form factor for electric and magnetic quantities sinusoidally varying with time. The EMF expressed in terms of the inductance L and current i in the coil

$$e = L\frac{di}{dt} \tag{1.17}$$

For a sinusoidal current

$$e = \omega L I_m \cos(\omega t) = E_m \cos(\omega t) \tag{1.18}$$

The effective value of the EMF expressed in complex form is

$$\mathbf{E} = jX_m\mathbf{I} \tag{1.19}$$

where $X_m = \omega L$ is the magnetizing reactance. For an ideal inductor, the induced voltage (EMF, E) is equal to the supply voltage $E = V$. The equivalent circuit and phasor diagram of an ideal inductor is shown in Fig. 1.13.

1.4.2 Practical inductor

A practical inductor has a real coil with its resistance R and a real ferromagnetic core, in which the hysteresis losses (1.7) and eddy-current losses (1.8) are produced.

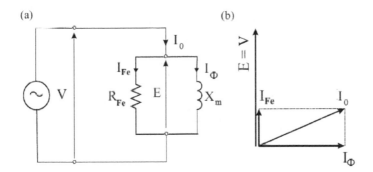

Fig. 1.14. Inductor with core losses: (a) equivalent circuit; (b) phasor diagram.

To draw the equivalent circuit of a practical inductor, an equivalent resistance R_{Fe}, in parallel with X_m, representing the core losses ΔP_{Fe} is introduced. Thus, the total power losses in the core at constant frequency can be expressed as

$$\Delta P_{Fe} = \frac{E^2}{R_{Fe}} \tag{1.20}$$

The power losses ΔP_{Fe} are proportional to the EMF E square, which appears across the resistance R_{Fe} that in turn represents the core losses. This allows building the equivalent circuit of a practical inductor (Fig. 1.14). The exciting current I_0 is split into two components:

- The magnetizing current I_Φ;
- The core loss current I_{Fe} proportional to the core power losses.

The currents I_Φ and I_{Fe} are displaced from each other by an angle $\pi/2$, i.e.,

$$\mathbf{I}_0 = I_{Fe} + jI_\Phi \qquad (1.21)$$

So far, the resistance R of the coil and the leakage flux Φ_σ have not been taken into account. The leakage flux Φ_σ is the flux, which goes through the air. It induces the voltage e_σ in the coil, which is equal

$$e_\sigma = L_\sigma \frac{di}{dt} = \omega L_\delta \cos(\omega t) \qquad (1.22)$$

or in complex notation

$$\mathbf{E}_\sigma = jX_\sigma \mathbf{I} \qquad (1.23)$$

where $X_\sigma = \omega L_\sigma$ is the leakage reactance and L_σ is the leakage inductance. The voltage balance equation for the equivalent circuit is

$$\mathbf{V} = \mathbf{E} + R\mathbf{I}_0 + \mathbf{E}_\sigma = \mathbf{E} + R\mathbf{I}_0 + jX_\sigma \mathbf{I}_0 = \mathbf{E} + (R + jX_\sigma)\mathbf{I}_0 \qquad (1.24)$$

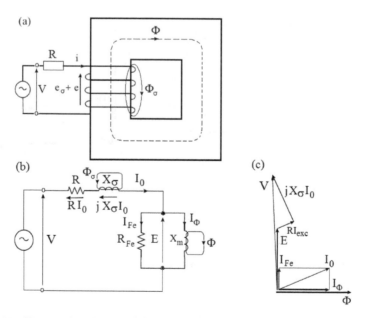

Fig. 1.15. Practical inductor: (a) coil with $R \neq 0$ wound on ferromagnetic core with losses and producing leakage flux; (b) equivalent circuit; (c) phasor diagram.

The equivalent circuit and phasor diagram of a practical inductor with $R \neq 0$, core losses and leakage flux taken into account is sketched in Fig. 1.15.

Example 1.1

Fig. 1.16 shows a magnetic circuit with air gap g. Two series connected windings are placed on external legs and fed with DC voltage $V = 48$ V. Such magnetic circuit can be used, for example, in magnetizers. The cross-section of laminated ferromagnertic core is $25 \times 20 = 500$ mm^2. The mean length of the magnetic flux path in each external section is $l_{Fe1} = l_{Fe2} = 170$ mm. The length of the magnetic flux path in the central leg is $l_{Fe} = 64$ mm. For 0.5-mm thick cold-rolled electrical sheet steel the stacking factor $k_i = 0.96$ and the relative magnetic permeablity $\mu_r = 1000$ (this is a rough simplification, why?).

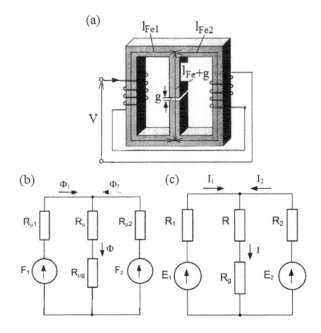

Fig. 1.16. Magnetic circuit with air gap: (a) 3D image; (b) equivalent magnetic circuit with concentrated parameters; (c) equivalent electric circuit with concentrated parameters.

The air gap $g = 6$ mm. Each of the two windings consists of $N = 2000$ turns wound with copper wire with its diameter $d = 0.5$ mm (without insulation). Find magnetic fluxes in each column and the air gap.

Solution

For the equivalent electric circuit (Fig. 1.16c), the following Kirchhoff's equations can be written:

$$E_1 - I_1 R_1 - I(R + R_g) = 0 \qquad (1.25)$$

$$E_2 - I_2 R_2 - I(R + R_g) = 0 \qquad (1.26)$$

$$I = I_1 + I_2 \qquad (1.27)$$

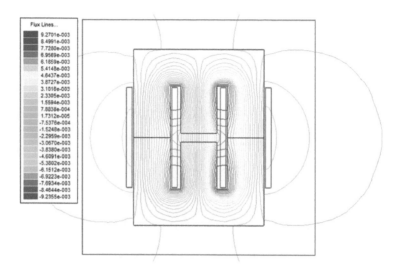

Fig. 1.17. Magnetic flux lines as obtained from the 2D FEM.

Solving the set of the above equations, the currents I_1, I_2 and I are

$$I_1 = \frac{E_1(R_2 + R + R_g) - E_2(R + R_g)}{R_1 + R + R_g) - (R + R_g)^2} \qquad (1.28)$$

$$I_2 = \frac{E_2(R_2 + R + R_g) - E_2(R + R_g)}{R_1 + R + R_g) - (R + R_g)^2} \qquad (1.29)$$

$$I_2 = \frac{E_1 R_2 + E_2 R_1}{R_1 + R + R_g) - (R + R_g)^2} \qquad (1.30)$$

Replacing in eqns (1.28), (1.29) and (1.30) the EMF with MMF and resistances with reluctances, the magnetic fluxes are

$$\Phi_1 = \frac{F_1(R_{\mu 2} + R_\mu + R_{\mu g}) - F_2(R_\mu + R_{\mu g})}{R_{\mu 1} + R_\mu + R_{\mu g}) - (R_\mu + R_{\mu g})^2} \qquad (1.31)$$

$$\Phi_2 = \frac{F_2(R_{\mu 2} + R_\mu + R_{\mu g}) - F_1(R_\mu + R_{\mu g})}{R_{\mu 1} + R_\mu + R_{\mu g}) - (R_\mu + R_{\mu g})^2} \qquad (1.32)$$

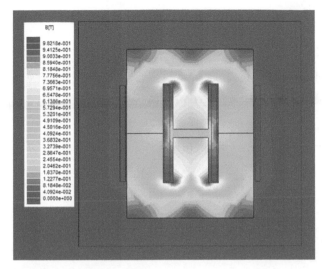

Fig. 1.18. Magnetic flux density distribution as obtained from the 2D FEM. The air gap magnetic flux density $B_g = 0.47$ T obtained analytically is almost the same as that obtained using the FEM (green color).

$$\Phi = \frac{F_1 R_{\mu 2} - F_2 R_{\mu 1}}{(R_{\mu 1} + R_\mu + R_{\mu g}) - (R_\mu + R_{\mu g})^2} \tag{1.33}$$

The values of reluctances have been calculated as follows:

$$R_{\mu 1} = \frac{l_{Fe1}}{\mu_0 \mu_r ab} = \frac{0.17}{0.4\pi \times 10^{-6} \times 1000 \times 0.25 \times 0.20} = 2.706 \times 10^5 \ 1/\text{H}$$

$$R_{\mu 2} = \frac{l_{Fe2}}{\mu_0 \mu_r ab} = \frac{0.17}{0.4\pi \times 10^{-6} \times 1000 \times 0.25 \times 0.20} = 2.706 \times 10^5 \ 1/\text{H}$$

$$R_\mu = \frac{l_{Fe}}{\mu_0 \mu_r ab} = \frac{0.064}{0.4\pi \times 10^{-6} \times 1000 \times 0.25 \times 0.20} = 1.019 \times 10^5 \ 1/\text{H}$$

$$R_{\mu g} = \frac{g}{\mu_0 \mu_r ab} = \frac{0.006}{0.4\pi \times 10^{-6} \times 1000 \times 0.25 \times 0.20} = 95.49 \times 10^5 \ 1/\text{H}$$

The mean length of turn of the winding

$$l_{mean} = 2(a + \Delta i + b + \Delta i) = 2(0.025 + 0.0015 + 0.020 + 0.0015) = 0.096 \text{ m}$$

where $\Delta i = 1.5$ mm is the assumed thickness of the winding–to–core insulation. The cross-section of the round conductor

$$s_w = \frac{\pi d^2}{4} = \frac{\pi \times 0.0005^2}{4} = 0.2 \times 10^{-6} \text{ m}^2$$

The electric conductivity of copper at 75° (hot winding) is 47×10^6 S/m. The resistance of two windings in series

$$R_w = 2\frac{N_c l_{mean}}{\sigma_{75} s_w} = \frac{2000 \times 0.096}{47 \times 10^6 \times 0.2 \times 10^{-6}} = 41.85 \; \Omega$$

The current in the winding and current density

$$I_w = \frac{V}{R_w} = \frac{48.0}{41.85} = 1.175 \text{ A} \qquad J_w = \frac{I_w}{s_w} = \frac{1.175}{0.2 \times 10^{-6}} = 5.85 \times 10^6 \text{ A/m}^2$$

The MMFs of windings

$$F_1 = N_c I_w = 2000 \times 1.175 = 2340 \text{ A} \qquad F_2 = N_c I_w = 2000 \times 1.175 = 2340 \text{ A}$$

Magnetic fluxes are calculated using eqns (1.31), (1.32), and (1.33)

$$\Phi_1 = 1.175 \times 10^{-4} \text{ Wb} \qquad \Phi_2 = 1.175 \times 10^{-4} \text{ Wb} \qquad \Phi = 2.35 \times 10^{-4} \text{ Wb}$$

The air gap magnetic flux density

$$B_g = \frac{\Phi}{ab} = \frac{2.35 \times 10^{-4}}{0.025 \times 0.020} = 0.47 \text{ Wb}$$

The magnetic flux lines and magnetic flux density distribution as obtained from the 2D finite element method (FEM) are plotted in Figs 1.17 and 1.18.

1.5 Two magnetically coupled electric circuits

Fig. 1.19 shows two electrical circuits, which are magnetically coupled. These two circuits can create, for example, a single-phase, two-winding transformer.

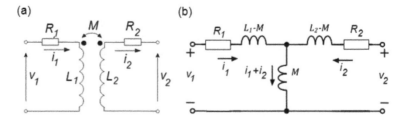

Fig. 1.19. Two magnetically coupled electric circuits: (a) magnetic coupling; (b) T-type electric equivalent circuit.

On the basis of Kirchhoff's voltage law (Fig. 1.19a)

$$v_1 - R_1 i_1 - L_1 \frac{di_1}{dt} - M \frac{di_2}{dt} = 0$$

$$v_2 - R_2 i_2 - L_2 \frac{di_2}{dt} - M \frac{di_1}{dt} = 0 \qquad (1.34)$$

where v_1, i_1, R_1, L_1 are the voltage, current, resistance and self-inductance of the primary winding, respectively, v_2, i_2, R_2, L_2 are the same quantities for the secondary winding and M is the mutual inductance between the primary and secondary windings. Eqns (1.34) can be brought to the following form (Fig. 1.19b)

$$v_1 - R_1 i_1 - (L_1 - M) \frac{di_1}{dt} - M \frac{di_1}{dt} - M \frac{di_2}{dt} = 0$$

$$v_2 - R_2 i_2 - (L_2 - M) \frac{di_2}{dt} - M \frac{di_2}{dt} - M \frac{di_1}{dt} = 0 \qquad (1.35)$$

Thus, the two magnetically coupled electric circuits can be replaced with the T-type electric equivalent circuit (four-terminal network).

1.6 Doubly-excited rotary device

Fig. 1.20 shows a rotary electromechanical energy conversion device with two electrical inputs (gates). It is shortly called a "doubly-excited device".

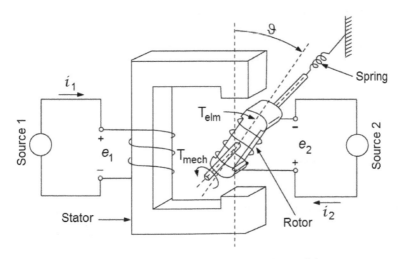

Fig. 1.20. Doubly-excited rotary device [8].

For a doubly-excited device, the relationships between linkage magnetic fluxes and currents are described by the following equations:

$$\psi_1 = L_{11}(\vartheta)i_1 + L_{12}(\vartheta)i_2$$

$$\psi_2 = L_{21}(\vartheta)i_1 + L_{22}(\vartheta)i_2 \tag{1.36}$$

The self-inductances L_{11}, L_{22} and the mutual inductances $L_{12} = L_{21}$ can be expressed as functions of the angle ϑ, i.e.,

$$L_{11}(\vartheta) = \frac{1}{2}(L_{1d} + L_{1q}) + \frac{1}{2}(L_{1d} - L_{1q}) = L_1 + \Delta L_1 \cos(2\vartheta) \tag{1.37}$$

$$L_{22}(\vartheta) = \frac{1}{2}(L_{2d} + L_{2q}) + \frac{1}{2}(L_{2d} - L_{2q}) = L_2 + \Delta L_2 \cos(2\vartheta) \tag{1.38}$$

$$L_{12}(\vartheta) = L_0 \cos(\vartheta) \tag{1.39}$$

where

$$L_1 = \frac{1}{2}(L_{1d} + L_{1q}) \qquad\qquad L_2 = \frac{1}{2}(L_{2d} + L_{2q}) \tag{1.40}$$

$$\Delta L_1 = \frac{1}{2}(L_{1d} - L_{1q}) \qquad\qquad \Delta L_2 = \frac{1}{2}(L_{2d} - L_{2q}) \tag{1.41}$$

L_{1d}, L_{2d} are the direct axis (d-axis) inductances, and L_{1q}, L_{2q} are the quadrature axis (q-axis) inductances. The d-axis is the axis of the magnetic flux (center axis of a pole). The q-axis is orthogonal to the d-axis. The following trigonometric identities have been used:

$$\sin^2 x = \frac{1}{2}(1 - \cos 2x) \qquad\qquad \cos^2 x = \frac{1}{2}(1 + \cos 2x) \tag{1.42}$$

For $\vartheta = 0$, the mutual inductances $L_{11} = L_d$ and $L_{12} = L_0$. For $\vartheta = \pi/2$, the mutual inductances $L_{11} = L_q$ and $L_{12} = 0$. The inductances $L_{11}(\vartheta)$ and $L_{12}(\vartheta)$ for $L_d = 0.01$ H and $L_q = 0.002$ H have been plotted against the angle ϑ in Fig. 1.21.

The electromagnetic torque can be found, for example, as a sum of first derivatives of inductances with respect to the rotation angle ϑ (magnetic energy stored)

$$T_{elm} = \frac{1}{2}i_1^2 \frac{dL_{11}(\vartheta)}{d\vartheta} + \frac{1}{2}i_2^2 \frac{dL_{22}(\vartheta)}{d\vartheta} + i_1 i_2 \frac{dL_{12}(\vartheta)}{d\vartheta} \tag{1.43}$$

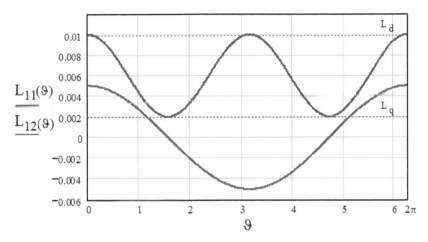

Fig. 1.21. Inductances $L_{11}(\vartheta)$ and $L_{12}(\vartheta)$ for $L_d = 0.01$ H and $L_q = 0.002$H as functions of the angle ϑ.

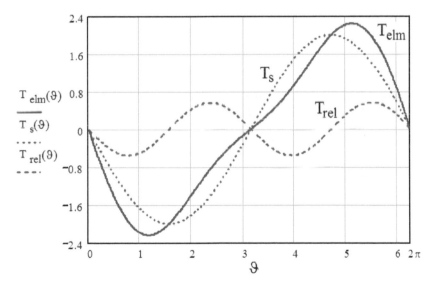

Fig. 1.22. Torques T_{elm}, T_s and T_{rel} as functions of the rotation angle ϑ.

The derivatives of inductances

$$\frac{dL_{11}(\vartheta)}{d\vartheta} = -2\Delta L_1 \sin(2\vartheta)$$

$$\frac{dL_{22}(\vartheta)}{d\vartheta} = -2\Delta L_2 \sin(2\vartheta) \tag{1.44}$$

$$\frac{dL_{12}(\vartheta)}{d\vartheta} = -L_0 \sin(\vartheta)$$

Thus, the electromagnetic torque is

$$T_{elm} = -(i_1^2 \Delta L_1 + i_2^2 \Delta L_2) \sin(2\vartheta) - i_1 i_2 L_0 \sin(\vartheta) \qquad (1.45)$$

The first term at the right-hand side of eqn (1.45) is the electromagnetic torque produced due to differences in inductances (reluctances) in the direct (d) and quadrature (q) axis. The magnitude of the *reluctance torque* is

$$T_{mrel} = i_1^2 \Delta L_1 + i_2^2 \Delta L_2$$

The second term at the right-hand side of eqn (1.45) depends on currents in the stator and rotor windings. The torque is called the *synchronous torque* with the magnitude

$$T_{ms} = i_1 i_2 L_0$$

Thus, the resultant electromagnetic torque can be expressed with the aid of a simple equation

$$T_{elm} = -[T_{ms} \sin(\vartheta) + T_{mrel} \sin(2\vartheta)]$$

The torques T_{elm}, T_s and T_{rel} are plotted against the rotation angle ϑ in Fig. 1.22. The following values of inductances and currents have been assumed: $L_{1d} = 1.0$ H, $L_{1q} = 0.2$ H, $L_{2d} = 0.5$ H, $L_{2q} = 0.1$ H, $i_1 = 1.0$ A, $i_2 = 2.0$ A.

Example 1.2.

The magnetic circuit shown in Fig. 1.20 is made of high-permeability electrical steel. The rotor does not have any winding and is free to turn about a vertical axis. The leakage flux and fringing effect are neglected.

(a) Derive an expression for the torque acting on the rotor in terms of the dimensions and the magnetic field in the two air gaps. Assume the reluctance of the steel to be negligible, i.e., $\mu \to \infty$ and neglect the effect of fringing.

(b) The maximum flux density in the overlapping portions of the air gaps is to be limited to approximately 1.6 T to avoid excessive saturation of the steel. Calculate the maximum torque for the radius of the rotor $r = 28$ mm, axial length (perpendicular to the page) of the magnetic circuit $L = 20$ mm and the air gap between the rotor and stator $g = 1.5$ mm.

Solution

(a) Derive an expression for the torque

There are two air gaps g in series, so the air gap magnetic field intensity H_g is

$$H_g = \frac{Ni}{2g}$$

Because the permeability of steel is assumed infinite and B_{Fe} must remain finite, $H_{Fe} = B_{Fe}/\mu = 0$ and the coenergy density in the steel is zero, i.e., $(\mu H_{Fe}^2/2 = B_{Fe}^2/(2\mu) = 0)$. Hence the system coenergy is equal to that of the air gaps, in which the coenergy density in the air gap is $\mu_0 H_g^2/2$. The volume of the two overlapping air gaps is $2gL(r_1 + 0.5g)\theta$. Consequently, the coenergy is equal to the product of the air gap coenergy density and the air gap volume, i.e.,

$$W_g' = \left(\frac{\mu_0 H_g^2}{2}\right)[2gL(r_1 + 0.5g)\theta] = \frac{\mu_0(Ni)^2 L(r_1 + 0.5g)\theta}{4g}$$

and the torque is given by eqns (1.43) and (3.18), i.e.,

$$T_{elm} = \left[\frac{\partial W_g'(i, \theta)}{\partial \theta}\right]_{i=const} = \frac{\mu_0(Ni)^2 L(r_1 + 0.5g)}{4g}$$

The sign of the torque is positive, hence acting in the direction to increase the overlap angle θ and thus to align the rotor with the stator poles.

(b) Calculate the maximum torque for $B_g = 1.6$ T, $r_1 = 28$ mm, $L = 20$ mm and $g = 1.5$ mm

$$H_g = \frac{B_g}{\mu_0} = \frac{1.6}{04\pi \times 10^{-6}} = 1.273 \times 10^6 \text{ A/m}$$

Thus, the MMF

$$Ni = 2gH_g = 2 \times 0.0015 \times 1.31 \times 10^{-6} = 3820 \text{ A}$$

The torque

$$T_{elm} = \frac{0.4\pi 10^{-6} \times 8000^2 \times 0.02(0.028 + 0.5 \times 0.0015)}{4 \times 0.0015} = 0.217 \text{ Nm}$$

Example 1.3

In the doubly-excited device shown in Fig. 1.20, the inductances are $L_{11} = 2.0 + 2.0\cos(2\theta) \times 10^{-3}$ H, axial length (perpendicular to the page) $L_{12} = 0.5\cos(\theta)$ H, $L_{22} = 25.0 + 15.0\cos(2\theta)$ H. Find the torque $T(\theta)$ for current $i_1 = 1.2$ A and $i_2 = 0.04$ A. Calculate the torque for the above values of currents and $\theta = 10°$, $\theta = 60°$ and $\theta = 120°$.

Solution

$$\frac{dL_{11}(\theta)}{d\theta} = \frac{2.0 + 2.0\cos(2\theta) \times 10^{-3}}{d\theta} = -4.0\sin(2\theta) \times 10^{-3}$$

$$\frac{dL_{22}(\theta)}{d\theta} = \frac{25.0 + 15.0\cos(2\theta)}{d\theta} = -30\sin(2\theta)$$

$$\frac{dL_{12}(\theta)}{d\theta} = \frac{0.5\cos(\theta)}{d\theta} = -0.5\sin(\theta)$$

The electromagnetic torque is given by eqn (1.43)

$$T_{elm} = \frac{i_1^2}{2}\frac{dL_{11}(\theta)}{d\theta} + \frac{i_2^2}{2}\frac{dL_{22}(\theta)}{d\theta} + i_1 i_2 \frac{dL_{12}(\theta)}{d\theta}$$

$$= \frac{1.2^2}{2}[-4.0\sin(2\theta) \times 10^{-3}] + \frac{0.4^2}{2}[-30\sin(2\theta)] - 1.2 \times 0.04[-0.5\sin(\theta)]$$

$$= -1.4029\sin(2\theta) - 0.24\sin(\theta)$$

For $\theta = 10°$ the torque $T = -0.013$ Nm, for $\theta = 60°$ the torque $T = -0.044$ Nm and for $\theta = 120°$ the torque $T = 2.494 \times 10^{-3}$ Nm.

Notice that the torque expression consists of terms of two types. One term, proportional to $i_1 i_2 \sin(\theta)$, is due to the mutual interaction between the rotor and stator currents. It acts in a direction to align the rotor and stator so as to maximize their mutual inductance.

The torque expression also has two terms each proportional to $\sin(2\theta)$ and to the square of one of the coil currents. These terms are due to the action of the individual winding currents alone and correspond to the torques one sees in singly-excited systems. Here the torque is due to the fact that the self inductances are a function of rotor position and the corresponding torque acts in a direction to maximize each inductance so as to maximize the coenergy. The 2θ variation is due to the corresponding variation of the self indutances, which in turn is due to the variation of the air gap inductance.

1.7 Basic coordinates and parameters of systems

The following forms of energy occur in the process of energy conversion in *electromechanical systems* (e.g., in electrical machines):

(a) *External energy* absorbed or produced by the machine;
(b) *Energy stored* in the machine or elements combined with the machine;
(c) *Dissipated energy*, e.g., in the form of heat.

Basic coordinates of processes of conversion can relate to electrical and mechanical parts. Basic coordinates of electric systems can be found under assumption, that all systems consist of *concentrated-parameter elements*, *conservative elements* (in which the energy is stored) and *dissipative elements* (in which the energy is dissipated).

Capacitive elements and *inductive elements* are conservative elements. *Resistive elements* are dissipative elements. The theory of electromagnetic field deals with *distributed-parameter elements and systems*.

1.7.1 Capacitive element

The basic coordinate that describes a *capacitive element* C is the electric charge q, because its quantity stored in this element determines a measure of the energy stored. The equation describing a linear capacitive element has the following form:

$$v(t) = \frac{1}{C} q(t) \tag{1.46}$$

When the capacitance C is a nonlinear function of the voltage $v(t)$, the function or their variability (variation) must be known in order to describe the state of the element. If the characteristic $q(t)$ is a straight line, the characteristic $v(t)$ in the case of linear capacitance C is a straight line too.

The electric current in the capacitive element is the first derivative of the electric charge q with respect to the time t. This quantity (electric current i) can be easily measured and is commonly used in electrical engineering, i.e.,

$$\frac{dq}{dt} = \dot{q} = i \tag{1.47}$$

1.7.2 Inductive element

Similar to capacitive element, the *inductive element* L can also store the energy. The basic coordinate that describes this element is coupled magnetic flux ψ (linkage flux). The equation describing a linear inductive element has the form:

$$i(t) = \frac{1}{L} \psi(t) \tag{1.48}$$

When the inductance L is a nonlinear function of the electric current $i(t)$ (element with a ferromagnetic core), the function or its graph must be known to describe the state of the element. The electric voltage v is the first derivative of the magnetic flux ψ with respect to the time t, i.e.,

$$\frac{d\psi}{dt} = \dot{\psi} = v \tag{1.49}$$

Similar to the current i, the voltage v can be easily measured.

1.7.3 Resistive element

A *resistive element* R is a dissipative element. It can be described using q and Ψ as basic coordinates.

The following relationship exists for a concentrated dissipative element with its conductance G

$$\dot{q} = i = G\dot{\psi} = \frac{1}{R} \tag{1.50}$$

where the current i is according to eqn (1.47) and the electric induced voltage v is according to eqn (1.49). The element is linear if the conductance G of the medium is constant. The element is nonlinear if the conductance G is a function of current or voltage. In this case the conductance $G(\dot{\psi})$ can be found on the basis of the characteristic $\dot{q} = G(\dot{\psi})$ for a given voltage $\dot{\psi}$.

1.7.4 Mass in translatory motion

The momentum p is the basic coordinate of the *mass* m in translatory motion, because the momentum is a measure of the energy stored by an element of mass. The equation describing a *linear element of mass* has the following form:

$$v(t) = \frac{1}{m}p(t) \tag{1.51}$$

If the mass m was dependent on the velocity v (as it is in relativistic physics) or dependent on the momentum p, then the characteristic of the element of mass $p = f(v)$ would be curvilinear. For a given velocity v the point corresponding to the velocity on this curve would describe the mass of the element. The first derivative of the momentum with respect to time

$$\frac{dp}{dt} = \dot{p} = f \tag{1.52}$$

is equal to the force f acting on the element of mass.

1.7.5 Elastic element in translatory motion

In most cases the *elastic element* is described by the x coordinate. The equation describing this element has the form

$$f(t) = \frac{1}{K}\tau(t) \tag{1.53}$$

where f is the force acting on the spring, N, K is the *compliance* of the elastic element, m/N. The inverse of compliance is the *stiffness* of the spring, N/m. Stiffness is the resistance of an elastic body to deformation by an applied force.

For full description of a nonlinear elastic element, the characteristic $x = f(F)$ is required. This characteristic defines the compliance in each point of the element and for each state of the element.

The momentum p and position x are the basic mechanical coordinates in translatory motion. Their first derivatives with respect to time are force and velocity, i.e.,

$$f = \frac{dp}{dt} = \dot{p} = \frac{d}{dt}(mv) \tag{1.54}$$

and velocity

$$v = \frac{dx}{dt} = \dot{x} \tag{1.55}$$

1.7.6 Dissipative element in translatory motion

For a clear description of a *dissipative element* D (viscous damping), basic coordinates for translatory motion are sufficient because a linear damping element is described by the following equation:

$$\dot{p} = D\dot{x} \tag{1.56}$$

where D is the damping coefficient, N/(m/s), which is numerically equal to the slope of the characteristic of the element $\dot{p} = D\dot{x}$. In general, this element can also be nonlinear, similar to all other elements previously described.

1.7.7 Concentrated-parameter elements in rotary motion

All *concentrated-parameter elements* in rotary motion, i.e.,

- Inertia J,
- Torsional compliance K_ϑ,
- Torsional damping D_ϑ

can be defined in a similar way as in translatory motion. In the case of rotary motion, the basic coordinates are

- Angular momentum

$$l = mr^2 \Omega(t) = J\Omega(t) \tag{1.57}$$

- Angular position ϑ.

Their first derivatives with respect to the time t are

- Torque

$$T = \frac{dl}{dt} = i \tag{1.58}$$

- Angular velocity

$$\Omega = \frac{d\vartheta}{dt} = \dot{\vartheta} \tag{1.59}$$

The moment of inertia of a material point is

$$J = mr^2 \tag{1.60}$$

while the *centrifugal force*

$$F_r = mr\dot{\vartheta}^2 = mr\Omega^2 \tag{1.61}$$

Equations describing concentrated elements in rotary motion are as follows:

- Inertial element

$$\Omega(t) = \frac{1}{J}l(t) \tag{1.62}$$

- Elastic element

$$T(t) = \frac{1}{K_\vartheta}\vartheta(t) \tag{1.63}$$

- Torsional damping element

$$i = D_\vartheta \dot{\vartheta} \tag{1.64}$$

All basic coordinates of energy conversion, their first derivatives and parameters describing each element are set up in Table 1.4.

Table 1.4. Basic coordinates of energy conversion, their first derivatives and parameters describing each element.

Form of energy	Mechanical translatory motion	Mechanical rotary motion	Electrical
Coordinates	Momentum, p, Linear position, x	Angular momentum, l, Angular position, ϑ	Electric charge, q Linkage flux, ψ
Their first derivative with respect to time	Force, $f = \dot{p}$ Velocity, $v = \dot{x}$	Torque, $T = \dot{l}$ Angular velocity, $\Omega = \dot{\vartheta}$	Electric current, $i = \dot{q}$ Electric voltage, $v = \dot{\psi}$
Mathematical description	Translatory motion	Rotary motion	Electric system
Conservative element	Mass $m(\dot{x}) = dp/d\dot{x}$	Moment of inertia $J(\dot{\vartheta}) = dl/d\dot{\vartheta}$	Capacitance $C(\dot{\psi}) = dq/d\dot{\psi}$
Conservative element	Compliance $K(\dot{p}) = dx/d\dot{x}$	Torsional compliance $K(\dot{\vartheta}) = d\vartheta/d\dot{l}$	Inductance $L(\dot{q}) = d\psi/d\dot{q}$
Dissipative element	Damping (friction) $D(\dot{x}) = d\dot{p}/d\dot{x}$	Torsional damping $D_\vartheta(\dot{\vartheta}) = d\dot{l}/d\dot{\vartheta}$	Conductance $G(\dot{\psi}) = d\dot{q}/d\dot{\psi}$

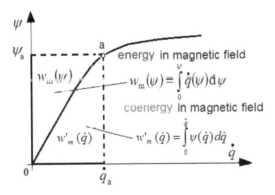

Fig. 1.23. Magnetic flux–current characteristic of a nonlinear inductive element.

1.8 Energy and coenergy

1.8.1 Energy and coenergy of a nonlinear inductive element

If basic coordinates and their first derivatives are known, equations describing energies in the process of conversion can be obtained. A nonlinear inductive element with flux–current characteristic sketched in Fig. 1.23 will be considered.

Coenergy (a second state function of the energy) is an auxiliary function necessary for calculations of the force or torque at constant current. In the linear case, energy and coenergy are numerically equal.

The instantaneous value of the power in a single coil is

$$P(t) = \dot{q}\dot{\psi} \tag{1.65}$$

The energy influx per unit of time to the inductive element is the energy growth stored in the magnetic field and described by the formula

$$dW_m(t) = P(t)dt = \dot{q}\frac{d\psi}{dt}dt \tag{1.66}$$

The total energy delivered in the time interval $t \in (t_0, t)$ is calculated as an integral of eqn (1.66), i.e.,

$$W_m(t) - W_m(t_0) = \int_{t_0}^{t} \dot{q}\frac{d\psi}{dt}dt \tag{1.67}$$

or in equivalent form

$$W_m(\psi) - W_m(\psi_0) = \int_{\psi_0}^{\psi} \dot{q}(\psi)d\psi \tag{1.68}$$

Assuming that at the time instant t_0 the linkage flux $\psi = 0$, eqn (1.68) can be simplified to the form

$$W_m(\psi) = \int_{0}^{\psi} \dot{q}(\psi)d\psi \tag{1.69}$$

Eqn (1.69) represents the total energy of the magnetic field stored in an inductive element. This is the surface area above the curve $\psi(\dot{q})$ in Fig. 1.23. In the case of linear element this energy is expressed by the formula

$$W_m = \frac{1}{2}\dot{q}\psi = \frac{1}{2}i\psi \tag{1.70}$$

Putting $\psi = Li$ to eqn (1.70), the total energy of the magnetic field is

$$W_m = \frac{1}{2}Li^2 \tag{1.71}$$

This is the well-known energy formula in the theory of electric circuits.

The area below the curve (Fig. 1.23) described by the equation

$$W'_m(\dot{q}) = \int_{0}^{\dot{q}} \psi(\dot{q})d\dot{q} = \tag{1.72}$$

is called the *coenergy* (in this case the magnetic coenergy). The following relationship exists between the energy and coenergy:

$$W_m + W'_m = \dot{q}\psi \tag{1.73}$$

The last eqn (1.73) says that the stored energy is equal to cooenergy (Fig. 1.23). This relationship is true both for linear and nonlinear elements. In the case of linear elements, the energy W_m is equal to the coenergy W'_m.

1.8.2 Energy and coenergy of a nonlinear capacitive element

In a similar way relationships describing the energy and coenergy in a capacitive element can be derived, i.e.,

- Electrical energy

$$W_e(q) = \int_0^q \dot{\psi}(q)dq \tag{1.74}$$

- Electrical coenergy

$$W'_e(\dot{\psi}) = \int_0^{\dot{\psi}} q(\dot{\psi})d\dot{\psi} \tag{1.75}$$

In the case of a linear element

$$W_e = \frac{1}{2}\dot{\psi}q = \frac{1}{2}vq = \frac{1}{2}Cv^2 = W'_e \tag{1.76}$$

This is the triangle area below the straight line $v = \dot{\psi} = f(\dot{q})$.

1.8.3 Energy and coenergy of mechanical systems

For mechanical systems with translatory or rotary motion the kinetic energy and coenergy can be found in a similar way. Appropriate relationships for instantaneous powers delivered to mass or inertia elements have the following forms:

- For translatory motion

$$P(t) = \dot{x}(t)\dot{p}(t) = v(t)F(t) \tag{1.77}$$

- For rotary motion

$$P(t) = \dot{\vartheta}(t)\dot{l}(t) = \Omega(t)T(t) \tag{1.78}$$

Kinetic energy and kinetic coenergy in translatory motion are described by the following equations (Fig. 1.24a):

$$W_k(p) = \int_0^p \dot{x}(p)dp \qquad \text{and} \qquad W'_k(\dot{x}) = \int_0^{\dot{x}} p(\dot{x})d\dot{x} \tag{1.79}$$

In rotary motion it is necessary to replace the linear velocity $\dot{x} = v$ with the angular velocity $\dot{\vartheta} = \Omega$ and the force $\dot{p}$ with the torque $\dot{l} = T$. Potential energy and potential coenergy, which describe the state of elastic elements, per analogy, are expressed by the following equations (Fig. 1.24b):

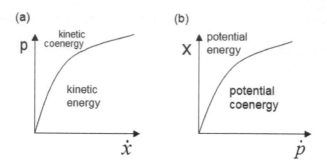

Fig. 1.24. Energy and coenergy in mechanical systems: (a) kinetic energy and coenergy; (b) potential energy and coenergy.

$$W_p(x) = \int_0^x \dot{p}(x)dx \qquad \text{and} \qquad W_p'(\dot{p}) = \int_0^{\dot{p}} x(\dot{p})d\dot{p} \qquad (1.80)$$

In this case, similar to conservative electrical elements, the following relationships result (Fig. 1.24):

$$W_k + W_k' = p\dot{x} \qquad \text{and} \qquad W_p + W_p' = x\dot{p} \qquad (1.81)$$

To determine the state function of a system consisting of multiple elements, it is necessary to add the adequate energies. Then, eqns (1.66) to (1.80) will become functions of many coordinates or generalized velocities. Tables 1.5 and 1.6 contain all relationships describing the energy stored in linear conservative elements and discharged in dissipative elements.

Example 1.4.

Find the kinetic energy and momentum of

(a) Passenger aicraft with mass of 78.6 t that lands with speed of 270 km/h;
(b) Passenger car with mass 1.5 t moving with velocity of 100 km/h.

(a) Landing passenger aicraft

$$W_k = \frac{1}{2}mv^2 = \frac{1}{2}78,600\left(\frac{270}{3.6}\right)^2 = 221,062,500 \text{ J} = 221.06 \text{ MJ}$$

$$p = mv = 78,600\left(\frac{270}{3.6}\right) = 5,895,000 \text{ Ns} = 5895 \text{ MNs}$$

Table 1.5. Energy stored and dissipated in linear concentrated-parameter electrical elements.

Elements	Electrical elements	
	Parameter	Energy
Conservative	Capacitance $C = \frac{dq}{d\psi}$	$w_e = \frac{1}{2}Cv^2 = \frac{1}{2}C(\dot\psi)^2$
	Inductance $L = \frac{d\psi}{d\dot q}$	$w_m = \frac{1}{2}Li^2 = \frac{1}{2}L(\dot q)^2$
Dissipative	Resistance $L = \frac{d\psi}{d\dot q}$	$\frac{dw_R}{dt} = Ri^2 = R(\dot q)^2$
	Conductance $G(\dot\psi) = \frac{d\dot q}{d\psi}$	$\frac{dw_R}{dt} = Gv^2 = G(\dot\psi)^2$

Table 1.6. Energy stored and dissipated in linear concentrated-parameter mechanical elements.

Elements	Mechanical elements			
	Translatory motion		Rotary motion	
	Parameter	Energy	Parameter	Energy
Conservative	Mass $m(\dot x) = \frac{dp}{d\dot x}$	$w_k = \frac{1}{2}mv^2 = \frac{1}{2}m(\dot x)^2$	Inertia $J(\dot\vartheta) = \frac{dl}{d\dot\vartheta}$	$w_k = \frac{1}{2}J\Omega^2 = \frac{1}{2}J(\dot\vartheta)^2$
	Compliance $K(\dot p) = \frac{dx}{d\dot p}$	$w_p = \frac{1}{2}KF^2 = \frac{1}{2}K(\dot p)^2$	Torsional compliance $K_\vartheta(\dot l) = \frac{d\vartheta}{d\dot l}$	$w_p = \frac{1}{2}K_\vartheta T^2 = \frac{1}{2}K_\vartheta(\dot l)^2$
Dissipative	Damping (friction) $D(\dot x) = \frac{dp}{d\dot x}$	$\frac{dw_R}{dt} = Dv^2 = D(\dot x)^2$	Torsional friction $D_\vartheta(\dot\vartheta) = \frac{dl}{d\dot\vartheta}$	$\frac{dw_R}{dt} = D_\vartheta\Omega^2 = D_\vartheta(\dot\vartheta)^2$

(a) Passenger car

$$W_k = \frac{1}{2}mv^2 = \frac{1}{2}1500\left(\frac{100}{3.6}\right)^2 = 1,157,407 \text{ J} = 1.16 \text{ MJ}$$

$$p = mv = 1500\left(\frac{100}{3.6}\right) = 41,667 \text{ Ns} \approx 0.042 \text{ MNs}$$

The consequences of collison with an obstacle, e.g., tree or wall, will be much more serious for the passenger car because its kinetic energy and momentum are much smaller than those of a heavy aircraft.

1.9 Force and torque balance equations

For a system of n particles, d'Alembert principle can be expressed as follows:
The work performed by the sum of external forces and inertia forces on the distance that is a virtual displacement, or virtual work, is equal to zero. This principle can be expressed with the aid of the following equation:

$$\sum_{i=1}^{n} (\mathbf{F}_i + \mathbf{F}_{Ji})\, \delta \mathbf{r}_i \tag{1.82}$$

where $\mathbf{F}_i$ is the total applied force (excluding constraint forces) acting on the ith particle of the system, $\mathbf{F}_{Ji} = -m_i \mathbf{a}_i$ is the force of inertia acting on the ith particle with its mass m_i, $\mathbf{a}_i$ is the accelleration of the ith particle of the system, and $\delta \mathbf{r}_i$ is the virtual displacement of the ith particle of the system. The product $m_i \mathbf{a}_i$ represents the time derivative of the momentum of the ith particle

On the basis of the d'Alembert principle for translatory motion

$$m\frac{dv}{dt} + Dv + K \int v dt = F \tag{1.83}$$

where F is the sum of external and electromagnetic forces. Eqn (1.83) can also be written in the form

$$m\frac{d^2 x}{dt^2} + D\frac{dx}{dt} + Kx = F \tag{1.84}$$

or

$$m\ddot{x} + D\dot{x} + Kx = F \tag{1.85}$$

Similar equations can be written for a series electrical circuit RLC

$$L\frac{d^2 q}{dt^2} + R\frac{dq}{dt} + \frac{1}{C}q = v \tag{1.86}$$

or

$$L\ddot{q} + R\dot{q} + \frac{1}{C}q = v \tag{1.87}$$

For rotary motion the torque balance equation is

$$J\frac{d^2 \vartheta}{dt^2} + D_\vartheta \frac{d\vartheta}{dt} + K_\vartheta \vartheta = T \tag{1.88}$$

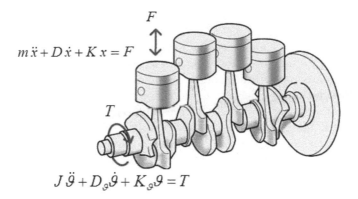

$$m\ddot{x} + D\dot{x} + K\,x = F$$

Fig. 1.25. Crankshaft: conversion of reciprocating motion $m\ddot{x} + D\dot{x} + Kx = F$ into rotary motion $J\ddot{\vartheta} + D_\vartheta\dot{\vartheta} + K_\vartheta\vartheta = T$.

or

$$J\ddot{\vartheta} + D_\vartheta\dot{\vartheta} + K_\vartheta\vartheta = T \qquad (1.89)$$

where T is the sum of external and electromagnetic torques, D_ϑ is the torsional damping, K_ϑ is the torsional compliance and T_ϑ is the external torque.

Fig. 1.25 shows a crankshaft of a combustion engine. A crankshaft is a basic part of combustion engine able to perform a conversion of reciprocating motion of a piston into rotational motion of shaft. The reciprocating motion of pistons can be described by eqns (1.83), (1.84), and (1.85), while the rotary motion of the shaft can be described by eqns (1.88) and (1.89).

Summary

Electromechanical energy conversion is a conversion of mechanical energy into electrical energy (generator) or vice versa (motor) with the aid of rotary motion (rotary machines) or translatory (linear, reciprocating) motion (linear machines and actuators).

Electrical machines, solenoid actuators and electromagnets are generally called *electromechanical energy conversion devices*.

Transformers and solid-state converters do not belong to the group of electromechanical energy conversion devices because they only convert one kind of electrical energy into another kind of electrical energy with different parameters (change in voltage, current, frequency, number of phases, conversion of DC into AC current, etc.) without any motion.

The corresponding electric circuit parameters and magnetic circuit parameters are: electric voltage, V – magnetic voltage drop, V_μ; SEM, E – MMF,

F; electric current, I – magnetic flux, Φ; resistance, R – reluctance, R_μ; conductance, $G = 1/R$ – permeance, Λ_μ; electric conductivity, σ – magnetic permeability, μ. Other analogies are given in Table 1.2.

Ferromagnetic materials for cores of electrical machines and electromagnetic devices are magnetically nonlinear (nonlinear variation of magnetic flux density B with magnetic field intensity H) and are described by magnetization curves $B(H)$ and specific core loss curves $\Delta p(B)$.

Electromagnetic torque of a rotary doubly-excited electromechanical energy conversion device (Fig. 1.20) has two components: synchronous torque $T_{ms}\sin(\vartheta)$ and reluctance torque $T_{mrel}\sin(2\vartheta)$, i.e.,

$$T_{elm} = -i_1 i_2 L_0 \sin(\vartheta) - (i_1^2 \Delta L_1 + i_2^2 \Delta L_2)\sin(2\vartheta)$$

$$= -[T_{ms}\sin(\vartheta) + T_{mrel}\sin(2\vartheta)]$$

All elements in electrical and mechanical systems can be described by the following equations:

- Capacitive element, inductive element and resistive element

$$v(t) = \frac{1}{C}q(t) \qquad i(t) = \frac{1}{L}\psi(t) \qquad \dot{q} = G\dot{\psi} = \frac{1}{R}\dot{\psi}$$

- Element of mass, elastic element and dissipative element (viscous friction) in translatory motion

$$v(t) = \frac{1}{m}p(t) \qquad F(t) = \frac{1}{K}x(t) \qquad \dot{p} = D\dot{x}$$

- Inertial element, elastic element and torsional damping element in rotary motion

$$\Omega(t) = \frac{1}{J}l(t) \qquad T(t) = \frac{1}{K_\vartheta}\vartheta(t) \qquad l = D_\vartheta \dot{\vartheta}$$

Characteristic $\psi(i)$ of a nonlinear inductive element is plotted in Fig. 1.23. The area above the curve is the energy stored in magnetic field

$$W_m(\psi) = \int_0^\psi \dot{q}(\psi)d\psi$$

The area below the curve is the so-called *coenergy*, i.e.,

$$W_m'(\dot{q}) = \int_0^{\dot{q}} \psi(\dot{q})d\dot{q}$$

Coenergy (a second state function of the energy) is an auxiliary function necessary for calculations of the force or torque at constant current. The sum of energy and coenergy, both for linear and nonlinear elements, is equal to the area of the rectangle $\dot{q}\psi$, or

$$W_m + W'_m = \dot{q}\psi$$

In the case of a linear element, the energy W_m is equal to coenergy W'_m, i.e.,

$$W_m = \frac{1}{2}\dot{q}\psi = Wm' = \frac{1}{2}\psi\dot{q}$$

For a capacitive element, the energy stored in electric field and coenergy are, respectively,

$$W_e(q) = \int_0^q \psi(q)dq \qquad\qquad W'_e(\dot{\psi}) = \int_0^{\dot{\psi}} q(\dot{\psi})d\dot{\psi}$$

In the case of a linear capacitive element

$$W_e = \frac{1}{2}\dot{\psi}q = W'_e = \frac{1}{2}q\dot{\psi}$$

Energy and coenergy in mechanical motion have the following forms:

- For translatory motion

$$W_k(p) = \int_0^p \dot{x}(p)dp \qquad\qquad W'_k(\dot{x}) = \int_0^{\dot{x}} p(\dot{x})d\dot{x}$$

- For rotary motion

$$W_k(l) = \int_0^l \dot{\vartheta}(l)dl \qquad\qquad W'_k(\dot{\vartheta}) = \int_0^{\dot{\vartheta}} l(\dot{\vartheta})d\dot{\vartheta}$$

Energies and coenergies are functions of generalized velocity, i.e., first derivatives with respect to time of basic coordinates, i.e., electric current, $i = \dot{q}$, electric voltage, $v = \dot{\psi}$, linear velocity, $v = \dot{x}$, momentum, p, angular velocity, $\Omega = \dot{\vartheta}$ and angular momentum, l.

On the basis of d'Alembert's principle, the force balance equation in translatory motion has the form

$$m\ddot{x} + D\dot{x} + Kx = F$$

A similar equation for torque balance in rotary motion:

$$J\ddot{\vartheta} + D_\vartheta\dot{\vartheta} + K_\vartheta\vartheta = T$$

Problems

1. A ferromagnetic ring made of M19 laminated silicon steel has a circular cross-section with diameter $d = 0.02$ m and mean diameter $D = 0.3$ m. The ring is uniformly wound with a coil of $N = 1000$ turns. The magnetization curve of M19 silicon steel is plotted in Fig. 1.8.

   (a) Find the current in the coil required to produce the magnetic flux of $\Phi = 0.0004$ Wb in the ring;
   (b) If an air gap $g = 0.002$ m is cut in the ring, find the air gap flux produced by the current found in (a);
   (c) Find the current that produces the same flux in the air gap as in (a).

   Answer: (a) $I = 0.304$ A; (b) $\Phi_g = 5.216 \times 10^{-5}$ Wb; (c) $I' = 2.33$ A.

2. A coil with the number of turns $N = 520$ and resistance $R = 12.0$ Ω is fed with DC voltage $V_{dc} = 3.0$ V. The coil is wound uniformly on a ring-shaped core with rectangular cross-section made of silicon steel tape M19 (Fig. 1.8). The inner diameter of the core $D_{in} = 0.15$ m, outer diameter $D_{out} = 0.2$ m, and its thickness $L = 0.08$ m. Find the magnetic flux density B_c and magnetic flux Φ in the core. The stacking factor $k_i = 0.96$.

   Answer: $B_c = 1.23$ T, $\Phi = 0.00236$ Wb.

3. The ring-shaped core with coil of Problem 1.2 is fed with AC voltage $V = 380$ V and frequency $f = 50$ Hz. The specific mass density of M19 silicon steel is $\rho_{Fe} = 7650$ W/kg. Find: (a) the magnetizing current I_Φ; (b) core losses ΔP_c; (c) total current I_0 drawn by the coil. The voltage drop across the coil resistance, additional losses and excess eddy current losses can be neglected.

   Answer: (a) $I_\Phi = 6.53$ A; (b) $\Delta P_c = 30.9$ W; (d) $I_0 = 6.53$ A.

4. For the ferromagnetic core with coil of Problems 1.2 and 1.3 calculate the self-inductance, if the coil is fed with DC voltage: (a) $V_1 = 1.0$ V; (b) $V_2 = 6.0$ V.
   For each voltage find the relative magnetic permeablity using the $B - H$ curve plotted in Fig. 1.8.

   Answer: (a) $\mu_{r1} = 8047$, $L_1 = 9.55$ H; (b) $\mu_{r2} = 1341$, $L_2 = 1.59$ H.

5. A simple electrical machine has the following dimensions of ferromagnetic core:
   - Stator core outer diameter $D_{1out} = 0.16$ m;
   - Stator core inner diameter $D_{1in} = 0.11$ m;
   - Rotor core outer diameter $D_{2out} = 0.108$ m;
   - Rotor core inner diameter $D_{2in} = 0.055$ m;

- Axial length of core $L = 0.1$ m.

The relative magnetic permeability of the stator and rotor core is $\mu_r = 500$ and the magnetic flux density in the air gap is $B_g = 1.0$ T. Find:

(a) The ferromagnetic core-to-air gap volume ratio k_V;

(b) The ferromagnetic core-to-air gap energy ratio k_W.

Answer: (a) $k_V = 50.78$; (b) $k_W = 0102$.

6. An electromagnetic relay has an exciting coil of $N = 1000$ turns. The coil has a cross-sectional area $a \times b = 50 \times 60$ mm^2. The reluctance of the magnetic circuit and fringing effect are neglected. Find:

(a) The coil inductance L_g if the air gap is $g = 3$ mm;

(b) The stored magnetic field energy W_m for a coil current of $i = 1.5$ A;

(c) The mechanical energy output based on field energy changes dW_m when the armature moves to position for which $g' = 1.5$ mm and coil current remains constant at $i = 1.5$ A. Assume slow movement of armature;

(d) Repeat (c) above based on force-calculations and mechanical displacement.

Answer: (a) $L_g = 1.257$ H; (b) $W_m = 1.414$ J; (c) $dW_m = 1.414$ J; (d) $W_{mech} = 1.414$ J.

7. The magnetic circuit shown in Fig. 1.20 is made of high-permeability electrical steel. The rotor does not have any winding and is free to turn about a vertical axis. The leakage flux and fringing effect are neglected.

(a) Derive an expression for the torque acting on the rotor in terms of the self-inductance L_s;

(b) Calculate the torque for the radius of the rotor $r = 28$ mm, axial length (perpendicular to the page) of the magnetic circuit $L = 80$ mm, number of the stator turns $N = 1100$, stator current $i = 2.0$ A and the air gap between the rotor and stator $g = 1.5$ mm;

(c) The maximum flux density in the overlapping portions of the air gaps is to be limited to approximately 1.6 T to avoid excessive saturation of the steel. Calculate the maximum current and torque for the dimensions and numbers of turns as above.

Answer: (a) $L_s = \mu_0 L(r+0.5g)N^2\theta/(2g)$, $T_{elm} = \mu_0 L(r+0.5g)(iN)^2/(4g)$; (b) $T_{elm} = 3.467$ Nm; (c) $i_{max} = 2.315$ A, $T_{elmmax} = 4.645$ Nm.

8. In the doubly-excited device shown in Fig. 1.20, the inductances are $L_{11} = 2.0 + 2.0\cos(2\theta) \times 10^{-3}$ H, $L_{12} = 0.5\cos(\theta)$ H, $L_{22} = 25.0 + 15.0\cos(2\theta)$ H. Find and plot the torque $T(\theta)$ for current $i_1 = 1.2$ A and $i_2 = 0.04$ A. Calculate the torque for the above values of currents and $\theta = 10°$, $\theta = 60°$ and $\theta = 120°$.

Answer: For $\theta = 10°$ $T = -0.013$ Nm; for $\theta = 60°$ $T = -0.044$ Nm and for $\theta = 120°$ $T = 2.494 \times 10^{-3}$ Nm.

9. Find the stored magnetic field energy and the associated power in a coil of inductance $L = 0.003$ H after $t = 0.001$ s, if the coil is fed with sinusoidal voltage $v = V_m \sin(\omega t)$, $V_m = 311.1$ V, $f = 50$ Hz.

Answer: $W_m = 147.8$ J, $P = -30\ 180$ W.

10. An AC voltage $v = V_m \sin(\omega t)$ is applied to the capacitor $C = 0.000020$ F. The frequency is $f = 50$ Hz, and the rms voltage is $V = 100$ kV. Find the stored electric field energy after $t = 0.001$ s and associated maximum power.

Answer: $W_e = 19\ 100$ J, $P_{max} = 62.83 \times 10^6$ W.

11. A ferromagnetic UI-core with two air gaps as shown in Fig. 2.3b has the following dimensions: width of leg $a = 0.05$ m, height of yoke $a = 0.05$ m, height of U-core $h = 0.25$ m, width of U-core (equal to the width of I-shaped moving armature) $b = 0.2$ m, thickness of stack $L = 0.07$ m. The core is made of a silicon electrical steel M19 (Fig. 1.8). The stacking factor is $k_i = 0.96$ and the specific mass density of the silicon steel is $\rho_{Fe} = 7650$ kg/m^3. The air gap depends on the position of moving armature (I-shaped). The number of turns on stationary U-section is $N = 200$. The winding is fed with AC voltage $V = 127$ V and frequency $f = 50$ Hz. Calculate:

(a) The magnetizing current for air gaps $g_0 = 0$, $g_1 = 0.001$ m and $g_2 = 0.002$ m;
(b) Total current including core loss current for the same air gaps;
(c) Attraction force between I-shaped armature and stationary U-shaped core and energy stored in magnetic field for the same air gaps.

Answer: (a) $I_{\Phi 0} = 0.229$ A for $g_0 = 0$, $I_{\Phi 1} = 5.016$ A for $g_0 = 1$ mm, $I_{\Phi 2} = 9.804$ A for $g_0 = 2$ mm; (b) the core loss current is only $I_{Fe} = 0.006$ A so the total currents are the same as magnetizing currents; (c) $F_0 \rightarrow \infty$ for $g_0 = 0$ if magnetic saturation is not included, $W_{m0} = 0.046$ J, $F_1 = 1062.5$ N, $W_{m1} = 1.063$ J for $g_1 = 1.0$ mm, $F_2 = 1015.0$ N $W_{m2} = 2.029$ J for $g_2 = 2.0$ mm.

12. The UI-core with two air gaps of Problem 1.11 as shown in Fig. 2.3b is fed with DC voltage $V_{dc} = 6.0$ V. The winding is wound with the same number of turns $N = 200$ using a round wire with diameter of bare wire $d_w = 1.0236$ mm (AWG 18). The thickness of core-to-coil insulation is $t_i = 1.0$ mm. Find the attractive force and stored magnetic energy for air gaps $g_1 = 1.0$ mm and $g_2 = 2.0$ mm. The electric conductivity of copper at 75°C is $\rho_{75} = 47 \times 10^6$ S/m/. The mean length of turn can be approximately calculated from the formula $l_{mean} = 2(a + L + 2t_i + 2d_w)$.

Answer: $F_1 = 923.5$ N, $W_{m1} = 0.924$ J at $g_1 = 1.0$ mm; $F_2 = 230.9$ N, $W_{m2} = 0.462$ J at $g_1 = 2.0$ mm.

13. A disk with its diameter $D = 1.5$ m, thickness $t_d = 0.16$ m and specific mass density $\rho_d = 7700$ kg/m^3 rotates with the speed of $n = 480$ rpm. Calculate: (a) moment of inertia J; (b) stored kinetic energy E_k; (c) input external torque, to obtain the rotational speed of $n = 480$ rpm in the time $t = 15$ s.

Answer: (a) $J = 612.3$ kgm^2; (b) $E_k = 773.5 \times 10^3$ J; (c) $T = 2052$ Nm.

2

TRANSFORMERS

A *transformer* is a static electromagnetic device for transforming electrical energy in an AC system from one (primary) circuit into another (secondary) circuit at the same frequency but with different values of voltages and currents. A simple single-phase transformer is represented by two electric circuits coupled magnetically (Fig. 1.19).

2.1 Single-phase transformer

2.1.1 Principle of operation and construction

A typical *single-phase transformer* consists of two windings, which are coupled electromagnetically by a ferromagnetic core (Fig. 2.1). The winding that absorbs electrical energy is called the *primary winding* and the winding that delivers electrical energy is called the *secondary winding*. The number of secondary windings can be more than one. The winding connected to the circuit with the higher voltage is called the *high voltage winding* (HV) and the winding connected to the circuit with lower voltage is called the *low voltage winding* (LV).

With respect to the construction of ferromagnetic core, there are two types of single-phase transformers: shell-type (Figs 2.2a, 2.3a) and core-type (Figs 2.2b, 2.3b). Nowadays, C-shaped wound cores are frequently used (Fig. 2.4). Transformers can be air-cooled, usally by natural convection (Fig. 2.5) or oil-cooled (Fig. 2.6). Oil-cooled transformers are immersed in oil tanks.

2.1.2 Ideal single-phase transformer

An ideal transformer is analyzed on the basis of the following assumptions:

(a) The primary winding resistance R_1 and secondary winding resistance R_2 are equal to zero ($R_1 = R_2 = 0$);

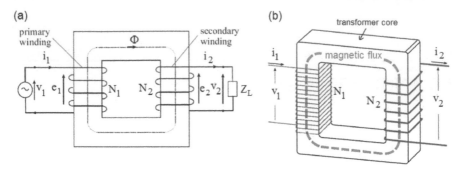

Fig. 2.1. Single-phase, two-winding transformer: (a) two electric circuits electromagnetically coupled; (b) 3D image.

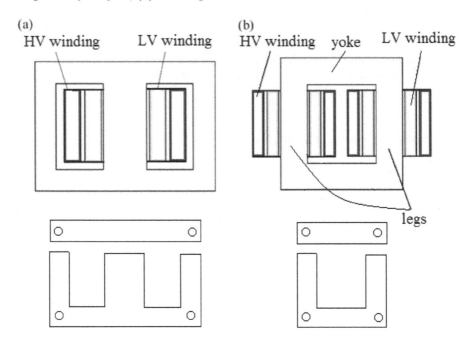

Fig. 2.2. Single-phase transformers: (a) shell-type; (b) core-type. HV – high-voltage winding, LV – low voltage winding. The LV winding is situated close to the core.

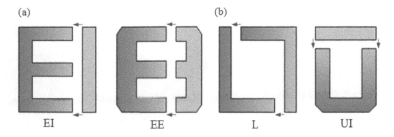

Fig. 2.3. Laminations for single-phase laminated cores: (a) shell-type; (b) core-type.

Fig. 2.4. Wound cores of single-phase transformers.

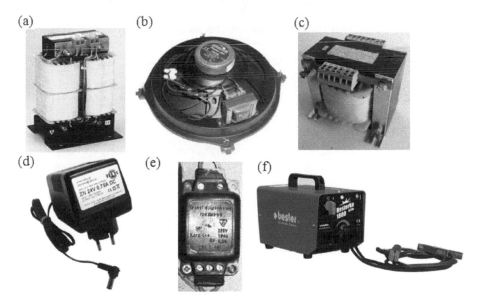

Fig. 2.5. Single-phase air cooled transformers: (a) separation transformer; (b) loud-speaker transformer; (c) radio transformer; (d) step-down transformer with rectifier; (e) door bell transformer; (f) welding transformer.

Fig. 2.6. Distribution transformers: (a) oil cooled distribution transformers with corrugated tank; (b) pole distribution transformer; (c) interior of a pole distribution transformer.

(b) The primary $\Phi_{\sigma 1}$ and secondary $\Phi_{\sigma 2}$ leakage fluxes are equal to zero ($\Phi_{\sigma 1} = \Phi_{\sigma 2} = 0$);

(c) The core losses ΔP_{Fe} are equal to zero ($\Delta P_{Fe} = 0$).

Assuming a sinusoidal time variation of the magnetic flux

$$\Phi = \Phi_m \sin(\omega t) \tag{2.1}$$

the instantaneous EMFs induced in the primary and secondary windings are, respectively,

$$e_1 = N_1 \frac{d\Phi}{dt} = N_1 \omega \Phi_m \cos(\omega t) = 2\pi f N_1 \Phi_m \cos(\omega t) = \sqrt{2} E_1 \cos(\omega t) \tag{2.2}$$

$$e_2 = N_2 \frac{d\Phi}{dt} = N_2 \omega \Phi_m \cos(\omega t) = 2\pi f N_2 \Phi_m \cos(\omega t) = \sqrt{2} E_2 \cos(\omega t) \tag{2.3}$$

The *rms* EMFs (induced voltages)

$$E_1 = \pi\sqrt{2} f N_1 \Phi_m = 4.44 f N_1 \Phi_m \qquad\qquad E_2 = \pi\sqrt{2} f N_2 \Phi_m = 4.44 f N_2 \Phi_m \tag{2.4}$$

For explanation, please see eqn (1.16). The ratio of induced voltages

$$\frac{e_1}{e_2} = \frac{E_1}{E_2} = \frac{N_1}{N_2} = \vartheta \tag{2.5}$$

For an ideal transformer $V_1 = E_1$ and $V_2 = E_2$, thus

$$\frac{V_1}{V_2} = \frac{N_1}{N_2} \tag{2.6}$$

There are no power losses, so that the apparent power of the primary side is equal to the apparent power of the secondary side, i.e.,

$$S_1 = S_2 \Rightarrow V_1 I_1 = V_2 I_2 \tag{2.7}$$

From eqns (2.6) and (2.7)

$$\frac{V_1}{V_2} = \frac{I_2}{I_1} = \vartheta \tag{2.8}$$

To eliminate the magnetic coupling, the secondary voltage and current is expressed with the aid of the voltage ratio ϑ as

$$V_1 = \vartheta V_2 = V_2' \tag{2.9}$$

$$I_1 = \frac{1}{\vartheta} I_2 = I_2' \tag{2.10}$$

where V_2' is the secondary voltage referred to as the primary winding and I_2' is the secondary current referred to as the primary winding.

2.1.3 Real transformer

In a real transformer the primary R_1 and secondary R_2 resistances, leakage fluxes $\Phi_{\sigma 1}$, $\Phi_{\sigma 2}$ and core losses ΔP_{Fe} are taken into account. The equation of balance of magnetomotive forces (MMFs) in complex notation is

$$\mathbf{F}_0 = \mathbf{F}_1 - \mathbf{F}_2 \tag{2.11}$$

where $\mathbf{F}_0$ is the MMF that generates the magnetic flux, $\mathbf{F}_1$ is the primary winding MMF and $\mathbf{F}_2$ is the secondary winding MMF. Eqn (2.11) also can be written as

$$N_1 \mathbf{I}_0 = N_1 \mathbf{I}_1 - N_2 \mathbf{F}_2 \tag{2.12}$$

Thus,

$$\mathbf{I}_0 = \mathbf{I}_1 - \mathbf{I}_2 \frac{N_2}{N_1} = \mathbf{I}_1 - \mathbf{I}_2' \tag{2.13}$$

where $\mathbf{I}_0$ is the exciting current (no-load current) and $\mathbf{I}_2'$ is the secondary current referred to as the primary side. The *excitating current*

$$\mathbf{I}_0 = \mathbf{I}_{Fe} + j\mathbf{I}_\Phi \tag{2.14}$$

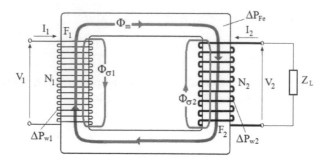

Fig. 2.7. A real single-phase two-winding transformer. Φ_m is the main flux, $\Phi_{\sigma1}$, $\Phi_{\sigma2}$ are leakage fluxes, $F_1 = I_1 N_1$ is the primary winding MMF, $F_2 = I_2 N_2$ is the secondary winding MMF, ΔP_{w1}, ΔP_{w2} are primary and secondary winding losses, respectively, ΔP_{Fe} are core losses, Z_L is the load impedance.

or

$$I_0 = \sqrt{I_{Fe}^2 + I_\Phi^2} \tag{2.15}$$

has two components:

- The active component I_{Fe}, which is the so-called *core loss current*;
- The reactive component I_Φ, which is the so-called *magnetizing current*.

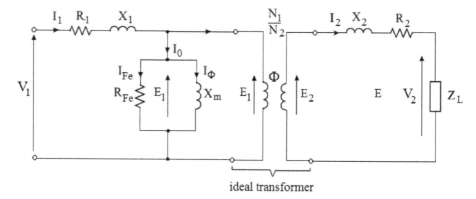

ideal transformer

Fig. 2.8. Equivalent circuit of a real transformer with magnetic coupling. R_1 – primary winding resistance, X_1 – primary winding leakage reactance, R_2 – secondary winding resistance, X_2 – secondary winding leakage reactance, R_{Fe} – resistance representing core losses, X_m – mutual reactance, I_1 – primary current, I_2 – secondary current, I_0 – exciting current, I_{Fe} – core loss current, I_Φ – magnetizing current, Φ – linkage magnetic flux, V_1 – primary voltage, V_2 – secondary voltage, E_1 – primary winding EMF, E_2 – secondary winding EMF, Z_L – load impedance.

The *equivalent circuit* with magnetic coupling and vertical branch containing the exciting current I_0 is shown in Fig. 2.8. From the equivalent circuit on the basis of Kirchhoff's voltage law:

$$\mathbf{V}_1 = \mathbf{E}_1 + (R_1 + jX_1)\mathbf{I}_1 \tag{2.16}$$

$$\mathbf{V}_2 = \mathbf{E}_2 - (R_2 + jX_2)\mathbf{I}_2 \tag{2.17}$$

Multiplying both sides of eqn (2.17) by the voltage ratio ϑ

$$\vartheta\mathbf{V}_2 = \vartheta\mathbf{E}_2 - \vartheta^2 R_2\frac{\mathbf{I}_2}{\vartheta} - j\vartheta^2 X_2\frac{\mathbf{I}_2}{\vartheta} \tag{2.18}$$

or

$$\mathbf{V}_2' = \mathbf{E}_2' - R_2'\mathbf{I}_2' - jX_2'\mathbf{I}_2' \tag{2.19}$$

where

$$\mathbf{V}_2' = \vartheta\mathbf{V}_2 \qquad \mathbf{E}_2' = \vartheta\mathbf{E}_2 \qquad \mathbf{I}_2' = \frac{1}{\vartheta}\mathbf{I}_2 \tag{2.20}$$

$$R_2' = \vartheta^2 R_2 \qquad\qquad X_2' = \vartheta^2 X_2 \tag{2.21}$$

Eqns (2.20) and (2.21) explain how to bring the secondary winding parameters to the primary winding using the voltage ratio ϑ. Since $E_2' = E_1$, both the left-hand side and right-hand side circuits in Fig. 2.8 can be electrically connected and the new equivalent circuit (Fig. 2.9a) does not contain any magnetic coupling.

Fig. 2.9b shows the construction of the phasor diagram on the basis of the equivalent circuit (Fig. 2.9a).

2.1.4 Open-circuit test

The purpose of the *open-circuit test* on a transformer is to find:

- Core losses ΔP_{Fe};
- Resistance R_{Fe} and reactance X_m of the vertical branch of the equivalent circuit (Fig. 2.9a);
- Magnetizing current I_Φ and core loss current I_{Fe};
- Voltage ratio ϑ.

The measured quantities are (Fig. 2.10):

- Input voltage V_1;
- No-load current I_0;
- Secondary voltage V_2;
- No-load power P_0.

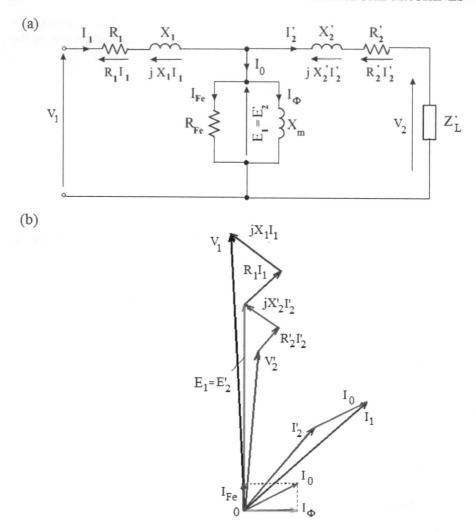

Fig. 2.9. Equivalent circuit and phasor diagram of a real transformer without magnetic coupling: (a) T-type equivalent circuit; (b) phase diagram.

The primary winding is fed with variable voltage using, for example, a variable autotransfomer. In the primary circuit there is a voltmeter, ammeter and wattmeter (Fig. 2.10). The secondary winding is open. A voltmeter with high internal resistance is connected across the secondary winding terminals, so the secondary current I_2 is practically equal to zero ($I_2 \approx 0$).

Since $I_o \ll I_{1n}$, where I_{1n} is the nominal (rated) primary winding current, $R_1 \ll R_{Fe}$ and $X_1 \ll X_m$, the secondary branch can be neglected. The modified equivalent circuit at no load is shown in Fig. 2.11c. From the

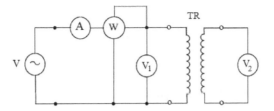

Fig. 2.10. Circuit diagram with instruments for open-circuit test on a transformer.

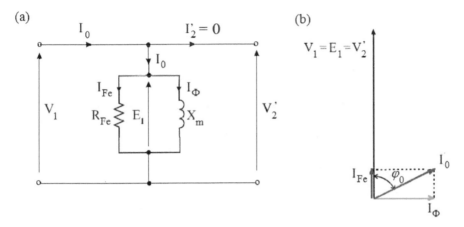

Fig. 2.11. Open-circuit test on a single-phase transformer: (a) equivalent circuit; (b) phasor diagram.

measured quantities, i.e., primary voltage V_1 and no-load power P_0, the following parameters can be calculated:

$$P_0 = \frac{E_1^2}{R_{Fe}} \approx \frac{V_1^2}{R_{Fe}} \Rightarrow R_{Fe} = \frac{V_1^2}{P_0} \qquad (2.22)$$

The vertical branch (mutual) reactance is

$$X_m = \frac{E_1}{I_\Phi} \approx \frac{V_1}{I_\Phi} \qquad (2.23)$$

According to the no-load phasor diagram (Fig. 2.11b), the magnetizing current is

$$I_\Phi = \sqrt{I_0^2 - I_{Fe}^2} \qquad (2.24)$$

where

$$I_{Fe} = \frac{P_0}{V_1} \qquad (2.25)$$

The voltage ratio

$$\vartheta = \frac{V_1}{V_2} \tag{2.26}$$

The *no-load current* is usually expressed as a percentage of the nominal (rated) current I_{1n} of the transformer, i.e.,

$$i_{0\%} = \frac{I_0}{I_{1n}} 100\% \tag{2.27}$$

In power transformer, the no-load current is in the interval $1 \leq i_{0\%} \leq 10\%$ of the nominal (rated) current I_{1n}. Rule: The higher the power, the lower the no-load current.

2.1.5 Short-circuit test

The porpose of a *short-circuit test* on a transformer is to find:

- Primary and secondary winding resistances R_1 and R_2;
- Primary and secondary leakage reactances X_1 and X_2;
- Primary and secondary winding losses ΔP_{w1} and ΔP_{w2}.

The measured quantities are (Fig. 2.12)

- Input voltage V_1;
- Short circuit current I_{sc};
- Short circuit power P_{sc}.

In the short-circuit test the secondary winding is shorted ($V_2 = 0$), while the primary winding is fed with a reduced voltage to obtain nominal currents in the primary and secondary windings, i.e., $I_{1sc} = I_{1n}$ and $I_{2sc} = I_{2n}$ (Fig. 2.12).

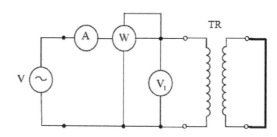

Fig. 2.12. Circuit diagram with instruments for short-circuit test on a transformer.

The *short-circuit voltage* is such a voltage across the primary winding terminals at which the primary and secondary currents take nominal values,

(a)   (b)

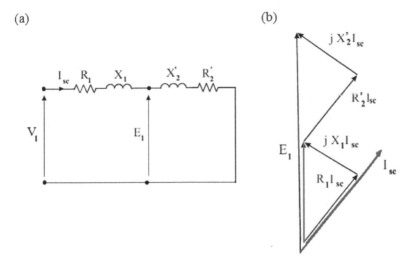

Fig. 2.13. Short-circuit test on a single-phase transformer: (a) equivalent circuit; (b) phasor diagram.

while the secondary winding is short-circuited. The short-circuit voltage $v_{sc\%}$ is given on the name plate of a transformer and it is expressed as a percentage of the nominal (rated) voltage V_{1n}, i.e.,

$$v_{sc\%} = \frac{V_{1sc}}{V_{1n}} 100\% \qquad (2.28)$$

For typical power transformers the short-circuit voltage $v_{sc\%}$ is in the range $3 \leq v_{sc\%} \leq 15$ % of the nominal voltage V_{1n}. Rule: The higher the power, the higher the short-circuit voltage $v_{sc\%}$.

Since $I_0 << I_{sc}$, the vertical branch in the equivalent circuit can be neglected. The short circuit power is

$$P_{sc} = R_{sc} I_{1sc} \qquad (2.29)$$

The short-circuit resistance R_{sc} and short-circuit impedance Z_{sc} can be determined from the measured quantities V_{1sc}, I_{1sc} and P_{sc}, i.e.,

$$R_{sc} = R_1 + R_2' = \frac{P_{sc}}{I_{1sc}^2} \qquad\qquad Z_{sc} = \frac{V_{1sc}}{I_{1sc}} = \sqrt{R_{sc}^2 + X_{sc}^2} \qquad (2.30)$$

The short-circuit reactance X_{sc} is calculated on the basis of short-circuit impedance Z_{sc} and short circuit resistance R_{sc}, i.e.,

$$X_{sc} = X_1 + X_2' = \sqrt{Z_{sc}^2 - R_{sc}^2} \qquad (2.31)$$

The primary winding resistance R_1 and secondary winding resistance R_2 are measured using, e.g., an ohmmeter. For most power transformers

$$R_1 \approx R_2' \approx \frac{R_{sc}}{2} \qquad\qquad X_1 \approx X_2' \approx \frac{X_{sc}}{2} \qquad (2.32)$$

The short-circuit power P_{sc} is equal to short circuit losses ΔP_{sc}. The short circuit losses are winding losses, i.e.,

$$P_{sc} = \Delta P_{sc} = \Delta P_{w1} + \Delta P_{w2} = R_1 I_1^2 + R_2 I_2^2 \qquad (2.33)$$

Power losses in the primary and secondary windings are

$$\Delta P_{w1} = R_1 I_1^2 \qquad\qquad \Delta P_{w2} = R_2 I_2^2 = R_2'(I_2')^2 \qquad (2.34)$$

Example 2.1

The ferromagnetic core of a single-phase transformer is made of cold-rolled electrical silicon steel M19 with magnetization curve plotted in Fig. 1.8. The mean path of magnetic flux in ferromagnetic core is $l_{Fe} = 1.04$ m and the cross-section of the core is $S_{Fe} = 0.0033$ m^2. Design the primary and secondary transformer windings, i.e., find the number of turns and cross-sections of conductors. The primary and secondary windings should meet the following requirements: nominal primary voltage $V_1 = 380$ V, nominal secondary voltage $V_2 = 42$ V, EMF per turn $e_t = 1.0$ V, nominal frequency $f = 50$ Hz, maximum current density in transformer windings $j_{max} = 3.5$ A/mm^2 (natural air cooling).

The leakage fluxes of the primary and secondary windings are neglected, there are no air gaps in ferromagnetic cores and the magnetizing current $i_{0\%} = 5$ % of the nominal primary current.

Solution

Number of primary and secondary turns

$$N_1 = \frac{V_1}{e_t} = \frac{380.0}{1.0} = 380 \qquad\qquad N_2 = \frac{V_2}{e_t} = \frac{42.0}{1.0} = 42$$

Magnetic flux in ferromagnetic core ($V_1 \approx E_1$ and $V_2 \approx E_2$)

$$\Phi \approx \frac{V_1}{\pi\sqrt{2}fN_1} = \frac{380.0}{\pi\sqrt{2} \times 50.0 \times 380} = 0.0045 \text{ Wb}$$

or

$$\Phi \approx \frac{V_2}{\pi\sqrt{2}fN_2} = \frac{42.0}{\pi\sqrt{2} \times 50.0 \times 42} = 0.0045 \text{ Wb}$$

Peak value of magnetic flux density in transformer core

$$B_m = \frac{\Phi}{S_{Fe}} = \frac{0.0045}{0.0033} = 1.364 \text{ T}$$

This value of magnetic flux density corresponds to magnetic field intensity of 680 A/m (Fig. 1.8). The magnetic field intensity at no load is

$$H_{m0} = \frac{I_{10}N_1}{l_{Fe}}$$

It can be assumed that $H_{m0} \approx H_m$, where $H_m = 680$ A/m is the magnetic field instensity under load. Thus, the magnetizing current

$$I_{10} = \frac{\frac{H_m}{\sqrt{2}} l_{Fe}}{N_1} = \frac{\frac{680.0}{\sqrt{2}} 1.04}{380} = 1.316 \text{ A}$$

Nominal primary current

$$I_1 = \frac{I_{10}}{i_{0\%}} 100\% = \frac{1.316}{5.0} \times 100\% = 26.3 \text{ A}$$

Approximate value of nominal current in the secondary winding

$$I_2 \approx I_1 \frac{N_1}{N_2} = 26.3 \frac{380}{42} = 238.1 \text{ A}$$

Cross-section of primary winding

$$s_1 = \frac{I_1}{j_{max}} = \frac{26.3}{3.5 \times 10^6} = 7.519 \times 10^{-6} \text{ m}^2 = 7.52 \text{ mm}^2$$

Cross-section of secondary winding

$$s_2 = \frac{I_2}{j_{max}} = \frac{238.1}{3.5 \times 10^6} = 53.68 \times 10^{-6} \text{ m}^2 = 68.0 \text{ mm}^2$$

Nominal apparent power of transformer

$$S = V_1 I_1 = 380.0 \times 26.3 = 10\ 000 \text{ VA} = 10 \text{ kVA}$$

Example 2.2

A single-phase transformer has the following nominal parameters:

- Apparent power $S_n = 15$ kVA
- Primary voltage $V_{1n} = 500$ V
- Secondary voltage $V_{2n} = 220$ V
- Frequency $f_n = 50$ Hz

- No-load current $i_{0\%} = 7.0\%$
- No-load power per unit $p_{0\%} = 1.0\%$
- Short-circuit voltage $v_{sc\%} = 4.5\%$
- Short-circuit power per unit $p_{sc\%} = 2.5\%$

Calculate

(a) Nominal primary and secondary current, no-load current in amps and short-circuit voltage in volts;
(b) Core losses and winding losses;
(c) Resistances and reactances of equivalent circuit.

Solution

(a) Nominal primary and secondary current, no-load current in amps and short-circuit voltage in volts

- Nominal primary current

$$I_{1n} = \frac{S_n}{V_{1n}} = \frac{15\ 000.0}{500.0} = 30 \text{ A}$$

- Nominal secondary current

$$I_{2n} = \frac{S_n}{V_{2n}} = \frac{15\ 000.0}{220.0} = 68.2$$

- No-load current

$$I_0 = i_{0\%} I_{1n} = 7.0 \times 30 = 2.1 \text{ A}$$

- Short-circuit voltage

$$V_{1sc} = v_{sc\%} V_{1n} = 4.5 \times 500 = 22.5 \text{ V}$$

(b) Core losses and winding losses

- The no-load power is equal to core losses $P_0 = \Delta P_{Fe}$, i.e.,

$$P_0 = p_{0\%} S_n = 1.0 \times 15\ 000 = 150 \text{ W}$$

- The short-circuit power is equal to winding losses $P_{sc} = \Delta P_w$, i.e.,

$$P = p_{sc\%} S_n = 2.5 \times 15\ 000 = 375 \text{ W}$$

(c) Resistances and reactances of equivalent circuit

- Active component of no-load current

$$I_{Fe} = \frac{P_0}{V_{1n}} = \frac{150}{500} = 0.3 \text{ A}$$

- Reactive component of no-load current

$$I_{\Phi} = \sqrt{I_0^2 - I_{Fe}^2} = \sqrt{2.1^2 - 0.3^2} = 2.1 \text{ A}$$

- Reactance in vertical branch of the equivalent circuit (at no-load R_1 and X_1 can be neglected)

$$X_m \approx \frac{V_{1n}}{I_{\Phi}} = \frac{500}{0.3} = 240.6 \text{ } \Omega$$

- Resistance in vertical branch representing core losess

$$R_{Fe} \approx \frac{V_{1n}}{I_{Fe}} = \frac{500.0}{0.3} = 1666.7 \text{ } \Omega$$

- Equivalent series resistance (6.63) and reactance (6.64) in vertical branch

$$R_0 = \frac{1666.7 \times 240.6^2}{1666.7^2 + 240.6^2} = 34.0 \text{ } \Omega$$

$$X_0 = \frac{1666.7^2 \times 240.6}{1666.7^2 + 240.6^2} = 235.7 \text{ } \Omega$$

- Short-circuit resistance

$$R_{sc} = \frac{P_{sc}}{I_{1n}^2} = \frac{375.0}{30.0} = 0.42 \text{ } \Omega$$

- Short-circuit impedance

$$Z_{sc} = \frac{V_{1sc}}{I_{1n}} = \frac{22.5}{30.0} = 0.75 \text{ } \Omega$$

- Short-circuit reactance

$$X_{sh} = \sqrt{Z_{sc}^2 - R_{sc}^2} = \sqrt{0.75^2 - 0.42^2} = 0.62 \text{ } \Omega$$

- Resistances and leakage reactances of primary and secondary windings

$$R_1 = \frac{0.42}{2} = 0.21 \text{ } \Omega \qquad\qquad R_2' = \frac{0.42}{2} = 0.21 \text{ } \Omega$$

$$X_1 = \frac{0.62}{2} = 0.31 \text{ } \Omega \qquad\qquad X_2' = \frac{0.62}{2} = 0.31 \text{ } \Omega$$

2.1.6 Voltage regulation (secondary voltage change)

As the current is drawn through a transformer, the secondary voltage changes because of voltage drop in the internal impedance of the transformer. The secondary voltage change $\Delta v_\%$ is used to identify this characteristic of *voltage change*. It is defined as

$$\Delta v_\% = \frac{V_{20} - V_2}{V_{20}} 100\% = \frac{V'_{20} - V'_2}{V'_{20}} 100\% = \frac{V_{1n} - V'_2}{V_{1n}} 100\% \qquad (2.35)$$

where V_{20} is the secondary voltage at no load and V_2 is the secondary voltage under load. In engineering practice, the *voltage regulation* is calculated on the basis of the following simplified formula:

$$\Delta v_\% \approx i_1 \left(v_{R\%} \cos \varphi_2 \pm v_{X\%} \sin \varphi_2 \right) \qquad (2.36)$$

where

$$i_1 = \frac{I_1}{I_{1n}} \qquad v_{R\%} = \frac{R_{sc}I_{1n}}{V_{1n}} 100\% \qquad v_{X\%} = \frac{X_{sc}I_{1n}}{V_{1n}} 100\% \quad (2.37)$$

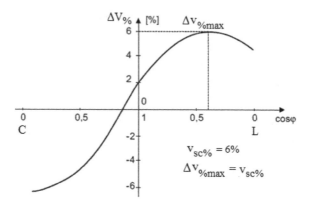

Fig. 2.14. Voltage change characteristic.

The maximum percentage voltage change $\Delta v_{\%max}$ is equal to percentage short-circuit voltage $v_{sc\%}$. The voltage regulation depends on power factor of the load.

To keep the output voltage unchanged, i.e., to adjust it to the required value, turns ratio is changed by means of tap-changing switch. The tap changing switch allows for adjustment of the secondary voltage ±5% of the nominal value (Fig. 2.15).

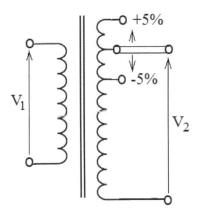

Fig. 2.15. Tap changing switch ±5%.

2.1.7 Efficiency

The *efficiency* of a transformer is the ratio of the output active power P_{out} to the input active power P_{in}, i.e.,

$$\eta = \frac{P_{out}}{P_{in}} \qquad \text{or} \qquad \eta = \frac{P_{out}}{P_{in}} 100\% \qquad (2.38)$$

If the transformer has more than one secondary winding, the output power P_{out} is the power absorbed by all secondary windings.

The nominal (rated) efficiency is defined under nominal operating conditions, power factor $\cos\varphi_2 = 1$ and temperature of windings equal to 75°C (348.2 K).

The efficiency of transformers is usually high: it is the largest efficiency of all electric apparatus and can achieve 99%. In terms of power losses

$$\eta = 1 - \frac{\sum \Delta P}{P_{out} + \sum \Delta P} \qquad (2.39)$$

The *power losses* consist of *basic losses* and *additional losses*. The basic losses are winding losses ΔP_w and core losses ΔP_{Fe}. The additional losses in the windings are due to skin effect and the additional losses in the core are due to stamping and metallurgical processing. The additional losses in metallic parts, such as bolts for tightening beams clamping the yoke, tank, tank cover, etc., are due to stray leakage fluxes.

The basic power losses mainly consist of *core losses* ΔP_{Fe} given by eqn (e1-dpfe3) and *winding losses* ΔP_w, i.e.,

$$\Delta P_w = R_{sc} I_2^2 = \left(\frac{I_2}{I_{2n}} \right)^2 \Delta P_{wn} = i_2^2 \Delta P_{wn} \qquad (2.40)$$

where ΔP_{wn} are winding losses at nominal secondary current I_{2n} (nominal load). Since the ouput power P_{out} is at $V_2 = V_{2n}$

$$P_{out} = V_2 I_2 \cos \varphi_2 = V_2 I_{2n} \cos \varphi_2 = S_n i_2 \cos \varphi_2 \qquad (2.41)$$

where

$$S_n = V_2 I_{2n} \qquad\qquad i_2 = \frac{I_2}{I_{2n}} \qquad (2.42)$$

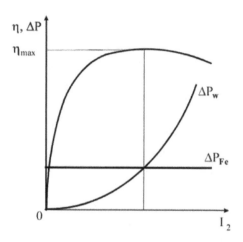

Fig. 2.16. Efficiency and power losses plotted against the secondary current at $f = const.$

The transformer efficiency as a function of i_2, S_n, $\cos \varphi_2$, ΔP_{wn} and ΔP_{Fe}

$$\eta = \frac{P_{out}}{P_{out} + \Delta P_w + \Delta P_{Fe}} = \frac{S_n i_2 \cos \varphi_2}{S_n i_2 \cos \varphi_2 + i_2^2 \Delta P_{wn} + \Delta P_{Fe}} \qquad (2.43)$$

The maximum efficiency is when the core losses ΔP_{Fe} are equal to the winding losses ΔP_w. Performing the first derivative test for maximum of function (2.43), i.e.,

$$\frac{d\eta}{di_2} = 0 \qquad (2.44)$$

at $S_n = const$, $\cos \varphi_2 = const$, $\Delta P_{wn} = const$ and $\Delta P_{Fe} = const$, the maximum efficiency is at

$$\cos \varphi_2 = 1 \qquad\qquad \text{and when} \qquad\qquad i_2^2 \Delta P_{wn} = \Delta P_{Fe} \qquad (2.45)$$

or

$$\Delta P_w = \Delta P_{Fe} \tag{2.46}$$

2.2 Three-phase transformes

2.2.1 Principle of operation and construction

Three-phase transformers are the most popular transformers in power system engineering. A three-phase transformer can be created of three single-phase transformers (Fig. 2.17). Construction of a three-phase transformer using a three-ray star core is impractical, so a typical three-phase transformer has a flat three-leg core (Fig. 2.18). The path for the magnetic flux in the center leg is shorter than in the two remaining side legs.

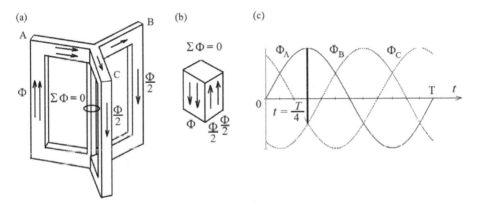

Fig. 2.17. Three single-phase transformers fed from a three-phase balanced (symmetrical) source: (a) symmetrical core with center leg; (b) $\sum \Phi = 0$ in center leg; (c) magnetic fluxes at $t = T/4$.

The input and output powers of a three-phase transformer are

- Input apparent power

$$S_{in} = 3V_1 I_1 \tag{2.47}$$

- Input active power

$$P_{in} = 3V_1 I_1 \cos \varphi_1 \tag{2.48}$$

- Output apparent power

$$S_{out} = 3V_2 I_2 \tag{2.49}$$

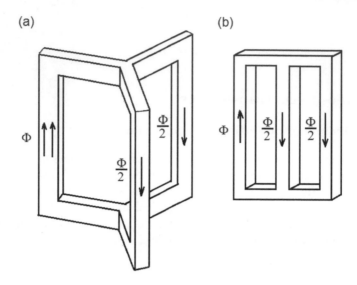

Fig. 2.18. Configuration of three-phase transformer cores: (a) with magnetic symmetry, (b) core of a real three-phase core-type transformer.

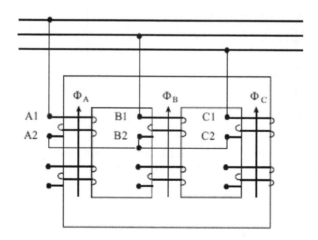

Fig. 2.19. Windings connections in legs of a three-phase transformer.

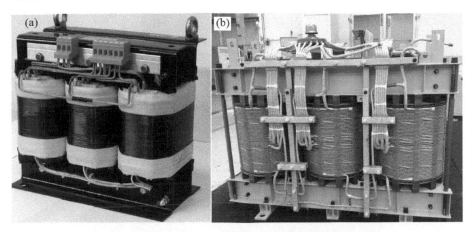

Fig. 2.20. Wound cores of three-phase transformers: (a) dry (air-cooled) separation transformer; (b) oil-cooled core with winding removed from the tank.

- Output active power

$$P_{out} = 3V_2 I_2 \cos\varphi_2 \tag{2.50}$$

where V_1, V_2 are the phase voltages, I_1, I_2 are the phase currents, and $\cos\varphi_1$, $\cos\varphi_2$ are power factors for the primary and secondary windings, respectively. The power losses in the three-phase windings are

- In the primary winding

$$\Delta P_{w1} = 3R_1 I_1^2 \tag{2.51}$$

- In the secondary winding

$$\Delta P_{w2} = 3R_2 I_2^2 = 3R_2'(I_2')^2 \tag{2.52}$$

The connection of primary windings of a three-phase transformer is shown in Fig. 2.19. In this diagram the primary windings are Y-connected; however, they can also be Δ-connected. Similar connections apply to the secondary windings.

Three-phase transformers are shown in Figs 2.20, 2.21, 2.22 and 2.23. Only small three-phase transformers are air-cooled (Fig. 2.20a). Transformers used in power engineering are oil-cooled.

Constructions of tanks of oil-cooled three-phase transformers are shown in Fig. 2.24. There are corrugated tanks (Fig. 2.24a,b), tubular tanks (Fig. 2.24c) and radiator tanks (Fig. 2.24d,e).

Fig. 2.21. Three-phase distribution transformer.

Fig. 2.22. Large three-phase oil cooled power transformer with external air blowers.

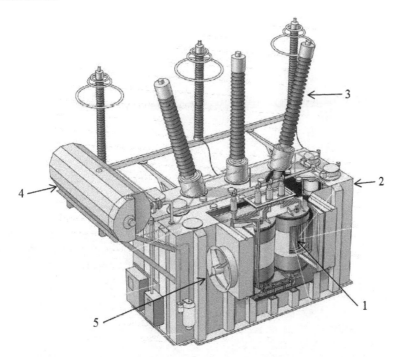

Fig. 2.23. Major components of power transformer: 1 – windings and core, 2 – oil tank, 3 – bushing, 4 – oil conservator, 5 – radiator and fan. Source: www.cbsa-asfc.gc.ca

2.2.2 Name plate

The name plate (Fig. 2.25) of a three-phase transformer usually contains:

- Name or logo of the manufacturer
- Name and type of the product
- Serial number
- Year of manufacture
- Nominal (rated) frequency and number of phases
- Nominal (rated) power (kVA), rated voltages and currents of all windings
- Measured short-circuit voltage
- Measured no-load and short-circuit losses
- Nominal (rated) duty
- Group (and diagram) of winding connection
- Class of insulation
- Degree of protection
- Total mass

Fig. 2.24. Tanks of three-phase oil-cooled transformers: (a), (b) corrugated tanks; (c) tubular tank; (d), (e) radiator tanks. In Fig. 2.24e cooling fans for radiators are visible.

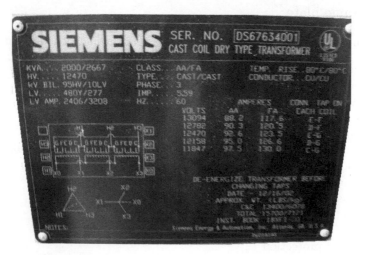

Fig. 2.25. Name plate of a three-phase transformer.

2.2.3 Voltage ratio of three-phase transformers

The voltage ratio of a three-phase transformer is the high voltage (HV) to low voltage (LV) ratio of line-to-line voltages, i.e.,

$$\vartheta_v = \frac{V_{HV}}{V_{LV}} \tag{2.53}$$

Windings of three-phase transformers can be star (Y), delta (Δ) or zig-zag connected. The number after connection symbol, e.g., Dy5 denotes that the line voltages of the primary and secondary windings are shifted by 150° ≡ 5 hours (1 hour = 30°.

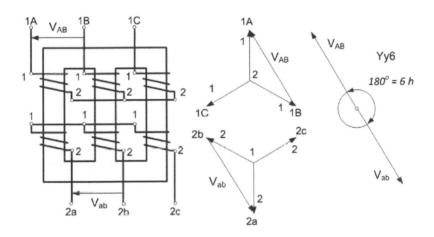

Fig. 2.26. Yy6 connection.

The relationships between voltage ratio and turn ratio for some winding connections are

(a) Yy6 connection (Fig. 2.26)

$$\vartheta_{Yy} = \frac{V_{HV}}{V_{LV}} = \frac{\sqrt{3}\pi\sqrt{2}N_{HV}f\Phi}{\sqrt{3}\pi\sqrt{2}N_{LV}f\Phi} = \frac{N_{HV}}{N_{LV}} = \vartheta \tag{2.54}$$

(b) Dy5 connection (Fig. 2.27)

$$\vartheta_{Dy} = \frac{V_{HV}}{V_{LV}} = \frac{\pi\sqrt{2}N_{HV}f\Phi}{\sqrt{3}\pi\sqrt{2}N_{LV}f\Phi} = \frac{1}{\sqrt{3}}\frac{N_{HV}}{N_{LV}} = \frac{1}{\sqrt{3}}\vartheta \tag{2.55}$$

(c) Yd5 connection (Fig. 2.28)

$$\vartheta_{Yd} = \frac{V_{HV}}{V_{LV}} = \frac{\sqrt{3}\pi\sqrt{2}N_{HV}f\Phi}{\pi\sqrt{2}N_{LV}f\Phi} = \sqrt{3}\frac{N_{HV}}{N_{LV}} = \sqrt{3}\vartheta \tag{2.56}$$

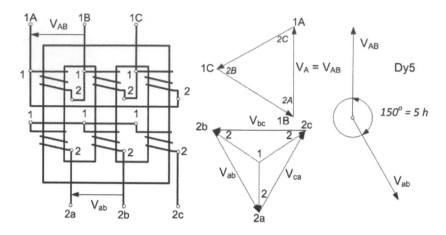

Fig. 2.27. Dy5 connection.

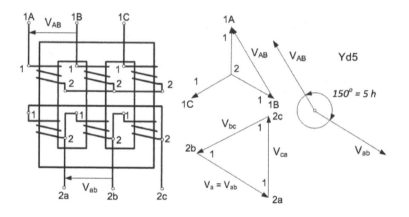

Fig. 2.28. Yd5 connection.

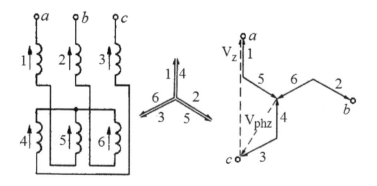

Fig. 2.29. Zig-zag connection.

The phase voltage of a zig-zag connected winding (Fig. 2.29) is smaller than it would be for a Y connection of the same halves of the winding. The phase voltage

$$V_{phz} = 2\frac{V_y}{2}\cos 30° = \frac{\sqrt{3}}{2}V_y \approx 0.86V_y \qquad (2.57)$$

Example 2.3

The nominal power of a three-phase transformer is $S_n = 630$ kVA, cross-section of its center leg $S_{Fe} = 0.024$ m^2, magnetic flux density in legs is $B_{Fe} = 1.58$ T. The primary winding is connected to a 50-Hz utility grid, so the primary line voltage is $V_{HV} = 6.3$ kV. The secondary winding on each leg consits of two halves with $N_{LV}/2 = 14$ turns each so that the secondary voltage can be Wye, Δ or zig-zag connected.
(1) For Dy and Yd connections calculate

(a) Number of turns of the primary (HV) winding and line voltage of the secondary (LV) winding;
(b) Phase currents in primary (HV) and secondary (LV) windings;
(c) Power losses in windings if cross-sections of conductors are $s_{HV} = 10$ mm^2, $s_{LV} = 265$ mm^2, mean lengths of turns (MLT) $l_{HV} = 0.98$ m, $l_{LV} = 0.76$ m and coefficients of additional losses for both windings are $k_{HV} \approx k_{LV} = 1.04$.

(2) How will the secondary (LV) voltage change if the secondary winding is zig-zag connected?
The electric conductivity of copper at 75° is $\sigma = 47 \times 10^6$ S/m.

Solution

The EMF per one turn

$$e_t = \pi\sqrt{2}fB_{Fe}S_{Fe} = \pi\sqrt{2}50 \times 1.58 \times 0.024 = 8.418 \text{ V}$$

(1a) Number of turns of the primary (HV) winding and line voltage of the secondary (LV) winding for Dy and Yd connections

- For Dy connection

$$N_{HV} = \frac{V_{HV}}{e_t} = \frac{6300.0}{8.418} \approx 748$$

$$V_{LV} = N_{HV}\sqrt{3}e_t = 748\sqrt{3} \times 8.418 = 408.0 \text{ V}$$

or

$$V_{LV} = V_{HV} \frac{N_{LV}}{N_{HV}} \sqrt{3} = 6300.0 \frac{28}{748} \sqrt{3} = 408.0 \text{ V}$$

- For Yd connection

$$N_{HVYd} = \frac{V_{HV}}{\sqrt{3} e_t} = \frac{6300.0}{\sqrt{3} \times 748} \approx 432$$

$$V_{LVYd} = N_{LV} e_t = 28 \times 8.418 = 236 \text{ V}$$

or

$$V_{LVYd} = V_{HV} \frac{N_{LV}}{\sqrt{3} N_{HVYd}} = 6300.0 \frac{28}{\sqrt{3} \times 432} = 236.0 \text{ V}$$

(1b) Phase currents in primary (HV) and secondary (LV) windings

- For Dy connection

$$I_{HV} = \frac{S_n}{3 V_{HV}} = \frac{630\,000.0}{3 \times 6300.0} = 33.33 \text{ A}$$

$$I_{LV} = \frac{S_n}{3 V_{LV}/\sqrt{3}} = \frac{630\,000}{3 \times 408.0/\sqrt{3}} = 891.5 \text{ A}$$

- For Yd connection

$$I_{HVYd} = \frac{S_n}{3 V_{HV}/\sqrt{3}} = \frac{630\,000}{3 \times 6300.0/\sqrt{3}} = 57.74 \text{ A}$$

$$I_{LVYd} = \frac{S_n}{3 V_{LVYd}} = \frac{630\,000}{3 \times 236.0} = 889.83 \text{ A}$$

(1c) Power losses in windings

- For Dy connection
  Phase resistance of HV winding

$$R_{HV} = \frac{l_{HV} N_{HV}}{\sigma s_{HV}}$$

$$= \frac{0.98 \times 748}{47 \times 10^6 \times 10.0 \times 10^{-6}} = 1.56 \text{ }\Omega$$

Phase resistance of LV winding

$$R_{LV} = \frac{l_{LV}N_{LV}}{\sigma s_{LV}} = \frac{0.76 \times 28}{47 \times 10^6 \times 265.0 \times 10^{-6}} = 0.00171\ \Omega$$

Losses in HV winding

$$\Delta P_{wHV} = 3I_{HV}^2 R_{HV}k_{HV} = 3 \times 33.3^2 \times 1.56 \times 1.04 = 5458.8\ \text{W}$$

Losses in LV winding

$$\Delta P_{wLV} = 3I_{LV}^2 R_{LV}k_{LV} = 3 \times 891.5^2 \times 0.00171 \times 1.04 = 4277.4\ \text{W}$$

- For Yd connection

  Phase resistance of HV winding

$$R_{HVYd} = \frac{l_{HV}N_{HVYd}}{\sigma s_{HV}} = \frac{0.98 \times 432}{47 \times 10^6 \times 10.0 \times 10^{-6}} = 0.901\ \Omega$$

  Phase resistance of LV winding

$$R_{LVYd} = \frac{l_{LV}N_{LVYd}}{\sigma s_{LV}} = \frac{0.76 \times 28}{47 \times 10^6 \times 265.0 \times 10^{-6}} = 0.00171\ \Omega$$

  Losses in HV winding

$$\Delta P_{wHVYd} = 3I_{HVYd}^2 R_{HVYd}k_{HV} = 3 \times 57.7^2 \times 0.901 \times 1.04 = 9458.0\ \text{W}$$

  Losses in LV winding

$$\Delta P_{wLVYd} = 3I_{LVYd}^2 R_{LVYd}k_{LV} = 3 \times 889.8^2 \times 0.00171 \times 1.04 = 4261.4\ \text{W}$$

(2) How will the secondary (LV) voltage change if the secondary winding is zig-zag connected?

- For Dz (Delta–zig-zg) connection ($V_{HV} = 6300.0$ V, $V_{LV} = 408.0$ V, $N_{HV} = 748$, $N_{LV} = 28$)

  Voltage ratio

$$\vartheta_{Dz} = \frac{V_{HV}}{0.5V_{LV}\sqrt{3}} = \frac{6300.0}{0.5 \times 408.0\sqrt{3}} = 17.83$$

or

$$\vartheta_{Dz} = \frac{2N_{HV}}{3N_{LV}} = \frac{2 \times 748}{3 \times 28} = 17.81$$

  Secondary voltage

$$V_{LV} = \frac{V_{HV}}{\vartheta_{Dz}} = \frac{6300.0}{17.83} = 353.7\ \text{V}$$

- For Yz conenction ($V_{HV} = 6300.0$ V, $V_{LV} = 353.7$ V, $N_{HVYd} = 432$, $N_{LV} = 28$)

  Voltage ratio

$$\vartheta_{Yz} = \frac{V_{HV}}{V_{LV}} = \frac{6300.0}{353.7} = 17.81$$

or

$$\vartheta_{Yz} = \frac{2N_{HVYd}}{\sqrt{3}N_{LV}} = \frac{2 \times 432}{\sqrt{328}} = 17.81$$

Secondary voltage

$$V_{LV} = \frac{V_{HV}}{\vartheta_{Yz}} = \frac{6300.0}{17.81} = 353.6 \text{ V}$$

2.2.4 Parallel operation

Given below are demands put upon the operation of *transformers connected in parallel* (Fig. 2.30), which must be fulfilled to avoid wrong operation at no-load and load conditions:

(a) There must be no currents in the secondary windings at no-load conditions;
(b) The transformers must load themselves accordingly to their rated powers under load;
(c) The phase angles of the secondary line currents of all in parallel connected transformers must be the same.

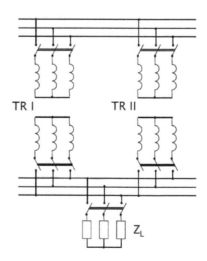

Fig. 2.30. Parallel operation of three-phase transformes.

To meet these demands the transformers must satisfy the following require-ments:

(a) Transformers must have the same voltage ratio ϑ;
(b) The connection group of transformers must be identical;
(c) The rated short-circuit voltages of transformers must be the same;
(d) The ratio of rated apparent powers (S_I/S_{II}) should not exceed $1/3$, i.e.,

$$\frac{S_I}{S_{II}} \leq \frac{1}{3} \tag{2.58}$$

The *goodness factor* of parallel connection is defined as

$$k_g = \frac{S_{par}}{S_{In} + S_{IIn}} \tag{2.59}$$

where S_{par} is the apparent power that can be transformed by two parallel transformers, while at least one of them reaches the nominal (rated) power, S_{In} is the nominal (rated) power of transformer I and S_{IIn} is the nominal (rated) power of transformer II.

2.3 Autotransformer

An *autotransformer* is any single-coil transformer in which the primary and secondary circuits are connected to the same common winding (Fig. 2.31). Part of the primary winding is part of the secondary winding and vice versa. Similar to the two-winding transformer, the turn ratio is defined as

$$\vartheta = \frac{N_1}{N_2} \tag{2.60}$$

and it is approximately equal to the voltage ratio

$$\vartheta = \frac{V_1}{V_2} \tag{2.61}$$

The power is transferred from the primary side to the secondary side in two ways: by *conduction* and *induction* (Fig. 2.32). Ignoring the power losses, the total volt-ampere power is the sum of conduction power S_c and induction power S_i, i.e.,

$$S = S_c + S_i \tag{2.62}$$

where

$$S_c = V_2 I_1 = V_1 I_1 \frac{1}{\vartheta} \tag{2.63}$$

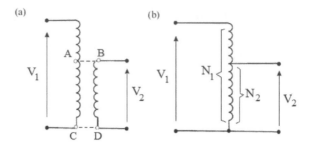

Fig. 2.31. From transformer to autotransformer.

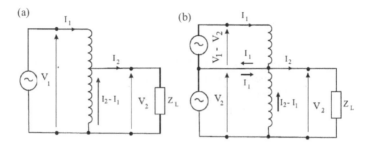

Fig. 2.32. Power transfer in an autotransformer.

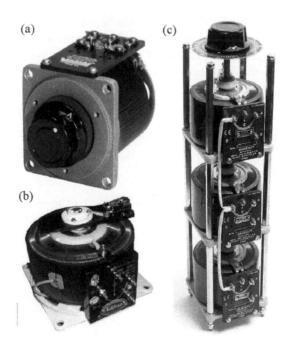

Fig. 2.33. Laboratory variable autotransformers: (a), (b) single-phase; (c) three-phase.

$$S_i = (V_1 - V_2)I_1 = V_1 I_1 \left(1 - \frac{1}{\vartheta}\right) \tag{2.64}$$

Sometimes the power transferred by conduction S_c is called "own power" and the power transferred by induction S_i is called "transitive power". Since $S = V_1 I_1$, the two power components expressed in terms of the total power are

$$S_c = S\frac{1}{\vartheta} \tag{2.65}$$

$$S_i = S\left(1 - \frac{1}{\vartheta}\right) \tag{2.66}$$

Variable autotransformers used in laboratories are shown in Fig. 2.33. Three-phase variable autotransformers consist of three single-phase variable auto-transformers.

2.4 Scott transformer

A *Scott connected transformer* (Fig. 2.34) is a special electromagnetic device used to convert a three-phase system to a two-phase system, or vice-versa. It consists of two single-phase transformers. The first single-phase transformer is called the *main transformer*. It has a center-tapped primary winding connected to the three-phase circuit with the secondary winding connected to the two-phase circuit. The ends of the center-tapped main primary winding are connected to two of the phases of the three-phase circuit.

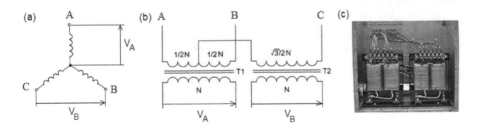

Fig. 2.34. Scott transformer: (a) principle of operation; (b) connection diagram of two single-phase transformers; (c) practical Scott transformer.

The second single phase transformer is called the *teaser transformer*. It has one end of its primary winding connected to the third phase of the three-phase circuit and the other end connected to the center tap of the primary winding of the main. The Scott connection requires no primary neutral connection, so zero-sequence currents are blocked.

In order to neutralize the ampturns of the center tapped primary winding of the main transformer, the two halves are wound on the same core leg. Also the center tapped winding of the main transformer may be interleaved to cancel the magnetic effect of current entering at the center point.

The secondary windings of both the main and teaser transformers are connected to the two-phase circuit. If the main transformer has a turns ratio of 1:1, then the teaser transformer requires a turns ratio of 0.866:1 for balanced operation.

Summary

A *transformer* is a static electromagnetic device for transforming electrical energy in an AC system from one (primary) circuit into another (secondary) circuit at the same frequency but with different values of voltage and current.

A *single-phase transformer* is represented by two electric circuits magnetically coupled (Fig. 1.19).

A typical single-phase transformer consists of two windings, which are coupled electromagnetically by a ferromagnetic core. The winding that absorbs electrical energy is called the *primary winding* and the winding that delivers electrical energy is called the *secondary winding*. The number of secondary windings can be more than one. The winding connected to the circuit with the higher voltage is called the *high voltage winding* (HV) and the winding connected to the circuit with lower voltage is called the *low voltage winding* (LV).

There are two types of single-phase transformes: *shell-type* (Figs 2.2a, 2.3a) and *core type* (Figs 2.2b, 2.3b). Frequently used are nowadays also *C-shaped cores* wound of a ferromagnetic tape (Fig. 2.4).

In an *ideal transformer* the primary winding resistance $R_1 = 0$, the secondary winding resistance $R_2 = 0$, leakage fluxes of the primary and secondary windings $\Phi_{\sigma 1} = 0$, $\Phi_{\sigma 2} = 0$ and core losses $\Delta P_{Fe} = 0$. The phase voltages and phase *rms* EMFs are equal, i.e.,

$$V_1 = E_1 = \pi\sqrt{2}fN_1\Phi_m \qquad\qquad V_2 = E_2 = \pi\sqrt{2}fN_2\Phi_m$$

$$\frac{V_1}{V_2} = \frac{N_1}{N_2} = \vartheta \qquad \frac{V_1}{V_2} = \frac{I_2}{I_1} = \vartheta \qquad V_1 = \vartheta V_2 = V_2'$$

where V_2' is the secondary voltage referred to as the primary side.

In a *real transformer* the primary R_1 and secondary R_2 resistances, leakage fluxes $\Phi_{\sigma 1}$, $\Phi_{\sigma 2}$ and core losses ΔP_{Fe} are taken into account. Eqns (2.11) and (2.12) are equations of balance of MMFs. The exciting (no-load) current I_0 has two components: a real component I_{Fe}, which is the core loss current and imaginary component I_Φ, which is the magnetizing current. The equivalent circuits and phasor diagram of a real transformer are shown in Figs 2.8 and

2.9. The secondary voltage V_2', EMF E_2' and current I_2' referred to as the primary side are

$$V_2' = \vartheta V_2 \qquad\qquad E_2' = \vartheta E_2 \qquad\qquad I_2' = \frac{1}{\vartheta} I_2$$

while the secondary resistance R_2' and secondary leakage reactance X_2' referred to as the primary winding are calculated by multiplying R_2 and X_2 by the voltage ratio ϑ^2 square, i.e.,

$$R_2' = \vartheta^2 R_2 \qquad\qquad\qquad X_2' = \vartheta^2 X_2$$

The purpose of the *open circuit test* is to find

- Core losses ΔP_{Fe};
- Resistance R_{Fe} and reactance X_m of the vertical branch of the equivalent circuit (Fig. 2.9a);
- Magnetizing current I_Φ and core loss current I_{Fe};
- Voltage ratio ϑ.

The measured quantities are the input voltage V_1, no-load current I_0, secondary voltage V_2, and no-load power P_0. The no-load current $i_{0\%}$ is usually expressed as a percentage of the nominal (rated) current I_{1n} of the transformer, as given by eqn (2.27). In power transformer, the *no-load current* is in the interval $1.0 \le i_{0\%} \le 10.0\%$ of the nominal (rated) current I_{1n}. The higher the power, the lower the no-load current.

The purpose of a *short-circuit* test on a transformer is to find:

- Winding resistances R_1 and R_2;
- Leakage reactances X_1 and X_2;
- Winding losses ΔP_{w1} and ΔP_{w2}.

The measured quantities are the input voltage V_1, short circuit current I_{sc} and short-circuit losses ΔP_{sc}. In the short-circuit test the secondary winding is shorted ($V_2 = 0$), while the primary winding is fed with a reduced voltage to obtain nominal currents in the primary and secondary windings, i.e., $I_{1sc} = I_{1n}$ and $I_{2sc} = I_{2n}$. The short-circuit voltage is such a voltage across the primary winding terminals at which the primary and secondary currents take rated values, while the secondary winding is short-circuited. The *short-circuit voltage* $v_{sc\%}$ is given on the name plate of a transformer and it is expressed as a percentage of the nominal (rated) voltage V_{1n}, as given by eqn (2.28). For typical power transformers the short-circuit voltage $v_{sc\%}$ is in the range $3 \le v_{sc\%} \le 15\%$ of the nominal voltage V_{1n}. The higher the power, the higher the short-circuit voltage $v_{sc\%}$.

The *secondary voltage change* or *voltage regulation* $\Delta v_{\%}$ is used to identify the behavior of voltage change as the current is drawn through the transformer. It is defined by eqn (2.35). The maximum percentage voltage change is equal to a percentage of the short-circuit voltage. The voltage regulation depends on power factor of the load. To keep the output voltage unchanged, i.e., to adjust it to the required value, the turns ratio is changed by means of a tap-changing switch. The *tap-changing switch* allows for adjustment of the secondary voltage ±5% of the rated value.

The *power losses* consist of *basic losses* and *additional losses*. The basic losses are winding losses ΔP_w and core losses ΔP_{Fe}. The additional losses in the windings are due to skin effect and additional losses in the core are due to stamping and metallurgical processing. The additional losses in metallic parts, such as bolts for tightening beams clamping the yoke, tank, tank cover, etc., are due to stray leakage fluxes. The efficiency of a transformer is defined as the output power to input power ratio. The *maximum efficiency* is when the winding losses ΔP_w are equal to the core losses ΔP_{Fe}. The efficiency of transformers is usually high: it is the largest effciency of all electric apparatus and can achieve 99%.

Three-phase transformers are the most popular transformers in power system engineering. A typical three-phase transformer has a flat three-leg core (Fig. 2.18b). The path for the magnetic flux in the center leg is shorter than in side legs. The input and output powers of a three-phase transformer are expressed by eqns (2.47) to (2.50). Only small three-phase transformers are air-cooled. Transformes used in power electrical engineering are oil-cooled.

The *name plate* (Fig. 2.25) of a three-phase transformer usually contains:

- Name or logo of the manufacturer
- Name and type of the product
- Serial number
- Year of manufacture
- Rated frequency and number of phases
- Rated power (kVA), rated voltages and currents of all windings
- Measured short-circuit voltage
- Measured no-load and short-circuit losses
- Rated duty
- Group (and diagram) of winding connection
- Class of insulation
- Degree of protection
- Total mass

Windings of three-phase transformers can be star (Y), delta (Δ or D) or zig-zag connected. The *voltage ratio of a three-phase transformer* is the high voltage (HV) to low voltage (LV) ratio of line-to-line voltages, i.e.,

$$\vartheta_v = \frac{V_{HV}}{V_{LV}}$$

The relationships between voltage ratio and turn ratio for some winding connections are given by eqns (2.54) to (2.57).

The following conditions must be met to *connect transformers in parallel* (Fig. 2.30) in order to avoid wrong operation:

(a) There must be no currents in the secondary windings at no-load conditions;
(b) The transformers must load themselves according to their rated powers under load;
(c) The phase angles of the secondary line currents of all in parallel connected transformers must be the same.

To meet the above demands the transformers must satisfy the following requirements:

(a) Transformers must have the same voltage ratio ϑ;
(b) The connection group of transformers must be identical;
(c) The rated short-circuit voltages of transformers must be the same;
(d) The ratio of rated apparent powers ($S_I = S_{II}$) should not exceed 1/3.

An *autotransformer* is any single-coil transformer in which the primary and secondary circuits are connected to the same common winding (Fig. 2.31). Part of the primary winding is part of the secondary winding and vice-versa. The power is transferred from the primary side to the secondary side in two ways: by conduction and induction (Fig. 2.32). Ignoring the power losses, the total apparent power is the sum of conduction power S_c and induction power S_i, i.e., $S = S_c + S_i$. The *conduction power* is given by eqn (2.65) and the induction power is given by eqn (2.66). Variable autotransformers are used in laboratories.

A *Scott transformer* (Fig. 2.34) is a special electromagnetic device used to convert a three-phase system to a two-phase system, or vice-versa. It consists of two single-phase transformers. The first single-phase transformer is called the *main transformer*. It has a center-tapped primary winding connected to the three-phase circuit with the secondary winding connected to the two-phase circuit. The ends of the center-tapped main primary winding are connected to two of the phases of the three-phase circuit. The second single-phase transformer is called the *teaser transformer*. It has one end of its primary winding connected to the third phase of the three-phase circuit and the other end connected to the center tap of the primary winding of the main. The Scott connection requires no primary neutral connection, so zero-sequence currents are blocked. The secondary windings of both the main and teaser transformers are connected to the two-phase circuit.

Problems

1. Calculate the no-load parameters of a three-phase transformer with the following nominal parameters: apparent power $S_n = 2.0$ MVA, frequency $f_n = 50$ Hz, primary voltage (HV) $V_{1n} = 30$ kV, no-load current $i_{0\%} =$

5.5%. The winding connection is YD11 and the no load power is $P_0 = 8.6$ kW.

Answer: $I_{10} = 2.12$ A, $I_\Phi = 2.11$ A, $I_{Fe} = 0.17$ A, $R_{Fe} = 104.6$ kΩ, $X_0 = 8.2$ kΩ.

2. Calculate the short-circuit parameters of a three-phase transformer with the following nominal parameters: apparent power $S_n = 2.4$ MVA, $f_n = 50$ Hz, primary voltage (HV) $V_{1n} = 35$ kV, no-load current $i_{0\%} = 5.0\%$, short-circuit voltage $v_{sc\%} = 6.5\%$. The winding connection is Yd11 and the short-circuit power is $P_{sc} = 26$ kW.

Answer: $Z_{sh} = 33.18$ Ω, $R_1 = 2.76$ Ω, $X_1 = 16.36$ Ω, $R_2' = 2.76$ Ω, $X_2' = 16.36$ Ω.

3. The parameters of the equivalent circuit of a 150-kVA, 2400-V/240-V transformers are: $R_1 = 0.2$ Ω, $R_2 = 0.002$ Ω, $X_1 = 0.45$ Ω, $R_{Fe} = 10$ kΩ, $X_m = 1555$ Ω. On the basis of the equivalent circuit shown in Fig. 2.9 calculate: voltage regulation $\Delta v_\%$; efficiency of the transformer operating at rated load with 0.8 lagging power factor.

Answer: (a) $V_{10} = 2454.7$ V, $V_2' = 2400$ V, $\Delta v_\% = 2.23\%$; (b) $\eta = 98.2\%$.

4. A single-phase transformer has the following nominal parameters: apparent power $S_n = 15$ kVA, primary voltage $V_{1n} = 600$ V, secondary voltage $V_{2n} = 240$ V, no-load power $p_\% = 1.2\%$, no-load current $i_{0\%} = 6.2\%$, short-circuit power $p_{sc\%} = 3.5\%$, short-circuit voltage $v_{sc\%} = 5.5\%$, frequency $f_n = 50$ Hz. The hysteresis losses $\Delta P_h = 3\Delta P_e$, where ΔP_e are eddy-current losses. Calculate: (a) parameters of the equivalent circuit; (b) no-load losses, no-load current and power factor when the primary winding is fed with the voltage of 720 V and frequency 60 Hz.

Answer: (a) $R_1 = 0.42$ Ω, $X_1 = 0.51$ Ω, $R_2' = 0.42$ Ω, $X_2' = 0.51$ Ω, $R_{Fe} = 2.0$ kΩ, $X_m = 394.6$ Ω; (b) $\Delta P_{060} = 226.8$ W, $I_{060} = 1.55$ A, $\cos\phi_0 = 0.203$.

5. A single-phase transformer has the following nominal parameters: apparent power $S_n = 25$ kVA, primary voltage $V_{1n} = 540$ V, secondary voltage $V_{2n} = 115$ V, frequency $f_n = 60$ Hz, no-load power $p_{0\%} = 1.1\%$, short-circuit power $p_{sc\%} = 3.5\%$, no-load current $i_{0\%} = 5.5\%$, short-circuit voltage $v_{sc\%} = 4.0\%$. The hysteresis losses $\Delta P_h = 2.5\Delta P_e$ where ΔP_e are eddy-current losses. Find:

(a) Parameters of equivalent circuit.

(b) If the transformer is fed with the voltage $V_1 = 380$ V at frequency $f_1 = 50$ Hz, calculate the primary current I_1 at which the total winding and core losses are the same as under nominal conditions.

(c) If the copper windings are replaced with aluminum windings with the same number of turns and cross sections, calculate the new nominal current assuming the power losses in windings are the same, $\sigma_{Cu} = 57 \times 10^6$ S/m, $\sigma_{Al} = 32. \times 10^6$ S/m.

Answer: (a) $R_1 = 0.2$ Ω, $X_1 = 0.11$ Ω, $R'_2 = 0.2$ Ω, $X'_2 = 0.11$ Ω, $R_{Fe} = 1060.4$ Ω, $X_m = 216.4$ Ω; (b) $\Delta P_{h60} = 162$ W, $\Delta P_{e60} = 136.2$ W, $I_1 = 49.8$ A; (c) $I_{1nAl} = 34.7$ A.

6. A single-phase transformer has the following nominal parameters: apparent power $S_n = 45$ kVA, primary voltage $V_{1n} = 460$ V, secondary voltage $V_{2n} = 6300$ V, frequency $f_n = 60$ Hz. Calculated voltage drops across resistance and leakage reactances of windings at nominal current are: voltage drop across primary winding resistance $V_{R1n} = 8.0$ V, voltage drop across primary winding leakage reactance $V_{X1n} = 18.2$ V, voltage drop across secondary winding resistance $V_{R2n} = 102.0$ V, voltage drop across secondary leakage reactance $V_{X2n} = 358.0$ V. When the primary winding is fed with nominal voltage V_{1n} at open secondary winding terminals, the current and no load power are $I_0 = 9.2$ V, $P_0 = 384$ V, respectively.
For the transformer fed with nominal voltage V_{1n} from the primary side at open secondary terminals, calculate:
(a) Voltage drops across resistances and leakage reactances of windings;
(b) EMF E_1 induced in the primary winding;
(c) Percentage difference between the terminal voltage and SEM induced in the primary winding.

Answer: (a) $V_{R1} = 0.752$ V, $V_{X1} - 1.712$ V, $V_{R2} = V_{X2} = 0$; (b) $E_1 = 458.2$ V; (c) 0.387 V.

7. A single-phase transformer has the following nominal parameters: apparent power $S_n = 20$ kVA, primary voltage $V_{1n} = 630$ V, secondary voltage $V_{2n} = 230$ V, frequency $f_n = 50$ Hz, no-load power $p_{0\%} = 0.9\%$, short-circuit power $p_{sc\%} = 3.2\%$, no-load current $i_{0\%} = 6.1\%$, short-circuit voltage $v_{sc\%} = 5.5\%$. Calculate:
(a) Resistances and reactances of equivalent circuit and input voltage to obtain maximum efficiency at primary current $I_1 = 14.5$ A;
(b) The input voltage, if the transformer operates at short circuit and the short circuit current $I_{sc} = I_{1n}$ and half of the primary winding is fed with frequency of $f_{60} = 60$ Hz.

Answer: (a) $R_1 = 0.318$ Ω, $R'_2 = 0.318$ Ω, $X_1 = 0.444$ Ω, $R_{Fe} = 2205$ Ω, $X_m = 328.9$ Ω, $V_1 = 380$ V; (b) $V_{sc60} = 14.76$ V.

8. A three-phase transformer has the following nominal parameters: $S_n = 10$ MVA, high voltage $V_{1n} = 220$ kV, low voltage $V_{2n} = 60$ kV, frequency $f_n = 50$ Hz, connection Yy0, winding losses $\Delta P_w = 60$ kW, core losses

$\Delta P_{Fe} = 28$ kW, short circuit voltage $v_{sc\%} = 7.5\%$, no load current $i_{0\%} = 3.0\%$. Calculate: (a) per unit parameters of equivalent circuit; (b) equivalent circuit parameters referred to HV winding.

Answer: (a) $v_{Rsc\%} = 0.6\%$, $v_{Xsc\%} = 7.476\%$; (b) $R_{sc} = 29.04$ Ω, $X_{sc} = 361.84$ Ω, $R'_2 = 0.5R_{sc} = 14.52$ Ω, $X'_2 = 0.5X_{sc} = 188.922$ Ω.

9. Two three-phase transformers A and B have the following nominal parameters: apparent power $S_{nA} = 3.0$ MVA, $S_{nB} = 2.5$ MVA, frequency $f_n = 50$ Hz, primary (HV) voltage $V_{1n} = 35$ kV, secondary (LV) voltage $V_{2n} = 6600$ V, short circuit voltage $v_{sh\%A} = 5.0\%$, $v_{sh\%B} = 6.5\%$, winding losses $\Delta P_{wA} = 26$ kW, $\Delta P_{wB} = 31$ kW, connection YD11 (both transformers) and operate in parallel. The HV bus bar has the voltage $V_1 = 30$ kV and frequency $f = 50$ Hz. Find:

   (a) Distribution of load between transformers, if the LV bus bars are loaded with current corresponding to arithmetic sum of nominal powers of both transformers at power factor of loads (secondary winding) $\cos \varphi_2 = 0.8$ lagging;

   (b) Maximum power the system can deliver without overloading the transformers;

   (c) Equalizing current I_{2e} when the primary winding tap changer -5% of transformer B and nominal tap changer of transformer A are conencted to HV bus bars.

   Answer: (a) $I_A = 293.2$ A, $I_B = 187.9$ A, $I_A/I_{nA} = 1.12$, $I_B/I_{nB} = 0.86$; (b) $S_{max} = 4.923$ MVA; (c) $I_{2e} = 102.5$ A.

10. In a single-phase autotransformer the total number of turns is $N_1 = 420$. The primary winding is connected across the voltage of $V_1 = 400$ V. It has been measured that the secondary voltage is $V_2 = 115$ V and the secondary current is $I_2 = 8.5$ A. Calculate the primary current I_1, conduction power S_c, induction power S_i and total apparent power S.

   Answer: $I_1 = 2.44$ A, $S_c = 281$ W, $S_i = 696.5$ W, $S = 977.5$ W.

3

SWITCHED-RELUCTANCE MACHINES

A *switched reluctance machine* (SRM) is the simplest electrical machine. The technology of SRMs was derived from variable reluctance (VR) stepping motors. A SRM was first used probably by Robert Davidson, a Scotish inventor, as a traction motor for an electric locomotive.

The first reference to the term SRM was made by S.A. Nasar in 1969 [31]. The "re-invention" of SRMs in the 1980s was possible due to progress in solid state switches. Nowadays, the SRM drive has become the commercial adjustable speed motor drive due to simple construction, low mass, potentially low manufacture cost, high efficiency, excellent torque-speed characteristics and operation in harsh environment.

3.1 What is a switched reluctance machine?

The *switched reluctance machine* (SRM) is a doubly-salient, singly-excited electrical machine. The stator and rotor have salient poles and the winding excited with current is only in the stator. The stator winding is usually a polyphase winding.

In an SRM the electromagnetic torque is produced by the magnetic attraction of a steel rotor to stator electromagnets. The normal component of the attraction force between the stator electromagnet (pole) and the rotor pole can be expressed as

$$F \approx \mu_0 \frac{(Ni)^2}{4g^2} S_p \tag{3.1}$$

where N is the number of turns per pole, i is the stator current, S_p is the overlaping area of the stator and rotor poles, $\mu_0 = 0.4\pi \times 10^{-6}$ H/m is the magnetic permeability of free space, and g is the air gap between the stator and rotor pole shoe.

No permanent magnets (PMs) are needed, and the rotor carries no windings. If the stator has, say, 12 poles, and rotor has, say, 8 poles, the SRM is shortly described as 12-8 SRM. In a SRM with 12 stator poles the stator can be designed as a 2-phase (6 poles per phase), 3-phase (4 poles per phase), or 6-phase (2-poles per phase). The least number of stator poles is 2.

The rotor position relative to the stator is detected using a simple hardware sensor or by electronic sensorless means. The controller then energizes each stator winding only when it can produce useful torque. By suitable timing of the stator excitation, the machine can operate as a motor or generator.

The fundamental frequency in one phase is

$$f = nN_r \tag{3.2}$$

where n is the speed in rev/s and N_r is the number of rotor poles. The fundamental frequency does not depend on the number of stator poles N_s.

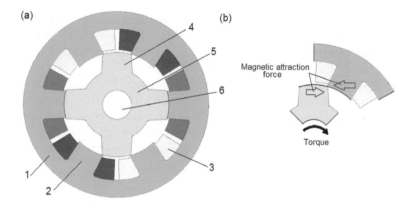

Fig. 3.1. Switched reluctance machine: (a) 6-stator pole, 4-rotor pole machine; (b) principle of operation. 1 – stator core, 2 – stator pole, 3 – stator winding, 4 – rotor pole, 5 – rotor core, 6 – shaft.

The unique features of SRMs are

- An SRM requires a controllable solid state converter and cannot be operated directly from a utility grid.
- Mutual inductance between stator phases is negligible. Each phase is electromagnetically independent of other phases. Owing to this feature, a short circuit fault in one phase winding has no effect on other phases. Operation of other healthy phases is possible. This is a feature unique to SRMs only.

- Since the current only needs to be unidirectional for all quadrants of operation, all power converter configurations have a switch (switches) in series with a stator phase winding facing the DC source voltage. This is also a distinctive feature of SRMs.
- No rotor excitation system (no PMs, no winding).

Stepping motors with reactive rotors may have similar constructions as SRMs. However, the fundamental difference between the stepping motor and the SRM is that the first one operates in an open control loop while the SRM operates in a closed control loop with the rotor position feedback.

3.2 Construction

Fig. 3.2 shows how a three-phase 6-4 SRM is designed. The stator has $N_s = 6$ poles (2 poles per phase) and the rotor has $N_r = 4$ poles. The coils of two poles belonging to the same phase can be either series or parallel connected creating a concentrated-parameter phase winding. Since the coils are wound on salient poles, the end turns are short. Short end turns means low leakage flux about the end turns and low winding losses. Both the stator and rotor cores are laminated to reduce the pulsating losses in pole shoes and simplify the manufacturing process.

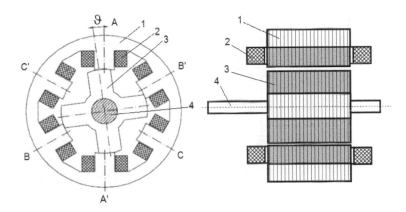

Fig. 3.2. Construction of a three-phase, 6-4 SRM. 1 – stator laminated stack, 2 – stator winding, 3 – rotor laminated stack, 4 – shaft.

The magnetic flux lines and magnetic flux density distribution in a 6-4 SRM are plotted in Fig. 3.3. The poles with the highest magnetic flux density carry the winding of the phase being energized.

Fig. 3.4 shows the power circuit of three-phase 6-4 SRM operating as a motor. The three-phase solid state converter contains six solid state switches, e.g., IGBTs [1]. The windings of the SRM are between upper and lower switches and connected in series with them. The six diodes are necessary to provide a return path for return current from phases being disconneted. The solid state converter is fed from a DC source with the voltage V_{dc}. The phase A in Fig.

---

[1] Integrated gate bipolar transistor.

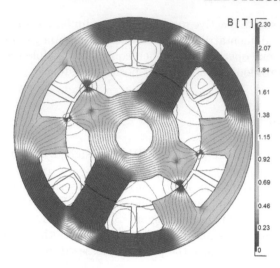

Fig. 3.3. Magnetic flux lines and magnetic flux density distribution in a 6-4 SRM as obtained from the 2D FEM.

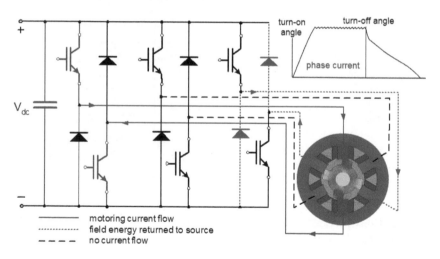

Fig. 3.4. Power circuit of a three-phase 6-4 SRM operating as a motor.

3.4 is energized. The current has been switched-off from the phase C and this phase winding returns the stored field energy (inductance of the winding) to the DC source.

The winding diagram of a three-phase 12-8 SRM is sketched in Fig. 3.5. Dots denote beginnings of coils. Groups of two coils, e.g., A1 and A2 can be connected in series, parallel or bifilarly. For two channel design, the set of coils A1,B1,C1 creates one channel, and the set of coils A2,B2,C2 creates the second channel. Two-channel architecture means that the winding of each

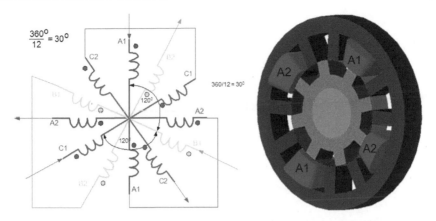

Fig. 3.5. Winding diagram of a three-phase 12-8 SRM. Dots denote beginnings of coils.

phase is doubled. This solution is used in applications where high reliability is the primary requirement, e.g., aircraft electric systems.

3.3 Aligned and unaligned positions

For the primitive two-pole reluctance machine as shown in Fig. 3.6, the *aligned* and *unaligned* positions are characterized by the properties summarized in Table 3.1.

Table 3.1. Properties in aligned and unaligned positions [27].

Aligned	Unaligned
$\vartheta = 0$ or $\vartheta = 180°$	$\vartheta = \pm 90°$
Maximum inductance	Minimum inductance
Magnetic circuit liable to saturate	Magnetic circuit unlikely to saturate
Zero torque: stable equilibrium	Zero torque: unstable equilibrium

The number of stator poles usually exceeds the number of rotor poles. Table 3.2 shows typical numbers of stator and rotor poles for three-phase SRMs.

The typical number of poles N_s, N_r, phases m and corresponding aligned positions n_a are shown in Fig. 3.6. For $N_s = 12$ and $m = 3$ or $m = 6$ the number of rotor poles $N_r = 8$ or $N_r = 10$. The number of rotor poles $N_r = 8$ is recommended because the number of aligned positions is greater, i.e., $n_a = 4$ instead of $n_a = 2$ for $N_r = 10$. The greater the number of aligned position

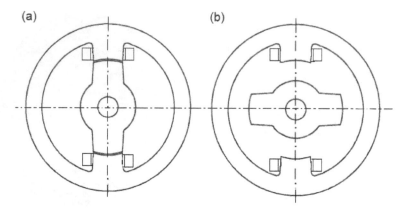

Fig. 3.6. Two-pole SRM: (a) stator and rotor aligned position; (b) stator and rotor unaligned position.

Table 3.2. Typical numbers of stator and rotor poles for SRMs.

Number of phases	Number of stator poles N_s	Number of rotor poles N_r
3	6	4
4	8	6
5	10	8
3 or 6	12	8 or 10
3	18	16, 14 or 12

n_a, the higher the electromagnetic torque developed by the SRM. The general rule for the number of poles is

$$n_a = GCD(N_s, \ N_r) \tag{3.3}$$

The number of aligned positions must be equal to the greatest common divisor GCD of the number of stator poles N_s and rotor poles N_r, e.g., $GCD(6,4) = 2$, $GCD(8,6) = 2$, $GCD(12,8) = 4$, $GCD(12,10) = 2$, $GCD(48,36) = 12$, etc. For most SRMs

$$\frac{N_s}{GCD(N_s, \ N_r)} = \frac{N_s}{n_a} = km, \qquad k = 1, 2, 3, \ldots \tag{3.4}$$

Also

$$\frac{N_s}{m} = 2k, \qquad k = 1, 2, 3, \ldots \tag{3.5}$$

where $k = N_s/2$ is the number of stator pole pairs.

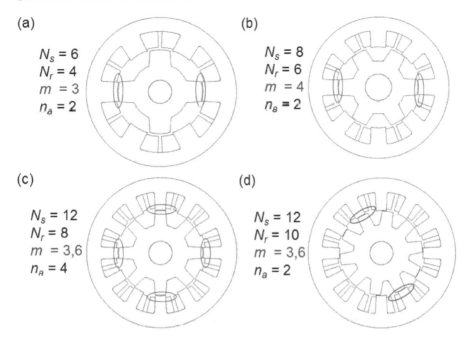

Fig. 3.7. Typical numbers of poles and phase in practical SRMs: (a) $N_s = 6$, $N_r = 4$, $m = 3$, $n_a = 2$; (b) $N_s = 8$, $N_r = 6$, $m = 4$, $n_a = 2$; (c) $N_s = 12$, $N_r = 8$, $m = 3, 6$, $n_a = 4$; (d) $N_s = 12$, $N_r = 10$, $m = 3, 6$, $n_a = 2$.

3.4 Electromagnetic torque

Fig. 3.8 shows graphs of inductance and electromagnetic torque as functions of the rotor position with respect to the stator [27]. The stator coil current is constant. When the rotor is in position U, and the current is switched on, the stator pole starts attracting the rotor pole. The inductance increases from its minimum value at point J until reaching maximum value in fully aligned position A. The corresponding linkage flux ψ is plotted in Fig. 3.9 [27]. The SRM produces attraction force and positive electromagnetic torque. This is motoring operation. The torque changes direction at the aligned position A. If the rotor continues past A, the attractive force between the poles produces a retarding (braking) torque. If the machine rotates with constant current in the coil, the negative and positive torque impulses cancel, and therefore the average torque over a complete cycle is zero. To eliminate the negative torque impulses, the current must be switched off while the poles are separating, i.e., during the intervals AK (Fig. 3.9).

The angle expressed in electrical degrees at which the current is switched-on to the given phase winding is called *turn-on angle*. The angle at which the current is switched-off from the given phase winding is called *turn-off angle*. Turn-on and turn-off angles, in general, are not the same as unaligned and

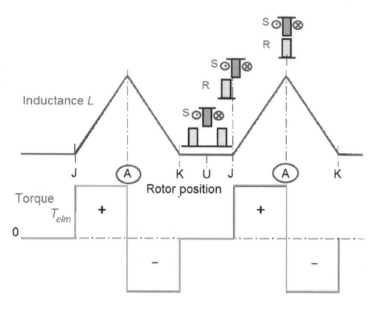

Fig. 3.8. Variation of inductance, electromagnetic torque and linkeage flux with rotor position. The small icons show the relative positions of the rotor and stator poles, with the rotor moving to the right. A – aligned position; U – unaligned position; J – start of overlap; K – end of overlap [27].

aligned positions. In practice, both the current and torque waveforms are not rectangular and distorted both by the inductance, commutation of phase and PWM modulation (Fig. 3.10).

The production of continuous unidirectional torque requires more than one phase, such that the gaps in the torque waveform are filled in by torques produced by currents flowing in the other phases (Fig. 3.11).

The ideal current waveform is therefore a series of pulses synchronized with the rising inductance intervals. The ideal torque waveform has the same waveform as the current.

The *stroke* is the cycle of torque production associated with one current pulse, i.e.,

$$\alpha_s = \frac{360^0}{mN_r} = \frac{360°}{S} \tag{3.6}$$

For m stator phases, there are $S = mN_r$ steps per revolution and the *step angle* or *stroke* is expressed by eqn (3.6). The maximum *torque zone*, i.e., the angle through which one phase can produce the torque is inversely proportional to the number of rotor poles

$$\tau_{max} = \frac{180^0}{N_r} \tag{3.7}$$

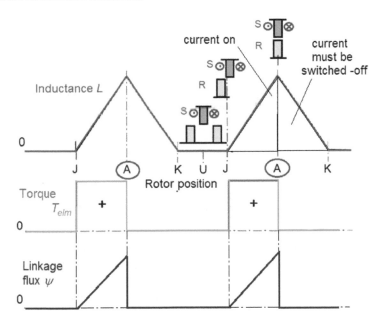

Fig. 3.9. Operation of an SRM machine as a motor. The attraction force and positive electromagnetic torque is in the interval JA. In the interval AK the attraction force retards the rotor, the electromagnetic troque is negative (braking, so the current must be switched off) [27].

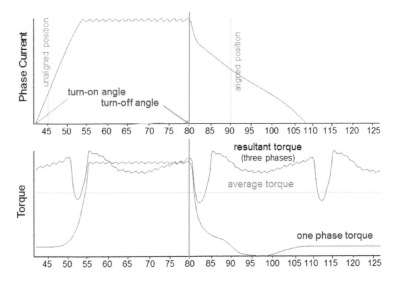

Fig. 3.10. Practical current and torque waveforms.

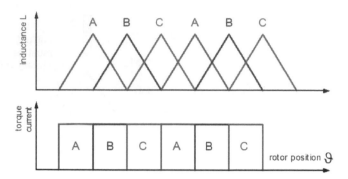

Fig. 3.11. Continuous torque production in a three-phase machine.

The smaller the number of rotor poles N_r, the larger the maximum torque zone. The maximum torque zone is affected by the saturation of magnetic circuit.

The terminal voltage equation for one stator phase

$$v = Ri + \frac{d\psi}{dt} = Ri + \Omega\frac{d\psi}{\vartheta} \qquad (3.8)$$

where $\Omega = d\vartheta/dt$. Suppose that the flux-linkage ψ is a function of both current i and rotor angle ϑ, i.e., $\psi = \psi(i, \vartheta)$. Then

$$\frac{d\psi}{dt} = \frac{\partial\psi}{\partial i}\frac{di}{dt} + \Omega\frac{\partial\psi}{\partial\vartheta} = L\frac{di}{dt} + e \qquad (3.9)$$

where $\psi = Li$. Thus, the terminal voltage equation takes the following form:

$$v = Ri + \frac{\partial\psi}{\partial i}\frac{di}{dt} + \Omega\frac{\partial\psi}{\partial\vartheta} = Ri + L\frac{di}{dt} + e \qquad (3.10)$$

where the EMF

$$e = \Omega\frac{\partial\psi}{\partial\vartheta} = \Omega i\frac{\partial L}{\partial\vartheta} \qquad (3.11)$$

Physically, $\partial L/\partial\vartheta$ is the slope of the inductance graph (Fig. 3.9). Multiplying both sides of voltage equation (3.10) by the current i

$$vi = Ri^2 + Li\frac{di}{dt} + \Omega i^2\frac{dL}{d\vartheta} \qquad (3.12)$$

The product vi is the input power, the first term Ri^2 on the right-hand side represents the resistive losses per phase in the stator winding, the second term $Li\,di/dt$ is the power corresponding to the magnetic energy stored in the inductance L and the third term $\Omega i^2 dL/d\vartheta$ is physically uninterpretable. Some researchers use the change of magnetic stored energy at any instant, i.e.,

$$\frac{d}{dt}\left(\frac{1}{2}Li^2\right) = \frac{1}{2}i^2\frac{dL}{dt} + Li\frac{di}{dt}$$

$$\frac{1}{2}i^2\frac{d\vartheta}{dt}\frac{dL}{d\vartheta} + Li\frac{di}{dt} = \frac{1}{2}i^2\Omega\frac{dL}{d\vartheta} + Li\frac{di}{dt} \qquad (3.13)$$

According to the law of conservation of energy, the mechanical power conversion

$$P = \Omega T_{elm} \qquad (3.14)$$

is what is left after the resistive loss Ri^2 and rate of change $d(0.5Li^2)/dt$ of magnetic stored energy are subtracted from the power input vi, i.e.,

$$P = vi - Ri^2 - \frac{d}{dt}\left(\frac{1}{2}Li^2\right) = \Omega\frac{1}{2}i^2\frac{dL}{d\vartheta} \qquad (3.15)$$

Thus, from eqns (3.14) and (3.15) the electromagnetic torque is

$$T_{elm} = \frac{P}{\Omega} = \frac{1}{2}i^2\frac{dL}{d\vartheta} \qquad (3.16)$$

The electromagnetic torque is proportional to the square of the current, hence the current can be unipolar to produce unidirectional torque. It is impossible to change the direction of rotation of an SRM by changing the direction of current. The direction of rotation can only be changed by changing the sequence of phases.

3.5 Electromagnetic torque derived from coenergy

The electromagnetic torque can also be found on the basis of magnetic coenergy. The magnetic coenergy $w'(i)$ (second energy-state function) of an inductive element is the area below the curve $\psi(i)$ in Fig. 1.23, i.e.,

$$W' = \int_0^i \psi di \qquad (3.17)$$

The electromagnetic torque in terms of coenergy is defined as the first partial derivative of the coenergy w' with respect to the angle ϑ

$$T_{elm} = \left[\frac{\partial w'}{\partial \vartheta}\right]_{i=const} = -\left[\frac{\partial w}{\partial \vartheta}\right]_{\psi=const} \qquad (3.18)$$

The energy stored in a coil (solenoid)

$$W(\psi) = \int_0^\psi i(\psi)d\psi = \frac{1}{2}\psi \qquad (3.19)$$

This is the area above the curve $\psi(i)$. Putting $\psi = Li$, the energy is $W = 0.5Li^2$ and the electromagnetic torque

$$T_{elm} = \frac{1}{2}i^2\frac{dL}{d\vartheta} \tag{3.20}$$

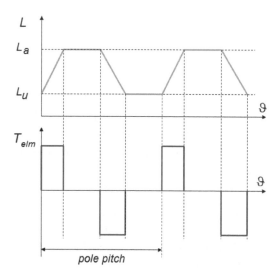

Fig. 3.12. Inductance and torque as functions of the rotor position angle ϑ for constant stator current and the width of the stator pole different than the width of the rotor pole.

The nonlinearity of the magnetic circuit has been neglected. The electromagnetic torque can also be expressed in a simplified way as

$$T_{elm} = \frac{1}{2}i^2\frac{dL}{d\vartheta} = \frac{1}{2}i^2\frac{L_a - L_u}{\beta_s} = \frac{1}{2}i^2\frac{L_a}{\beta_s}\left(1 - \frac{L_u}{L_a}\right) \tag{3.21}$$

where L_a is the inductance in aligned position, L_u is the inductance in unaligned position and β_s is the angle of stator pole in degrees. The above equation is useful in analytical calculations provided that L_a, L_u and β_s are known.

Inductance and torque waveforms for constant stator current and the width of the stator pole different than the width of the rotor pole are plotted in Fig. 3.12.

Torque equations (3.16) to (3.21) have the following implications:

- The torque is proportional to the square of the current and hence, the current can be unipolar to produce unidirectional torque. Only one power switch is required for the control of current in a phase winding (cost-effective solid state converter). This is a distinct advantage of a SR machine.

- Since the torque is proportional to the square of the current, it has a good starting torque.
- A generation action is made possible with unipolar current due to its operation on the negative slope of the inductance profile. As a result, this machine is suitable for four-quadrant operation with a converter.
- Because of its dependence on a power converter for its operation, this motor is an inherently variable-speed motor drive system.

It is necessary to mention that the electromagnetic torque of a SRM can be roughly estimated from eqn (1.2), in which $B = B_g$ is the air gap magnetic flux density under aligned stator and rotor poles, $i = 2N_p i$ is the total current per phase (two opposite stator poles active), $l = L$ is the stack length, i.e.,

$$T_{elm} = F\frac{D}{2} = 2B_g N_p i L \left(\frac{D}{2} - g\right) \tag{3.22}$$

The radius of the rotor is $0.5D - g$, where D is the stator core inner diameter.

Example 3.1

A three-phase SRM has $N_s = 6$ stator poles, $N_r = 4$ rotor poles and two coils per phase on opposite stator poles with $N_p = 50$ turns each. The overlap angle between these stator and rotor poles $\alpha = 12°$. The air gap is $g = 0.4$ mm, stator inner diameter $D = 60$ mm, and axial length of the stator stack $L = 55$ mm. The stator and rotor pole arcs are both $\beta = 24°$. A current of $i = 10$ A flows through the two coils in series. Neglect fringing, leakage, and saturation effects. Assume that the unaligned inductance $L_u = 0$. Calculate:

(a) Magnetic flux-density in the air gap between the active poles;
(b) Electromagnetic torque and angle of rotation at which the torque is essentially constant, if the current is constant.

Solution

(a) Magnetic flux-density in the air gap between the active poles

Magnetic field intensity in the air gap from Ampere's circuital law

$$H_g = \frac{2N_p i}{2g} = \frac{2 \times 50 \times 10.0}{2 \times 0.4 \times 10^{-3}} = 125,000 \text{ A}$$

Air gap magnetic flux density

$$B_g = \mu_0 H_g = 0.4\pi \times 10^{-6} \times 125,000 = 1.571 \text{ T}$$

(b) Electromagnetic torque and angle of rotation at which the torque essentially constant, if the current is constant

Area of arc pole

$$S_p = \beta \frac{D}{2} L = 24° \frac{\pi}{180°} \frac{0.06}{2} \times 0.055 = 0.00069 \text{ m}^2$$

Air gap permeance

$$\Lambda_g = \mu_0 \frac{S_p}{2g} = 0.4\pi \times 10^{-6} \frac{S_p}{2 \times 0.4 \times 10^{-3}} = 1.086 \times 10^{-6} \text{ H}$$

Maximum inductance (aligned position)

$$L_a = (2N_p)^2 \Lambda_g = (2 \times 50)^2 \times 1.086 \times 10^{-6} = 0.011 \text{ H}$$

The unaligned inductance $L_u \approx 0$. Thus, the electromagnetic torque according to eqn (3.21)

$$T_{elm} = \frac{1}{2} i^2 \frac{L_a - L_u}{\beta} = \frac{1}{2} \times 10.0^2 \times \frac{0.11 - 0.0}{24° \frac{\pi}{180°}} = 1.296 \text{ Nm}$$

The torque is essentially constant through the angle $\beta = 24°$.

Example 3.2

A three-phase, 2645-rpm, 12-8 SRM has the following dimensions of the magnetic circuit: stator inner diameter $D = 0.135$ m, length of stack $L = 0.142$ m, air gap $g = 0.57$ mm, radial height of rotor pole $h_{pr} = 29$ mm, stator pole arc $\beta_s = 17°$, rotor pole arc $\beta_r = 16.5°$. The stator winding is wound of round copper conductor with bare wire diameter $d_w = 0.6426$ mm (AWG 22). There are $a_p = 14$ parallel conductors (strands in hand). The number of turns per pole is $N_p = 34$. All coils per phase are connected in series. At nominal operating conditions the stator *rms* current is $I_{rms} = 53.8$ A. Assuming the electric conductivity of copper at 75° $\sigma_{Cu} = 47 \times 10^6$ S/m, rotational (mechanical) losses $\Delta P_{rot} = 0.25\Delta P_w$, where ΔP_w are stator winding losses, core losses $\Delta P_{Fe} = 0.33\Delta P_w$, calculate:

(a) The ratio of aligned–to–unaligned inductance L_a/L_u;
(b) Electromagnetic torque and electromagnetic power;
(c) Output power, input power and efficiency.

The magnetic saturation of the stator and rotor cores is to be neglected. The electromagnetic torque should be calculated according to eqn (3.21).

Solution

(a) The ratio of aligned–to–unaligned inductance L_a/L_u

For a three-phase motor the number of stator poles per phase is $n_{ph} = N_s/3 = 12/3 = 4$. The pole pitch is $\tau = \pi D/N_s = \pi \times 0.135/12 = 0.035$ m. The width of the stator and rotor poles are, respectively

$$b_s = \frac{D}{2}\beta_s\frac{\pi}{180} = \frac{0.135}{2} \times 17.0 \times \frac{\pi}{180} = 0.02 \text{ m}$$

$$b_r = \frac{D-2g}{2}\beta_r\frac{\pi}{180} = \frac{0.135 - 2 \times 0.00057}{2} \times 16.5 \times \frac{\pi}{180} = 0.019 \text{ m}$$

Mean area of pole

$$S_p = 0.5(b_s + b_r)L = 0.5(0.02 + 0.019) \times 0.142 = 0.00279 \text{ m}^2$$

Maximum permeance for fully aligned position

$$\Lambda_{gmax} = \mu_0\frac{S_p}{2g} = 0.4\pi \times 10^{-6}\frac{0.00279}{2 \times 0.00057} = 3.076 \times 10^{-6} \text{ H}$$

Maximum length of flux path in the air for unaligned position (from the stator pole face to the rotor yoke)

$$l_{\Phi max} \approx h_{pr} + g = 0.029 + 0.00057 \approx 0.03\text{m}$$

Minimum length of flux path in the air for unaligned position (between corners of the stator and rotor poles)

$$l_{\Phi min} \approx 0.5(\tau - b_s) + g = 0.5(0.035 - 0.02) + 0.00057 \approx 8.23 \times 10^{-3} \text{ m}$$

Mean path for magnetic flux for unaligned position

$$l_\Phi \approx \frac{l_{\Phi min}l_{\Phi max}}{l_{\Phi min} + l_{\Phi max}} = \frac{8.23 \times 10^{-3} \times 0.03}{8.23 \times 10^{-3} + 0.03} \approx 6.437 \times 10^{-6} \text{ m}$$

Minimum permeance for unaligned position

$$\Lambda_{gmin} = \mu_0\frac{S_p}{l_\Phi} = 0.4\pi \times 10^{-6}\frac{0.00279}{6.437 \times 10^{-6}} = 0.426 \times 10^{-6} \text{ H}$$

Inductance for fully aligned position under assumption that all pole coils per phase are series connected

$$L_a \approx (n_{ph}N_p)^2\frac{\Lambda_{gmax}}{2} = (4 \times 34)^2\frac{3.076 \times 10^{-6}}{2} = 0.028 \text{ H}$$

Inductance for unaligned position

$$L_u \approx (n_{ph}N_p)^2 \frac{\Lambda_{gmin}}{2} = (4 \times 34)^2 \frac{0.426 \times 10^{-6}}{2} = 0.0039 \text{ H}$$

Fully aligned–to–unaligned position ratio

$$\frac{L_a}{L_u} = \frac{0.028}{0.0039} \approx 7.22$$

The higher the ratio L_a/L_u, the better the performance and higher power density of a SRM.

(b) Electromagnetic torque and electromagnetic power

Electromagnetic torque

$$T_{elm} = \frac{1}{2}I_{rms}^2 \frac{L_a - L_u}{\beta_s} = \frac{1}{2} \times 53.8^2 \frac{0.028 - 0.0039}{17\frac{180}{\pi}} = 119.5 \text{ Nm}$$

Angular speed

$$\Omega = 2\pi \frac{2645}{60} = 277 \text{ rad/s}$$

Electromagnetic power

$$P_{elm} = 277 \times 119.5 = 33,106.7 \text{ W}$$

(c) Output power, input power and efficiency

Average distance of the stator turn from the pole core

$$t_{av} = \frac{1}{4}(\tau - b_s) = \frac{1}{4}(0.035 - 0.02) = 0.0038 \text{ m}$$

Mean length of turn (MLT)

$$l_{mean} = 2(L + 2t_{av}) + 2(b_s + 2t_{av})$$

$$= 2(0.142 + 2 \times 0.0038) + 2(0.02 + 2 \times 0.0038) = 0.355 \text{ m}$$

Cross-section of stator parallel conductors

$$s_{Cu} = \frac{\pi d_w^2}{4}a_p = \frac{\pi \times 0.6426 \times 10^{-3}}{4} \times 14 = 4.54 \times 10^{-6} \text{ m}^2$$

Resistance per phase

$$R = \frac{n_{ph}N_{p}l_{mean}}{\sigma_{Cu75}s_{Cu}} = \frac{4 \times 34 \times 0.355}{47 \times 10^6 \times 4.54 \times 10^{-6}} = 0.226 \ \Omega$$

Winding losses

$$\Delta P_w = 3I_{rms}^2 R = 3 \times 53.8^2 \times 0.226 = 1962.8 \text{ W}$$

Rotational losses including windage losses $\Delta P_{rot} = \Delta P_w/4 = 1962.8/4 = 490.7$ W and core losses $\Delta P_{Fe} = \Delta P_w/3 = 1962.8/3 = 654.3$ W. Total losses

$$\Delta P = \Delta P_w + \Delta P_{rot} + \Delta P_{Fe} = 1962.8 + 490.7 + 654.3 \approx 3108 \text{ W}$$

Output power

$$P_{out} = P_{elm} - \Delta P_{rot} = 33,106.7 - 490.7 = 32,616 \text{ W}$$

Input power

$$P_{in} = P_{out} + \Delta P = 32,616 + 3108.0 = 35,723.7 \text{ W}$$

Efficiency

$$\eta = \frac{P_{out}}{P_{in}} \times 100\% = \frac{32,616.0}{35,723.7} \times 100\% = 91.3 \text{ \%}$$

3.6 Power electronics converters for SRMs

The structure of SRM system is shown in Fig. 3.13. There are two main methods of SRM control:

- Current hysteresis control;
- Voltage pulse width modulation (PWM).

Information about the position of the rotor is a key issue to control the current and then the torque. A good torque control requires the knowledge of inductance profile.

3.6.1 Current hysteresis control

In the *hysteresis switching-base current controller* (Fig. 3.14), the current error is computed. The switching is generated from the current error depending on its relationship to the hysteresis current window. The switching logic of the hysteresis controller is summarized as:

$$\text{If} \quad i^* - i \geq \Delta i, \quad \text{then} \quad V = V_{dc}$$
$$\text{If} \quad i^* - i \leq -\Delta i, \quad \text{and} \quad i^* > 0, \quad \text{then} \quad V = V_{dc} \quad (3.23)$$
$$\text{If} \quad i^* - i \leq -\Delta i, \quad \text{and} \quad i^* \leq 0, \quad \text{then} \quad V = -V_{dc}$$

where Δi is the current hysteresis window, i^* is the current command and V_{dc} is the DC link voltage.

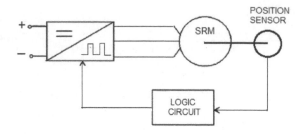

Fig. 3.13. Structure of SRM drive system.

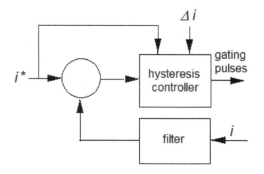

Fig. 3.14. Hysteresis controller.

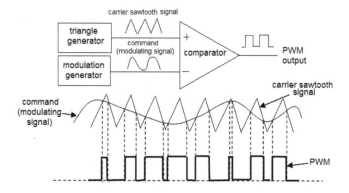

Fig. 3.15. Principles of pulse width modulation (PWM).

3.6.2 Voltage PWM control

Combining a triangle wave and a modulating signal generates the output PWM voltage waveform (Fig. 3.15). The triangle generator produces a saw-tooth wave that is the *carrier* or *switching frequency* of the solid state converter. The modulation generator produces a command signal or wave that determines the width of pulses and therefore the output voltage of the solid state converter. The larger the command signal, the wider the pulse.

3.6.3 Asymmetric bridge converter with freewheeling and regeneration capability

Fig. 3.16a shows a solid state converter for SRM. For comparison, a voltage source inverter for induction motor is shown in Fig. 3.16b.

In the *asymmetric bridge converter* (Fig. 3.16a) solid state switches are in series with each phase winding of SRM. Turning on the two power switches in each phase will circulate a current in that phase of SRM. If the current rises above the commanded value, the switches are turned off. The energy stored in the motor phase winding will keep the current in the same direction until it is depleted. The half-bridge phase leg circuit can supply current in only one direction, while it can supply a positive, negative (reversed) or zero voltage at the phase terminals. Utilization of the power switches in the asymmetric bridge converter is poor.

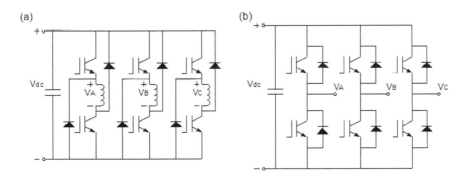

Fig. 3.16. Solid state converters for: (a) SRM; (b) induction motor.

For low-speed motoring, the voltage applied to the phase winding is:

(a) $+V$ if both switches are on (Fig. 3.17a);
(b) 0 if T2 is on and T1 is off (Fig. 3.17b);
(c) V if both power switches are off and the phase current is freewheeling through both diodes D1 and D2 (Fig. 3.17c).

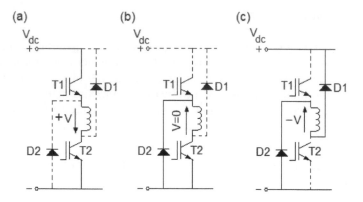

Fig. 3.17. Three conduction modes of asymmetric bridge converter: (a) both power switches T1 and T2 are on; (b) switch T2 is on and T1 is off; (c) both switches T1 and T2 are off and the phase current is freewheeling through both diodes D1 and D2.

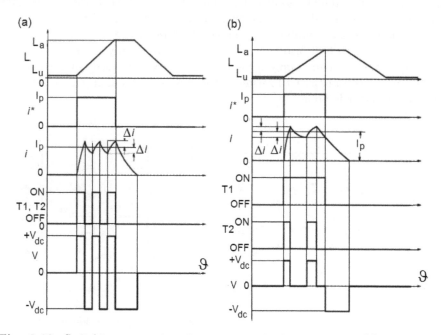

Fig. 3.18. Switching strategies of asymmetric bridge converter: (a) T1 and T2 turned-on and turned-off, (b) T2 turned-on and turned-off only. Alternating switches to balance thermal losses.

Fig. 3.18 shows switching strategies of an asymmetric bridge converter. In Fig. 3.18 I_p is the desired magnitude of current during the positive inductance slope for motoring action, $i*$ is the current command and Δi is the current window of hysteresis controller.

Hysteresis current controller is considered here due to its simplicity. The current command $i*$ is enforced with a current feedback loop where it is compared with the phase current i.

T1 and T2 turned-on and turned-off (Fig. 3.18a). When the current error exceeds Δi, the switches T1 and T2 are turned off simultaneously. At that time, diodes D1 and D2 take over the current and complete the path through the DC source. After the initial startup, during turn-on and turn-off of T1 and T2, the machine phase winding experiences twice the rate of change of DC link voltage, resulting in a higher deterioration of the insulation. More ripples are put into the DC link capacitor, thus reducing its life.

T2 turned-on and turned-off only (Fig. 3.18b). T1 is always turned-on. T2 is turned-on and turned-off.

3.6.4 $(m+1)$ converter

A more efficient converter topology is called $(m+1)$ switch and diode configuration (Fig. 3.19).

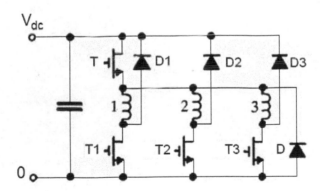

Fig. 3.19. Converter topology with $(m+1)$ switch and diode configuration.

When T and T1 are turned on, phase 1 is energized by applying the source voltage across the phase winding. The current can be limited to the set level by controlling either T or T1, or both. Similarly, phase 2 can be energized by T and T2. The circuit provides restricted current control during overlapping phase currents. In the $(m+1)$ converter utilization of power switches is better due to the shared switch operation. The disadvantage of $(m+1)$ configuration is that the current cannot be switched fast.

3.7 Advantages and disadvantages

SRMs have the following advantages:

- Simple construction (only laminations and stator coils);
- Highly robust machine both mechanically and electrically;
- No rotor windings, no rotor permanent magnets (PMs);
- Can operate in high temperatures;
- Shorter motor than induction motor of the same rating due to shorter overhangs of concentrated coils wound around salient poles (Fig. 3.20);
- The best performance-to-cost ratio (less expensive motor and converter in comparison with AC induction motor drive)
- Low energy losses in both the rotor and power electronics;
- Better efficiency than that of induction motors (Fig. 3.21);
- High efficiency over wide speed range;
- Negligible increase of *rms* current at low speed;
- High overload torque;
- Fault tolerant motor;
- Higher torque-to-current ratio as compared with induction motors;
- High torque-to-rotor moment of inertia ratio (excellent dynamics).

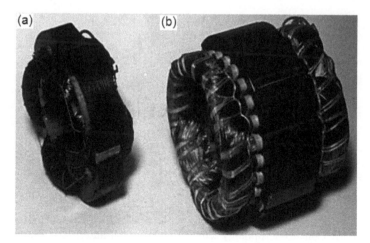

Fig. 3.20. Comparison of two motors for washers rated at 150 W, 1500 rpm: (a) SRM; (b) single-phase induction motor.

On the other hand, SRMs shows also serious drawbacks:

- High torque ripple;
- High level of acoustic noise;
- High level of stator vibration;

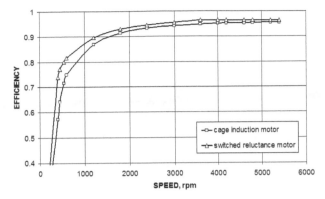

Fig. 3.21. Efficiency of large electric motors rated at 150 kW, 5400 rpm, 270 Nm for compressors versus speed and constant torque.

- High level of electromagnetic interference (EMI) noise due to nonsinusoidal curent waveforms;
- High windage losses due to rotor saliency in the case of high speed machines;
- Good performance only if the air gap is very small (in general, smaller than that in induction motors);
- Cannot be plugged in directly to the utility grid;
- Requires non-standard solid state converter (Fig. 3.16a);
- Not quite mature technology and limited applications thus far;
- Non-standard design.

These disadvantages can be minimized if the SRM is properly designed and optimized.

3.8 Applications of SRMs

Some possible applications of SRMs include, but are not limited to:

- Domestic life (washers and clothes dryers, vacuum cleaners)
- Compressors
- HEV and EV traction
- High speed turbine generators
- Process industry
- Servo systems.

Fig. 3.22 shows a Maytag Neptune washing machine with SRM. The low-noise, high-efficiency, maintenance-free motor with variable-speed control drives directly the drum of the washer without any belt transmission.

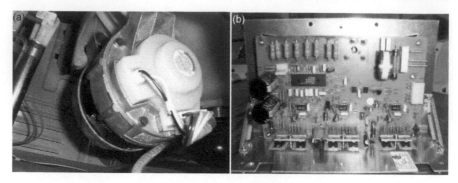

Fig. 3.22. Maytag Neptune front loading washing machine with direct SRM drive: (a) SRM; (b) power electronic circuit. An SRM is from Emerson Electric, St. Louis, MO, USA.

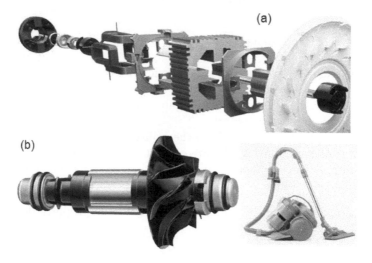

Fig. 3.23. High speed X020 SRM for vacuum cleaner: (a) expandend view of electromechanical drive; (b) rotor of turbine; (c) vacuum cleaner. Photo courtesy Dyson, Malmesbury, UK

Fig. 3.23 shows a high speed vacuum cleaner with an SRM spinning at 100,000 rpm. The turbine swirls air at high speed through a series of chambers to expel dirt and dust without trapping it in disposable filters that can clog.

3.9 Steady-state performance characteristics

Figs 3.24 and 3.25 show steady-state speed-torque and efficiency-torque characteristics at constant DC voltage (solid-state converter input voltage) of a high speed SRM for a vacuum cleaner with two rotor poles [3]. Both the speed and efficiency change minimally in wide range of the load torque.

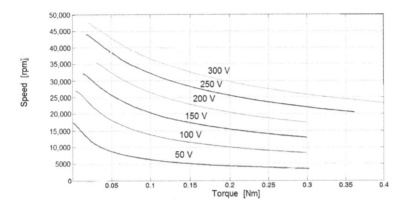

Fig. 3.24. Speed-torque characteristics at constant DC voltage $50 \leq V_{dc} \leq 300$ of a high speed SRM for a vacuum cleaner [3].

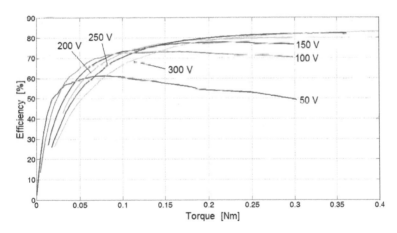

Fig. 3.25. Efficiency-torque characteristics at constant DC voltage $50 \leq V_{dc} \leq 300$ of a high speed SRM for a vacuum cleaner [3].

3.10 Design recommendation

Given below are some recommendations to design an SRM with good performance characteristics:

- The air gap should be as small as possible (attraction magnetic force is inversely proportional to the air gap square);
- The performance is very sensitive to the turn-on and turn-off angle;
- The performance is very sensitive to the width of the stator and rotor poles;
- The ratio L_a/L_u (aligned-to-unaligned inductance) should be as high as possible. This can be achieved by proper shaping of stator and rotor poles;
- The greater the number of aligned pole pairs $n_a = GCD(Ns, Nr)$, the higher the electromagnetic torque;
- The electromagnetic torque and efficiency are very sensitive to the saturation of the magnetic circuit.

Summary

The SRM is a doubly-salient, singly-excited electrical machine. In an SRM the electromagnetic torque is produced by the magnetic attraction of a steel rotor to stator electromagnets. No permanent magnets are needed, and the rotor carries no windings.

The *difference between a stepping motor* and an SRM is that a stepping motor operates in an open control loop, while SMRs require rotor position sensors and closed control loop.

The SRM cannot be plugged in directly to the utility grid. It must always operate with a solid state converter.

The unique features of SRMs are

- An SRM requires a controllable solid state converter and cannot be operated directly from a utility grid;
- Because the mutual inductance between phases is negligible, a short circuit fault in one phase winding has no effect on other phases;
- Since the current only needs to be unidirectional for all quadrants of operation, all power converter configurations have a switch (switches) in series with a stator phase winding facing the DC source voltage;
- No rotor excitation system.

A typical SRM has a laminated stator and rotor with salient poles. Only the stator has a winding with concentrated parameters.

The *number of rotor poles* is less than the number of stator poles. A general rule for the number of rotor poles is $GCD(N_s, N_r) = n_a$ where GCD is the greatest common divisor, N_s is the number of stator poles, N_r is the number of rotor poles and n_a is the number of aligned positions.

The *frequency of the phase current* only depends on the speed and the number of rotor poles, i.e., $f = nN_r$. It is independent of the number of stator poles.

Performance characteristics of SRMs are shaped electronically. The performance of an SRM is very sensitive to the turn-on and turn-off angle. The air gap should be as small as possible, even smaller than that in induction machines.

The *maximum torque zone* is the angle through which one phase can produce the torque. It is inversely proportional to the number of rotor poles. The smaller the number of rotor poles, the larger the maximum torque zone. The maximum torque zone is affected by the saturation of the magnetic circuit.

The *electromagnetic torque* of SRM can be calculated either on the basis of voltage balance equation or on the basis of coenergy. The torque is proportional to the current square and dL/dt, i.e.,

$$T_{elm} = \frac{1}{2}i^2\frac{dL}{dt}$$

The *direction of rotation* of an SRM can only be changed by changing the sequence of the phase current, i.e., from A,B,C,A,..., to A,C,B,A,C,.... . Change in current direction does not change the direction of rotation because the electromagnetic torque is proportional to i^2.

An SRM is a *reversible machine* and can operate as a generator. The generator operation is in the AK interval shown in Fig. 3.9.

The electromagnetic torque of a motor is produced only when the stator pole attracts the rotor pole, i.e., during the interval when the self-inductance increases from L_u to L_a. When the self-inductance decreases from L_a to L_u, the phase current should be switched-off. For decreasing inductance the SRM operates as a generator. The turn-on and turn-off angle is different for the motor and generator operation.

There are two *main methods of SRM control*:

- Current hysteresis control;
- Voltage pulse width modulation (PWM).

The most typical solid state converter for an SRM is an *asymmetric bridge converter* (Fig. 3.16a). It has two solid state switches in series with each phase winding and two freewheeling diodes. Turning on the two power switches in each phase will circulate a current in that phase of SRM. If the current rises above the commanded value, the switches are turned off. The energy stored in the motor phase winding will keep the current in the same direction until it is depleted. Utilization of the power devices in the asymmetric bridge converter is poor. A more efficient converter topology is called $(m+1)$ switch and diode configuration (Fig. 3.19).

The most important *advantages* of SRMs are

- Simple construction: no PMs and no windings on the rotor;
- Short end turns (concentrated parameter stator coils): low resistance of the stator winding, low winding losses, shorter axial length than that of an equivalent induction motor;
- High efficiency over a wide range of speed;
- Better efficiency than that of induction motors;

- Short-circuit fault in one phase winding has no effect on other phases;
- Fault tolerant machines;
- Low cost solution;
- Highly robust machine both mechanically and electrically;
- Can work in harsh environment (no PMs, no windings on the rotor);
- SRMs are the only electrical machines capable of operating at very high temperatures, 600°C and above;
- SRM is a reversible machine and can operate as a generator.

The most serious *disadvantages* of SRMs are (a) high torque ripple and (b) high level of noise.

Some possible *applications of SRMs* include, but are not limited to

- Domestic life (washers and clothes dryers, vacuum cleaners);
- Compressors;
- HEV and EV traction;
- High speed turbine generators;
- Process industry;
- Servo systems.

Problems

1. For a 4-phase SRM with the number of stator poles $N_s = 8$, 16, 24 and 32 find the number of rotor poles N_r.

   Answer: (1) $N_s = 8$, $N_r = 6$; (2) $N_s = 16$, $N_r = 14$, $N_r = 12$; (3) $N_s = 24$, $N_r = 22$, $N_r = 20$, $N_r = 18$; (4) $N_s = 32$, $N_r = 30$, $N_r = 28$, $N_r = 24$.

2. A three-phase SRM has $N_s = 12$ stator poles and $N_r = 10$ rotor poles. The nominal speed is 5000 rpm. Find: (a) the number of aligned positions; (b) stroke angle; (c) commutation frequency; (d) maximum torque zone.

   Answers: (a) $n_a = 2$; (b) $e = 12°$; (c) $f = 833.3$ Hz; (d) $\tau_{max} = 18°$.

3. A three-phase, 6-4 SRM has the air gap $g = 0.5$ mm, stator core inner diameter $D = 48$ mm, and axial length of stack $L = 42$ mm. There are $N_p = 80$ turns on each stator pole (160 turns per pole pair). The stator and rotor pole arcs are $\beta_s = \beta_r = 30°$. The rotor is in such a position that the "overlap angle" between the stator and rotor pole pair is 15°. The phase current that flows through two phase coils on opposite stator poles connected in series is $i = 7.4$ A.

   (a) Calculate the magnetic flux density in the air gap between the active poles;
   (b) Estimate the electromagnetic torque;
   (c) Through what angle of rotation is the torque essentially constant, if the current is constant?

Assumption: The fringing and leakage flux are to be neglected and the stator and rotor cores have infinitely large magnetic permeability.

Answer: (a) $B_g = 1.49$ T; (b) $T_{elm} = 0.888$ Nm; (c) $\theta = 30°$.

4. A four-phase SRM with $N_s = 8$ stator poles and $N_r = 6$ rotor poles has a stator pole arc $\beta_s = 28°$ and a rotor pole arc $\beta_r = 30°$. The unsaturated aligned inductance is $L_a = 11.8$ mH and the unaligned inductance is $L_u = 2.0$ mH. The fringing effect and magnetic saturation can be neglected. Find:

(a) Instantaneous electromagnetic torque when the rotor is 10° before the aligned position and the phase current is $i = 8.0$ A.

(b) Maximum energy conversion in one stroke if the current is limited to $i = 8.0$ A and the average torque corresponding to this energy conversion.

(c) Flux-linkage in the aligned position when phase current is $i = 8.0$ A. If this flux-linkage can be maintained constant while the rotor rotates from the unaligned position to the aligned position at low speed, determine the energy conversion per stroke and the average torque.

Answer: (a) $T_{elm} = 0.642$ Nm; (b) $W = 0.3136$ J, $T_{elmav} = 1.198$ Nm; (c) $\Psi_a = 0.094$ Wb, $W = 1.85$ J, $T_{elmav} = 7.067$ Nm

5. A five-phase, 3200-rpm, 10-8 SRM has the following dimensions of the magnetic circuit: stator inner diameter $D = 48$ mm, length of stack $L = 73$ mm, air gap $g = 0.254$ mm, radial height of rotor pole $h_{pr} = 4$ mm, stator pole arc $\beta_s = 16°$, rotor pole arc $\beta_r = 17°$. The stator winding is wound of round copper wire with wire diameter $d_w = 1.0236$ mm (AWG 18). There are $a_w = 4$ parallel conductors (strands in hand). The number of turns per pole is $N_p = 24$. All coils per phase are series connected. At nominal operating conditions the stator rms current is $I = 20.4$ A. Assuming that the electric conductivity of copper at 75° is $\sigma_{Cu} = 47 \times 10^6$ S/m, rotational losses are $\Delta P_{rot} = 0.1\Delta P_w$, where ΔP_w are stator winding losses, and core losses are $\Delta P_{Fe} = \frac{1}{12}\Delta P_w$, calculate: (a) the ratio of aligned-to-unaligned inductance L_a/L_u; (b) electromagnetic torque T_{elm} and electromagnetic power P_{elm}; (c) output power P_{out}, input power P_{in} and efficiency η.

Answer: (a) $L_a/L_u = 8.746$; (b) $T_{elm} = 0.944$ Nm, $P_{elm} = 316.2$ W; (c) $P_{out} = 287.8$ W, $P_{in} = 467.9$ W, $\eta = 61.5\%$.

6. The stator inner diameter of a SRM is $D = 62$ mm, the stack length is $L = 98$ mm, the air gap $g = 0.35$ mm and the number of turns per one pole is $N_p = 22$. Estimate: (a) the electromagnetic torque T_{elm} at phase current $i = 5.65$ A; (b) the alligned inductance L_a assuming that the stator pole arc $\beta_s = 16°$ and unaligned inductance $L_u \approx 0$ H.

Answer: (a) $T_{elm} = 0.333$ Nm; (b) $L_a = 5.8$ mH.

7. For the SRM of Problem 3.6 find: (a) the maximum (peak) electromagnetic torque T_{elmmax} produced at the onset of overlap (the maximum electromagnetic torque is when the ferromagnetic core is magnetically saturated); (b) number of turns N_p to obtain similar torque T_{elm} as in Problem 3.6 when the magnetic flux density in the core reaches the saturation value. Assume that the saturation magnetic flux density of ferromagnetic core is $B_{sat} = 1.75$ T.

Answer: (a) $T_{elmmax} = 5.124$ Nm; (b) the number of turns must be integer and be rounded to $N_p = 2$ to obtain $T_{elm} = 0.466$ Nm (the number of turns cannot be $N_p = 1$, because $T_{elm} < 0.333$ Nm).

4

DC MACHINES

In spite of intensive development of solid-state converter-fed electromechanical drives with induction and synchronous motors, the DC brush motors are still in use. This is due to the simple and cost-effective speed control and very good performance characteristics of DC motors.

4.1 Function and objective

The basic function of DC brush machines is conversion of the DC current energy into mechanical energy with controlled parameters, i.e., variable speed or variable torque (motors), or vice versa, mechanical energy into DC energy (generator).

The disadvantage of DC brush machines is higher cost of manufacturing and problems with the maintenance of brushes and commutator.

4.2 Prinicple of operation

Fig. 4.1 explains the principle of operation of a DC brush electrical motor. Electric current in the rectangular coil shown in Fig. 4.1a generates a magnetic field the N polarity of which is behind the rectangular coil and the S polarity is in front of the rectangular coil. The poles of magnets and rectangular coil of the same polarity repel each other and the poles of different polarity attract each other so that the coil will turn clockwise. There is no current in the rectangular coil in the position shown in Fig. 4.1b because semi-rings connected to the coil terminals do not touch brushes. The rectangular coil can pass this position due to its moment of inertia. When the semi-rings turn more than 90°, the current in the coil flows in the opposite direction and the polarity of the magnetic field generated by the rectangular coil is also in the opposite direction (Fig. 4.1c). Because the polarity of the magnetic field generated by the rectangular coil has been changed, the coil can turn further in the same direction. The semi-ring plays the role of an electromechanical inverter and creates the so-called "commutator". It is necessary to obtain continuous rotation of the rectangular coil. In practice, there is more than one coil, which creates the rotor winding and more than two segments of commutator to provide smooth movement of the rotor. A DC brush machine with multicoil rotor winding (armature winding) and two-pole stator field excitation winding (instead of PMs) is shown in Fig. 4.2.

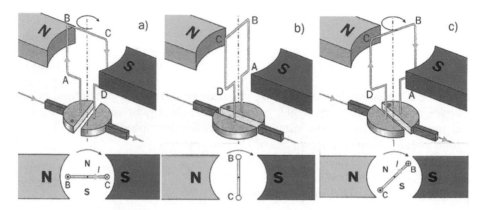

Fig. 4.1. Principle of operation of a DC motor: (a) rectangular coil turns clockwise; (b) neutral position (current-free state) of the rectangular coil; (c) current in the coil changes direction and the coil continues rotation.

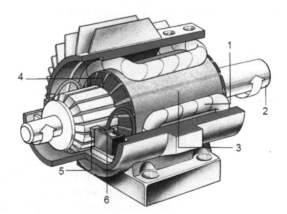

Fig. 4.2. A simple two-pole DC brush machine. 1 – stator field excitation winding, 2 – shaft, 3 – rotor with armature winding, 4 – commutator, 5 – brush, 6 – brush holder.

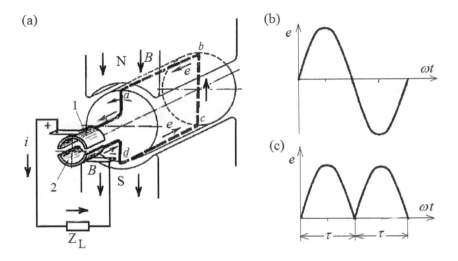

Fig. 4.3. Operation of a DC brush machine as a generator: (a) simple machine with single armature coil; (b) EMF induced in armature coil; (c) EMF after rectification by commutator.

If the brushes are connected to an electrical load, e.g., electric bulb and the rotor is spinning with the aid of an external prime mover, an EMF is induced in the rotor armature winding and the electric current makes the bulb light. The DC machine operates as a generator (Fig. 4.3). There is AC current in the rotor, but the commutator acts as a mechanical rectifier, so the output current (behind the brushess) is a DC current. In the case of a two-segment commutator, the rectified EMF contains large pulsations (Fig. 4.3c), two pulses per one period. An increase in the number of commutator segments reduces the amplitude of pulsations and increases their frequency.

4.3 Construction of DC brush machine

Fig. 4.4 shows the longitudinal and cross-sections of a typical *DC brush motor* with its electromagnetic excitation system and interpoles. The electromagnetic torque developed by this type of machine is created by two main windings: the *armature winding*, in which the EMF is induced, and the *field winding*, which produces the exciting magnetic flux. In typical designs, the armature winding is inserted into slots of a laminated rotor core (Fig. 4.5) while the field winding is located on salient stator poles (Fig. 4.6).

The armature coils are interconnected through the *commutator*, which consists of a number of insulated copper segments (Fig. 4.7). The commutator is located on the same shaft as the armature (rotor) and rotates together with the armature winding. The armature (rotor) core must be laminated to

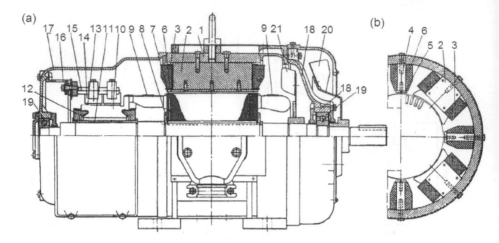

Fig. 4.4. A DC motor: (a) longitudinal section, (b) cross-section. 1 – frame, 2 – main pole core, 3 – field excitation winding, 4 – interpole core, 5 – interpole shoe, 6 – interpole coil, 7 – armature core, 8 – clamping rings, 9 – bands securing the end connections, 10 – commutator segment, 11 – commutator steel cylinder, 12 – commutator clamping cone, 13 – brush, 14 – brush holder, 15 – brush stud, 16 – brush rocker, 17,18 – end bells, 19 – bearing, 20,21 – fan blades.

Fig. 4.5. Rotor of a DC brush machine with armature winding inserted in slots, commutator and shaft.

reduce the core losses that may arise due to the flow of AC current. In DC motors the commutator serves as a mechanical inverter. In DC *generators*, the commutator serves as a mechanical rectifier. Current is fed to the commutator with the aid of *brushes*, often made of carbon or graphite (Fig. 4.8). The brushes are held in *brush holders* (Fig. 4.9) and they must be free to move radially in order to maintain contact with the commutator as they wear away. The current is conducted from the brush to the *brush stud* by means of a *brush pigtail*. The brush-commutator mechanism is shown in Fig. 4.10.

The *main pole* consists of a pole core with a concentrated-parameter coil and a pole shoe (Fig. 4.11). The pole shoes should also be laminated to reduce the additional iron losses due to any variable air gap reluctance caused by the rotor slots. The pole cores can be made either solid or laminated. The higher

Fig. 4.6. Stator of a 4-pole DC brush machine: 1 – main poles, 2 – interpoles (compoles).

Fig. 4.7. Commutators.

the voltage for a given diameter of armature, the fewer is the number of main poles, to provide space for the larger number of commutator segments. High-current motors require a large number of poles in order to carry currents which sometimes exceed 1 kA per brush set.

There are smaller poles, which are called *interpoles* or *commutating poles* (compoles) between the main poles (Fig. 4.6). The *interpole winding* may be called the *commutating winding*, and it is connected in series with the armature winding. Interpoles produce an MMF in opposition to that of the armature winding in order to achieve commutation with reduced sparking at

Fig. 4.8. Brushes of small DC machines.

Fig. 4.9. Brush holders of DC machines.

Fig. 4.10. Brush-commutator mechanism.

Fig. 4.11. Construction of main poles and field excitation coil: (a) punching (pole core with pole shoe); (b) field excitation coil; (c) main pole with field coil.

the brushes and to reduce the demagnetizing effect of the quadrature axis armature reaction MMF. Interpoles are generally omitted in small DC machines, such as fractional-horsepower motors.

In motors subject to heavy duties, the quadrature armature reaction is neutralized by means of a *compensating winding* embedded in the pole-shoe slots of the main poles (Fig. 4.12). Like the interpole winding, the compensating winding is connected in series with the armature winding.

Fig. 4.12. Stator of a 6-pole DC machine with compensating winding: 1 – field winding of main poles, 2 – interpole winding, 3 – compensating winding placed in slots made in pole shoes of main poles.

4.4 Armature winding

The armature winding of DC brush machines is distributud in axial slots made in the armature core (Fig. 4.5), although there are also slotless windings for small DC brush machines (Figs 4.35b, 4.38). Fig. 4.13 shows two basic armature windings of DC brush machines: (a) lap winding and (b) wave winding. There are the following relationships:

- Span of armature winding

$$y = y_1 \pm y_2 \qquad (4.1)$$

- Commutator span

$$y_c = y \qquad (4.2)$$

where y_1 is the partial first span, y_2 is the partial second span and the "+" sign is for the wave winding and the "−" sign is for the lap winding. Each coil is connected to the adequate segment of the commutator. The distribution of the armature winding in slosts is shown in Fig. 4.14. The armature slot is usually divided into two layers: upper layer and lower layer separated by an insulation. Each coil has its left side located in the upper layer of a slot and its right side located in the lower layer of the slots. The main feature of the armature winding of DC brush machines is that the armature winding is a closed winding (without the beginning and the end).

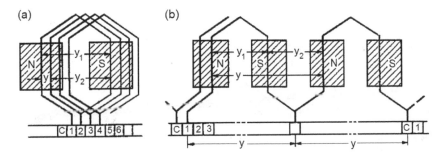

Fig. 4.13. Two basic windings of DC brush machines: (a) lap winding; (b) wave winding.

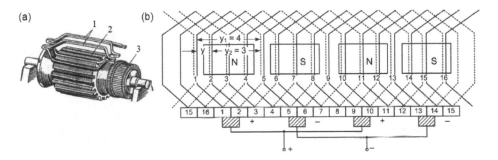

Fig. 4.14. Lap winding: (a) distribution of armature coils in slots; (b) example of a winding diagram with $s = 16$, $2p = 4$, $a = 2$, $C = 16$, $y_1 = 4$, $y_2 = 3$. 1 – armature coil, 2 – armature core, 3 – commutator.

The brushes divide the armature winding into parallel paths (Fig. 4.15). The number of pairs of parallel paths a is equal to the number of pairs of brushes. If the armature current is I_a, the current of parallel path is $I_a/(2a)$.

The number of parallel paths should be adjusted to the nominal armature current, which is divided uniformly between all parallel paths. For most DC machines

$$\frac{I_a}{2a} \leq 300 \text{ A} \qquad (4.3)$$

In the case of lap winding the number of parallel paths should be $a = p \geq I_a/600$.

The number of commutator segments

$$C = us \qquad (4.4)$$

where u is the number of coil sides in the upper layer of a slot and s is the number of armature slots. The following relationship exists between the

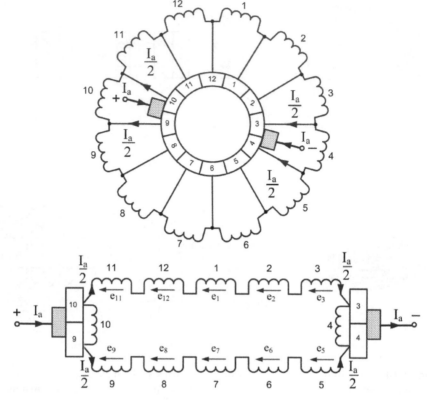

Fig. 4.15. Division of the lap winding into parallel paths for $p = 1$ and $a = 1$.

number of armature conductors N and the number of commutator segments C:

$$N = 2CN_c \tag{4.5}$$

where N_c is the number of turns per one armature coil.

The *absolute symmetry* of the armature winding is when at any position of the rotor and any waveform of the EMF, both the EMFs and resistances of all parallel paths are the same. In practice, it is enough to obtain a *relative symmetry*, when the following conditions are to be met:

$$\frac{C}{a} = integer; \qquad \frac{s}{a} = integer \qquad \frac{p}{a} = integer \tag{4.6}$$

If the first and second conditions are even numbers, the symmetry becomes absolute symmetry.

Example 4.1

Sketch the diagram of the lap winding with the following parameters:

- Number of slots $s = 16$
- Number of poles $2p = 4$
- Number of parallel paths $2a - 4$
- Number of coil sides in upper layer of a slot $u = 1$
- Number of turnes per coil $N_c = 1$

Solution

Number of commutator segments

$$C = us = 1 \times 16 = 16$$

Number of slots per pole

$$Q = \frac{s}{2p} = \frac{16}{4} = 4$$

First partial span

$$y_1 = Qu = 4 \times 1 = 4$$

The coil span has been assumed $y = +1$. Thus, the second partial span is

$$y_2 = y_1 - y = 4 - 1 = 3$$

The winding diagram is sketched in Fig. 4.14b. There are $2a = 4$ parallel paths.

4.5 Fundamental equations

4.5.1 Terminal voltage

From Kirchhoff's voltage law, the terminal (input) voltage of a DC machine is

$$V = E \pm I_a \sum R_a \pm \Delta V_{br} \tag{4.7}$$

where E is the voltage induced in the armature winding (called the EMF), I_a is the armature current, and ΔV_{br} is the brush voltage drop. The brush voltage drop is approximately constant, and for the majority of typical DC motors is practically independent of the armature current. For carbon brushes, $\Delta V_{br} \approx 2$ V. The "+" sign is for the motor and the "−" sign is for the generator. The armature circuit resistance is, in general:

$$\sum R_a = R_a + R_{int} + R_{comp} + R_{se} \tag{4.8}$$

where R_a is the resistance of the armature winding, R_{int} is the resistance of the commutation winding located on the interpoles, R_{comp} is the resistance of the compensating winding, and R_{se} is the resistance of the series winding. For small and medium-power shunt DC motors $\sum R_a = R_a + R_{int}$ because the compensating and series windings do not exist.

4.5.2 Armature winding EMF

The EMF induced in the armature winding by the main flux Φ is

$$E = \frac{N}{a}pn\Phi = c_E n\Phi \tag{4.9}$$

where N is the number of the armature conductors, a is the number of pairs of armature current parallel paths, p is the number of pole pairs, Φ is the main (useful) magnetic flux, and c_E is the so-called *EMF constant* or *armature constant*, which can be expressed as

$$c_E = \frac{Np}{a} \tag{4.10}$$

4.5.3 Magnetic flux

The magnetic flux in the airgap is

$$\Phi = b_p L_i B_g = \alpha_i \tau L_i B_g \tag{4.11}$$

where b_p is the pole shoe width, L_i is the effective length of the armature stack, B_g is the airgap magnetic flux density, $\alpha_i = b_p/\tau = 0.55\ldots0.75$ is the effective pole arc coefficient and $\tau = \pi D/(2p)$ is the pole pitch defined as the armature circumference πD divided by the number of poles $2p$.

4.5.4 MMF of the field winding

The MMF per pole pair of the main poles (field excitation winding) is

$$F_f = 2N_f I_f \tag{4.12}$$

where N_f is the number of turns of a single field coil (per one pole) and I_f is the current in the field winding. The magnetic circuit of a DC brush machine for a single pole pair is shown in Fig. 4.16.

According to Fig. 4.16, the MMF F_f per one pole pair can be calculated on the basis of Ampere's circuital law, i.e.,

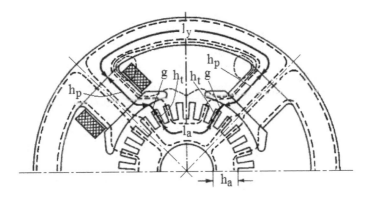

Fig. 4.16. Magnetic circuit of a DC brush machine per one pole pair.

$$F_f = \oint_l \mathbf{H} d\mathbf{l} = 2H_g g + 2H_m h_p + H_y l_y + 2H_t h_t + H_a l_a \qquad (4.13)$$

where the magnetic field intensity in the air gap is

$$H_g = \frac{B_g}{\mu_0} \qquad (4.14)$$

and g is the air gap between the rotor core and stator pole shoe, H_m is the magnetic field intensity in the main pole core, h_p is the height of main pole including pole shoe, H_y is the magnetic field intensity in the stator yoke, l_y is the circumferential length of the stator yoke segment per pole pair, H_t is the magnetic field intensity in the rotor tooth, h_t is the radial height of the rotor tooth, H_a is the magnetic field intensity in the rotor (armature) core, l_a is the circumferential length of the rotor (armature) core per pole pair and $\mu_0 = 0.4\pi \times 10^{-6}$ H/m is the magnetic permeability of free space.

4.5.5 Electromagnetic power

The electromagnetic power developed by the DC machine is

$$P_{elm} = \Omega T_{elm} \qquad (4.15)$$

where the rotor angular speed is

$$\Omega = 2\pi n \qquad (4.16)$$

On the other hand, the electromagnetic power may also be written as

$$P_{elm} = EI_a \qquad (4.17)$$

4.5.6 Electromagnetic (developed) torque

The electromagnetic torque developed by the DC machine is

$$T_{elm} = \frac{P_{elm}}{\Omega} = \frac{EI_a}{2\pi n} = \frac{VI_a \mp I_a^2 \sum R_a \mp \Delta V_{br} I_a}{2\pi n}$$

$$= \frac{N}{a}\frac{p}{2\pi} I_a \Phi = c_T \Phi I_a \qquad (4.18)$$

where

$$c_T = \frac{Np}{2\pi a} = \frac{c_E}{2\pi} \qquad (4.19)$$

is the *torque constant*. Note that the electromagnetic torque is proportional to the armature current.

4.5.7 Rotor and commutator linear speed

The rotor (armature) linear speed is

$$v = \pi D_a n \qquad (4.20)$$

where D_a is the external diameter of the rotor. Similarly, the commutator linear speed is

$$v_C = \pi D_C n \qquad (4.21)$$

where D_C is the external diameter of the commutator.

4.5.8 Input power, output power and efficiency

A motor converts an electrical input power

$$P_{in} = VI_a \qquad (4.22)$$

into a mechanical output power that is expressible as

$$P_{out} = \Omega T \qquad (4.23)$$

where T is the shaft (output) torque. The efficiency

$$\eta = \frac{P_{out}}{P_{in}} \qquad (4.24)$$

In generators, the input power $P_{in} = \Omega T$ and the output electrical power $P_{out} = VI_a$.

4.5.9 Losses

The DC machine losses are

$$\sum \Delta P = \Delta P_a + \Delta P_f + \Delta P_{se} + \Delta P_{Fe} + \Delta P_{br} + \Delta P_{rot} + \Delta P_{str} \quad (4.25)$$

where each component of the total power loss is listed below:

- The armature winding losses:

$$\Delta P_a = I_a^2 \sum R_a \quad (4.26)$$

- The shunt-field winding loss:

$$\Delta P_f = I_f^2 R_f \quad (4.27)$$

- The series-field winding loss:

$$\Delta P_{se} = I_{se}^2 R_{se} \quad (4.28)$$

- The armature core loss (see also Section 1.3):

$$\Delta P_{Fe} - \Delta P_{ht} + \Delta P_{et} + \Delta P_{hy} + \Delta P_{ey} \quad (4.29)$$

- The brush-drop loss:

$$\Delta P_{br} = I_a \Delta V_{br} \approx 2I_a \quad (4.30)$$

- The rotational losses:

$$\Delta P_{rot} = \Delta P_{fr} + \Delta P_{wind} + \Delta P_{vent} \quad (4.31)$$

- The stray load losses:

$$\Delta P_{str} \approx 0.01 P_{out} \quad (4.32)$$

In the above equations, I_f is the shunt-field current, I_{se} is the series-field current, $\Delta P_{ht} \propto f B_t^2$ are the hysteresis losses in the armature teeth, $\Delta P_{et} \propto f^2 B_t^2$ are the eddy-current losses in the armature teeth, $\Delta P_{hy} \propto f B_y^2$ are the hysteresis losses in the armature yoke, $\Delta P_{ey} \propto f^2 B_y^2$ are the eddy-current losses in the armature yoke, f is the frequency of the armature flux (current), B_t is the magnetic flux density in the armature teeth, B_y is the magnetic flux density in the armature yoke, ΔP_{fr} are the friction losses (due to the bearings and commutator–brushes), ΔP_{wind} are the windage losses, and ΔP_{vent} are the ventilation losses.

The frequency of the armature current is

$$f = pn \quad (4.33)$$

It is sometimes convenient to express the motor losses as a function of the motor's efficiency. Thus,

$$\sum \Delta P = P_{in} - P_{out} = \frac{P_{out}}{\eta} - P_{out} = P_{out}\frac{1-\eta}{\eta} \qquad (4.34)$$

4.5.10 Armature line current density

The *armature line current density*

$$A = \frac{NI_a}{2a\pi D_a} \qquad (4.35)$$

expresses the *armature electric loading*. Physically, this is the total armature current $0.5NI_a$ per armature periphery πD_a.

4.6 Armature reaction

The flux lines produced by the field winding alone are shown in Fig. 4.17a. The brushes are in neutral axis, i.e., in the q-axis. If the armature current $I_a = 0$, the lines of the magnetic flux are symmetrical with respect to the center axis of the main poles (d-axis). The normal component of the magnetic flux density B_g in the air gap as a function of the rotor circumferential distance has a trapezoidal shape, as shown in Fig. 4.17b.

Fig. 4.18a shows the flux lines produced by the armature winding alone when the brushes are in neutral axis. The armature MMF F_a distribution in the circumferential direction is a tooth-saw waveform (Fig. 4.18b) and the distribution of the normal component of the armature magnetic flux density corresponds to the distribution of the MMF except for the regions of brushes.

Superposition of the rotor and stator magnetic fields for neutral position of brushes is shown in Fig. 4.19. The shape of the normal component of the magnetic flux density distribution depends on how much the machine is loaded. Magnetic saturation of pole shoes distorts the resultant magnetic flux density waveform (curve 2 in Fig. 4.19).

The armature reaction on the field excitation winding has the following effects:

- Distortion of the resultant field (shift of the neutral axis);
- Weakening of the magnetic flux when the magnetic circuit is saturated (Fig. 4.20).

The armature reaction can be compensated by:

- Magnetic flux of the interpoles (compoles);
- Magnetic flux of the commutation winding that is placed in slots of the main poles.

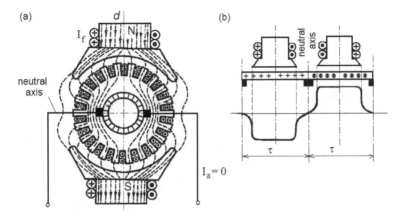

Fig. 4.17. Magnetic field of the stator field excitation winding: (a) magnetic flux lines; (b) magnetic flux density normal component distribution in circumferential direction.

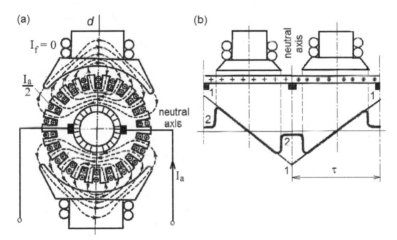

Fig. 4.18. Magnetic field of the rotor armature winding corresponding to brushes in neutral position: (a) magnetic flux lines; (b) MMF and magnetic flux density normal component distribution in circumferential direction. 1 – MMF, 2 – magnetic flux density.

4.7 Classification of DC machines according to armature and field winding connections

DC machines are reversible machines and can operate either as motors or generators. The motors absorb the electrical energy of DC current and convert it into mechanical energy. The generators are driven mechanically by a prime mover and convert mechanical energy into electrical DC energy.

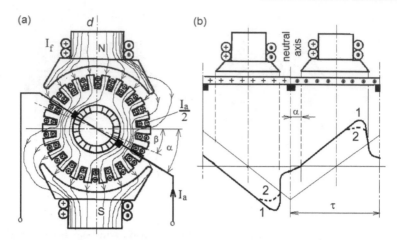

Fig. 4.19. Superposition of stator amd rotor magnetic fields: (a) magnetic flux lines; (b) magnetic flux density normal component distribution in circumferential direction. $\alpha-$ shift angle of brushes to obtain proper commutation without sparking in machines without interpoles, $\beta-$ shift angle (in the direction of rotation) of geometrical neutral axis under load, 1 – resultant waveform of normal component of magnetic flux density as a sum of field and armature magnetic flux density, 2 – resultant waveform of normal component of magnetic flux density deformed by magnetic saturation of pole shoes.

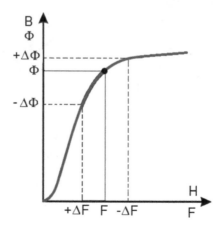

Fig. 4.20. Effect of magnetic saturation on the magnetic flux. $F = N_f I_f$ – MMF of the field winding, ΔF – increase in the armature MMF.

DC machines may be classified according to their armature and field winding connections as follows:

(a) *Separately-excited machines*, in which the field winding is fed from a source that is separate to that of the armature winding or the machine is excited by a system of PMs;

(b) *Shunt machines*, in which the armature and field windings are connected in parallel;

(c) *Series machines*, in which the armature and field windings are connected in series;

(d) *Compound machines* with two field windings, i.e., series and shunt (parallel).

Symbols for windings of DC brush machines are as follows:

- A1A2 – armature winding;
- B1B2 – interpole (compole) winding;
- C1C2 – compensating winding;
- D1D2 – series field excitation winding;
- E1E2 – shunt (parallel) field excitation winding;
- F1F2 – separate field excitation winding.

4.8 DC generators

4.8.1 Separately-excited generator

The circuit diagram of a separately-excited generator is shown in Fig. 4.21.

The *open-circuit (no-load) characteristic* (Fig. 4.22) is the variation of no-load voltage V_0 (EMF E) with the field excitation current I_f at constant speed $n = const$ and zero armature current $I_a = 0$.

The *external characteristic* (Fig. 4.23) is the variation of the terminal voltage with the armature current I_a at constant speed $n = const$ and constant excitation flux $\Phi_f = const$ ($I_f = const$). The no load voltage V_0 is at $I_a = 0$ and the short circuit armature current is when EMF $E = 0$, i.e.,

$$I_{ash} = -\frac{V + \Delta V_{br}}{\sum R_a} \tag{4.36}$$

The armature current of a separately-excited generator is

$$I_a = \frac{E - V - \Delta V_{br}}{\sum R_a} \tag{4.37}$$

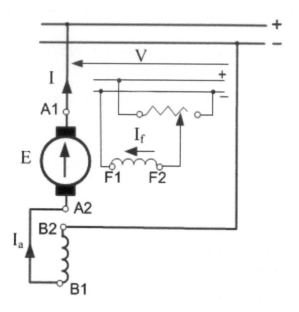

Fig. 4.21. Circuit diagram of a DC separately-excited generator.

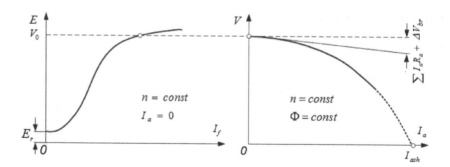

Fig. 4.22. Characteristics of a DC separately-excited generator: (a) open-circuit characteristic; (b) external characteristic.

4.8.2 Shunt generator

The circuit diagram of a shunt generator is shown in Fig. 4.23. Note that "shunt" is an early word that designated a parallel connection. The field excitation winding E1E2 is in parallel with the armature winding A1A2. The armature current I_a is the sum of the external current I delivered to the load and the field current, i.e.,

$$I_a = I + I_f \approx I \tag{4.38}$$

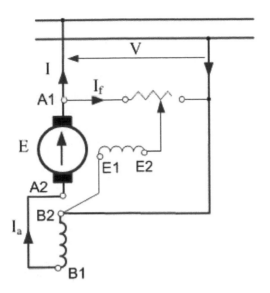

Fig. 4.23. Circuit diagram of a DC shunt generator. The EMF E_r corresponds to residual magnetic flux.

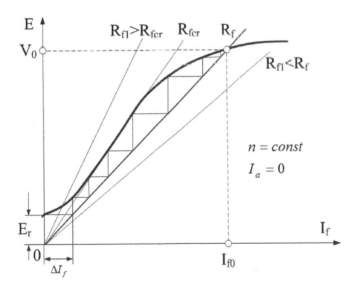

Fig. 4.24. Self-excitation of a DC shunt generator.

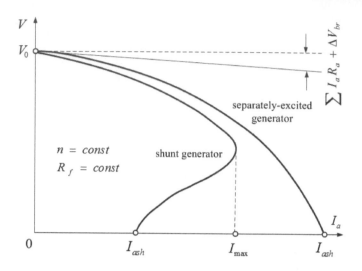

Fig. 4.25. External characteristic of a DC shunt generator.

Since $I_a \gg I_f$, it can be assumed that $I \approx I_a$.

A shunt generator is a self-excited generator if certain conditions are met. Conditions to be met for self-excitation are (Fig. 4.24):

- Residual magnetism must be present;
- The magnetic excitation flux must be in the same direction as the residual magnetic flux;
- The resistance of the field excitation circuit R_f must be less than the critical resistance R_{fcr}, i.e., $R_f < R_{fcr}$.

The external characteristic $V = f(I_a)$ at $n = const$ and $R_f = const$ is shown in Fig. 4.25. The critical resistance R_{fcr} is such a resistance above which the open-circuit nonlinear characteristic $E = f(I_f)$ will never crisscross the straight line $R_f I_f = f(I_f)$.

Example 4.2

A 60-kW, 960-rpm, 230-V, 260.2-A DC generator operates as a separately-excited generator at nominal speed and nominal voltage feeding an external circuit with such resistance that the current drawn by the load is equal to the nominal armature current. Resistances of winding at steady temperature and nominal operation are: armature winding resistance $R_a = 0.035\ \Omega$ and resistance of interpole winding $R_{int} = 0.015\ \Omega$. The brush voltage drop is $\Delta V_{br} = 2$ V. Assuming constant field excitation current, find the armature terminal voltage, current and delivered power to the external load for:

(a) The speed of prime mover decreases 15%;
(b) The speed of prime mover increases 10%.

Solution

(a) The speed of prime mover decreases 15%, i.e., $n = 0.85n_n$

The load resistance is

$$R_L = \frac{V_n}{I_{an}} = \frac{230.0}{260.2} = 0.885 \ \Omega$$

When a DC machine operates as a separately-excited generator, the load current is equal to the armature current. The nominal EMF is

$$E_n = V_n + I_{an}\sum R_a + \Delta V_{br} = V_n + I_{an}(R_a + R_{int}) + \Delta V_{br}$$

$$= 230.0 + 260.2(0.035 + 0.015) + 2.0 \approx 245.0 \ \text{V}$$

When the speed drops 15%, the EMF decreases also 15% because the magnetic flux Φ at constant field excitation current $I_f = const$ and armature reaction being neglected is also constant, i.e.,

$$E = 0.85E_n = 0.85 \times 245.0 = 208.25 \ \text{V}$$

The current in external circuit

$$I_a = \frac{E - \Delta V_{br}}{R_a + R_{int} + R_L} = \frac{208.25 - 2.0}{0.035 + 0.015 + 0.885} = 221.0 \ \text{A}$$

Armature terminal voltage

$$V = E - I_a c(R_a + R_{int}) - \Delta V_{br} = 208.25 - 221.0(0.035 + 0.015) - 2.0 = 195.2 \ \text{V}$$

Power delivered by generator to external circuit (load)

$$P = VI_a = 195.2 \times 221.0 = 43,140.0 \ \text{W} = 43.14 \ \text{kW}$$

In this case the 15% drop in speed causes the following drop in the power delivered to the load:

$$\frac{60.0 - 43.14}{60}100\% = 28.1\%$$

(b) The speed of prime mover increases 10%, i.e., $1.1n_n$

At constant field excitation current $I_f = const$ with armature reaction being neglected the magnetic flux $\Phi = const$. Thus, the EMF

$$E = 1.1E_n = 1.1 \times 245.0 = 269.5 \text{ V}$$

Current delivered to the load

$$I_a = \frac{269.5 - 2.0}{0.035 + 0.015 + 0.885} = 286.0 \text{ A}$$

Armature terminal voltage

$$V = 269.5 - 286.0(0.035 + 0.015) - 2.0 = 263.2 \text{ V}$$

Power delivered by generator to external circuit (load)

$$P = VI_a = 263.2 \times 286.0 = 75\ 275.0 \text{ W} = 75.27 \text{ kW}$$

In this case the 10% increase in speed causes the following increase in the power delivered to the load

$$\frac{75.27 - 60.0}{60} 100\% = 25.5\%$$

The generator is overloaded. Continuous operation is prohibited because of eccessive temparature rise of the armature winding.

Example 4.3

The nominal speed of a 4-pole DC shunt generator is $n_n = 1200$ rpm. The armature wave winding consists of single-turn coils ($N_c = 1$) and is distributed in $s = 31$ slots. The number of commutator segments is $C = 155$. The resistance of armature winding at 75°C is $R_a = 0.06$ Ω. The field excitation winding has its resistance at 75°C $R_f = 96.0$ Ω and produces magnetic flux $\Phi = 0.02$ Wb practically independent of the load. Since the generator is not equipped with interpoles, to improve the commutation, the brushes were shifted by an angle $\alpha = 10°$ with respect to neutral axis (Fig. 4.19). The generator feeds $k = 100$ electric bulbs rated at 250 W and 220 V each. Assuming that the brush voltage drop is $\Delta V_{br} = 2.0$ V and the torque corresponding to no-load losses is $T_0 = 0.005T_{elm}$, find:

(a) EMF of the generator;
(b) Armature terminal voltage at full load and no load;
(c) Electromagnetic torque, shaft torque and the torque of prime mover.

Solution

(a) EMF of the generator
   The number of armature conductors

$$N = 2CN_c = 2 \times 155 \times 1 = 310$$

The EMF when the brushes are in neutral line and assuming the armature winding consists of full pitch coils

$$E' = \frac{N}{a}pn\Phi = \frac{310}{1} \times 2\frac{1200}{60} \times 0.02 = 248.0 \text{ V}$$

The number of slots per pole $Q = s/(2p) = 31/4 = 7\frac{3}{4}$. It is likely that this machine has coils with extended span $Q' = 8$. The difference between practical and fractional coil span

$$\epsilon = |Q - Q'| = \left|\frac{31}{4} - 8\right| = \frac{1}{4} = 0.25$$

Electrical angle between neighboring slots

$$\alpha_{el} = p\frac{360°}{s} = 2\frac{360°}{31} = 23°15'$$

is equal to the electrical angle between phasors of slot voltages (Fig. 5.12). The angle between the EMF phasors induced in full pitch coil and extended coil is

$$\gamma = \frac{1}{2}\alpha_{el}\epsilon = \frac{1}{2}23°15' \times 0.25 = 2°55'$$

The real EMF of the DC generator

$$E = E'\cos(\gamma)\cos(\alpha) = 248.0\cos(2°55')\cos(10°) = 244.0 \text{ V}$$

(b) Armature terminal voltage at full load and no load

Resistance of a bulb

$$R_b = \frac{V_b^2}{P_b} = \frac{220.0^2}{250.0} = 194.0 \text{ }\Omega$$

Resistance of $k = 100$ bulbs in parallel (load resistance)

$$R_L = \frac{R_b}{k} = \frac{194.0}{100} = 1.94 \text{ }\Omega$$

In DC shunt generator the terminal voltage is across the load resistance R_L and field winding resistance R_f in parallel, i.e.,

$$V = I_a\frac{R_L R_f}{R_L + R_f}$$

The EMF of loaded generator

$$E = V + I_a R_a + \Delta V_{br} = I_a\frac{R_L R_f}{R_L + R_f} + I_a R_a + \Delta V_{br}$$

Hence, the armature current

$$I_a = \frac{E - \Delta V_{br}}{R_L R_f/(R_L + R_f) + R_a} = \frac{244.0 - 2.0}{1.94 \times 96.0/(1.94 + 96.0) + 0.06} = 123.5 \text{ A}$$

Voltage across the armature terminals of loaded generator

$$V = 123.5 \frac{1.94 \times 96.0}{1.94 + 96.0} = 234.5 \text{ V}$$

Field excitation current

$$I_f = \frac{V}{R_f} = \frac{234.5}{96.0} = 2.33 \text{ A}$$

At no load the armature current is equal to the field excitation current, i.e., $I_{a0} = I_f$ and the EMF is

$$E = I_f(R_f + R_a) + \Delta V_{br}$$

Hence, the field current at no load

$$I_f = \frac{E - \Delta V_{br}}{R_f + R_a} = \frac{244.0 - 2.0}{96.0 + 0.06} = 2.52 \text{ A}$$

Armature terminal voltage of unloaded generator

$$V_0 = I_f R_f = 2.52 \times 96.0 = 242.0 \text{ V}$$

At no load the armature terminal voltage $V_0 = 242.0$ V is very close to the EMF $E = 244.0$ V and higher than under load ($V = 234.5$ V).

(c) Electromagnetic torque, shaft torque and the torque of prime mover

Torque constant

$$c_T = \frac{1}{2\pi} \frac{N}{a} p = \frac{1}{2\pi} \frac{310}{1} \times 2 = 98.73$$

Electromagnetic torque when the brushes are in neutral line

$$T'_{elm} = c_T \Phi I_a = 98.73 \times 0.02 \times 123.5 = 244.0 \text{ Nm}$$

Electromagnetic torque when the brushes are shifted by $\alpha = 10°$

$$T_{elm} = T'_{elm} \cos(\alpha) = 244.0 \cos(10°) = 244.0 \times 0.985 \approx 240.0 \text{ Nm}$$

Shaft torque

$$T = T_{elm} + 0.05 T_{elm} = 1.05 T_{elm} = 1.05 \times 240 = 252.0 \text{ Nm}$$

Power of the prime mover

$$P_{in} \geq 2\pi n T = 2\pi \left(\frac{1200}{60}\right) \times 252.0 = 31,650 \text{ W} = 31.65 \text{ kW}$$

4.9 DC motors

4.9.1 DC shunt motor

In the DC shunt motor, the shunt field windings are connected in parallel with the armature winding. The circuit diagram of a shunt DC motor is shown in Fig. 4.26. The total line current, I, is given by

$$I = I_a + I_f \tag{4.39}$$

where I_a is the armature current and I_f is the shunt-field current.

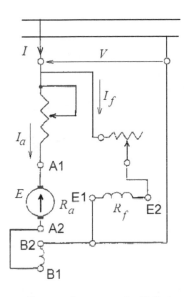

Fig. 4.26. Circuit diagram of a DC shunt motor.

The speed of the motor, as a function of the terminal voltage V, the magnetic flux Φ and the armature current I_a, can be expressed as follows:

$$n = \frac{E}{c_E \Phi} = \frac{V - I_a \sum R_a - \Delta V_{br}}{c_E \Phi} = \frac{V}{c_E \Phi} - \frac{\sum R_a}{c_E \Phi} I_a - \frac{\Delta V_{br}}{c_E \Phi} \tag{4.40}$$

If the shaft torque is zero, then $I_a \approx 0$ and the no-load speed is

$$n_o \approx \frac{V - \Delta V_{br}}{c_E \Phi} \tag{4.41}$$

Neglecting the armature reaction ($\Phi = const$) and neglecting any temperature fluctuation ($\sum R_a = const$), the speed equation can be brought to the simple form:

$$n = n_o - c_a I_a \tag{4.42}$$

in which $c_a = \sum R_a/(c_E \Phi)$ is a constant (if $\Phi = const$, i.e., there is no armature reaction). Since $I_a = T_{elm}/(c_T \Phi)$, the speed can be expressed as a function of the electromagnetic torque, i.e.:

$$n = n_o - K_a T_{elm} \tag{4.43}$$

where $K_a = \sum R_a/(c_E c_T \Phi^2)$ is also a constant (again, if $\Phi = const$).

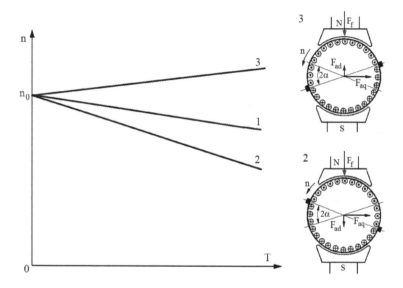

Fig. 4.27. Speed-torque characteristics of a DC shunt motor: 1 – brushes are on the geometrical neutral line, 2 – brushes are shifted ahead from the neutral line, 3 – brushes are shifted back from the neutral line.

The speed–torque characteristics are shown in Fig. 4.27. The curves in Fig. 4.27 are affected by the direct axis armature reaction (magnetizing or demagnetizing action). In general, the action of the armature MMF on the main MMF is termed the *armature reaction*.

4.9.2 DC series motor

In a DC series motor, the armature and series field windings are connected in series (see Fig. 4.28). The armature current and the field current are therefore the same:

$$I = I_a = I_f \tag{4.44}$$

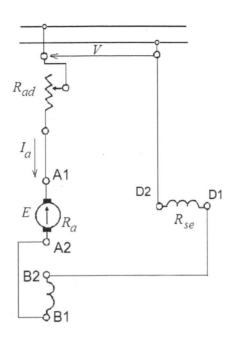

Fig. 4.28. Circuit diagram of a DC series motor.

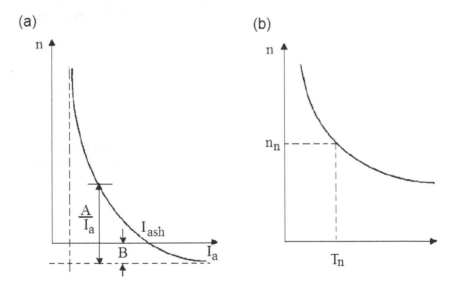

Fig. 4.29. Characteristics of a DC series motor: (a) speed–armature current, (b) speed–torque.

If the magnetic circuit is unsaturated, the magnetic flux $\Phi \propto I_a$ and

$$\Phi = c_I I_a \tag{4.45}$$

The electromagnetic torque of an unsaturated motor is proportional to the square of the armature current:

$$T_{elm} = c_T \Phi I_a \approx c_T c_I I_a^2 \tag{4.46}$$

The speed equation for such a motor is

$$n = \frac{E}{c_E \Phi} = \frac{V - I_a \sum R_a - \Delta V_{br}}{c_E c_I I_a} \approx \frac{V - \Delta V_{br}}{c_E c_I I_a} - \frac{\sum R_a}{c_E c_I} = \frac{A}{I_a} - B \tag{4.47}$$

where

$$A = \frac{V - \Delta V_{br}}{c_E c_I}; \qquad B = \frac{\sum R_a}{c_E c_I} \tag{4.48}$$

At no load ($\Phi \approx 0$, $I_a \approx 0$), the speed becomes very high. This is dangerous because mechanical damage will eventually ensue, such as rupture of the bandings, or damage to the armature winding, the commutator, or other important components. For this reason, a series motor should be operated so as to exclude the possibility of starting it without a load (by means of a permanent coupling, toothed gear, worm gear, etc.)

Since $T_{elm} = c_T c_I I_a^2$, the armature current can be expressed as a function of the torque:

$$I_a = \sqrt{\frac{T_{elm}}{c_T c_I}} = c\sqrt{T_{elm}} \tag{4.49}$$

where $c = 1/\sqrt{c_T c_I}$.

The speed-torque curve shown in Fig. 4.29b has a similar shape to the speed current relationship shown in Fig. 4.29a, and so the speed-torque relationship can be approximated by the equation

$$n = \frac{A}{I_a} - B = \frac{A}{c\sqrt{T_d}} - B = \frac{K}{\sqrt{T_d}} - B \tag{4.50}$$

where $K = A/c$.

4.10 Compound-wound motor

A compound-wound DC motor has two field windings: a shunt field winding and a series field winding (see Fig. 4.30). When the field windings are *cumulatively compounded*, their magnetizing forces are added, increasing the flux.

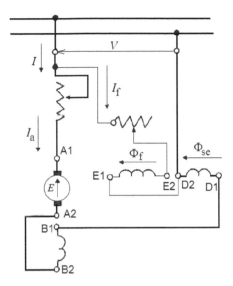

Fig. 4.30. Circuit diagram of a DC compound-wound motor: $\Phi = \Phi_f + \Phi_{se}$ for cumulative compound, $\Phi = \Phi_f - \Phi_{se}$ for differential compound.

When the field windings are *differentially connected*, the series field winding opposes the shunt winding. The resultant exciting flux is

$$\sum \Phi = \Phi_f \pm \Phi_{se} \tag{4.51}$$

where the "+" sign is for a cumulative compound motor and the "−" sign is for the differential case (which is almost never used). The load current is similar to that of a shunt motor and given by eqn (4.39). The speed of the compound-wound motor,

$$n = \frac{V - I_a \sum R_a - \Delta V_{br}}{c_E(\Phi_f \pm \Phi_{se})} \tag{4.52}$$

is inversely proportional to the total excitation flux.

Cumulative compound motors have characteristics resembling those of a series motor. The flux–armature current characteristic and the torque–armature current characteristic for a DC cumulative compound motor are shown in Fig. 4.31a. The shunt field winding limits the excessive speed increase when the load is removed, since in this case the flux Φ_f remains, and hence sets a limit on the no-load speed $n_o \approx (V - \Delta V_{br})/(c_E\Phi_f)$ (Fig. 4.31b).

4.10.1 Starting

To decrease the current inrush when starting a motor, a starting rheostat is inserted into the armature circuit (as in Fig. 4.32). By combining eqn (4.7) for

(a)   (b)

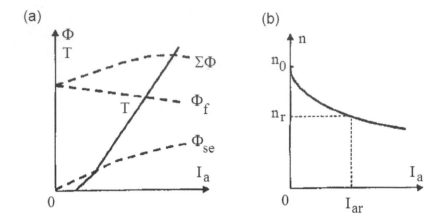

Fig. 4.31. Characteristics of a DC cumulative compound motor: (a) flux–armature current and torque–armature current curves, (b) speed–torque curve.

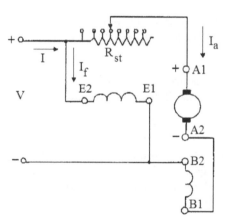

Fig. 4.32. Circuit diagram of a DC shunt motor with a starting rheostat R_{st}.

motoring mode (the "+" sign), in which $\sum R_a = \sum R_a + R_{st}$ (where R_{st} is the variable resistance of the starting rheostat) with eqn (4.9), the armature current can be expressed as a function of speed. Thus,

$$I_a = \frac{V - \Delta V_{br} - c_E n \Phi}{\sum R_a + R_{st}} \tag{4.53}$$

Suppose now that $R_{st} = 0$. At the first instant of starting the motor, the speed $n = 0$ and the EMF $E = 0$. Hence, at this instant, the starting current is equal to the blocked-rotor (short-circuit) current:

$$I_{ash} = \frac{V - \Delta V_{br}}{\sum R_a} \gg I_{an} \tag{4.54}$$

where I_{an} is the nominal (rated) armature current. To reduce the starting current I_{ash}, a starting rheostat is connected in series with the armature winding so that the resultant resistance of the armature circuit is $\sum R_a + R_{st}$. Note that the resistance R_{st} is greatest at the instant when the motor is started. Thereafter, the blocked-rotor armature current I_{ash} drops to the value of

$$I_{amax} = \frac{V - \Delta V_{br}}{\sum R_a + R_{st}} \qquad (4.55)$$

As the speed increases from 0 to n', the EMF increases too, and

$$I_{amin} = \frac{V - E' - \Delta V_{br}}{\sum R_a + R_{st}} \qquad (4.56)$$

where $E' = c_E n' \Phi < E$. When the speed reaches its nominal value n_n, the starting rheostat can be removed, since at that instant

$$I_a = I_{an} = \frac{V - c_E n_n \Phi - \Delta V_{br}}{\sum R_a} \qquad (4.57)$$

4.10.2 Speed control of DC motors

The relationship

$$n = \frac{1}{c_E \Phi}[V - I_a(\sum R_a + R_{rhe}) - \Delta V_{br}] \qquad (4.58)$$

summarizes all the important contributions to the speed of a DC motor. In eqn (4.58) R_{rhe} is the resistance of the speed control rheostat. In particular, eqn (4.58) shows that the speed of a DC motor can be controlled by changing:

- The supply mains voltage V;
- The armature-cirucit resistance $\sum R_a + R_{rhe}$;
- The field flux Φ.

Example 4.4

A 4-kW, 220-V, 20.6-A, 1200-rpm separately-excited DC motor has an armature circuit resistance $\sum R_a = 0.3\ \Omega$ and efficiency $\eta = 85.5\ \%$. If the terminal voltage is reduced to 50% of the rated voltage but the field excitation current is constant and the shaft torque is $T = 20\ \text{Nm} = const$ find:

(a) The speed n;
(b) The speed n' with additional armature series resistance $R_{rhe} = 5\Omega$;

(c) The speed–torque characteristics for $\sum R_a = 0.3\ \Omega$ and armature series resistance $\sum R_a + R_{rhe} = 5.3\ \Omega$.

The armature reaction is neglected and the brush voltage drop is $\Delta V_{br} = 2$ V.

Solution

(a) The speed n

The nominal (rated) EMF

$$E = V - I_a \sum R_a - \Delta V_{br} = 220.0 - 20.6 \times 0.3 - 2.0 = 211.82\ \text{V}$$

The nominal (rated) angular speed

$$\Omega_n = 2\pi n = 2\pi\frac{1200}{60} = 125.6\ \text{rad/s}$$

The armature constant c_E multiplied by the magnetic flux Φ

$$c_E\Phi = \frac{E}{n} = \frac{211.82 \times 60}{1200} = 10.591\ \text{Vs}$$

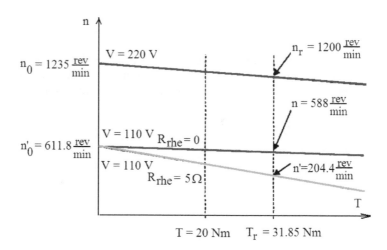

Fig. 4.33. Speed–torque characteristics for $V = 220$ V ($R_{rhe} = 0$), $V = 110$ V ($R_{rhe} = 0$), and $V = 110$ V and with the armature circuit rheostat $R_{rhe} = 5.0\ \Omega$.

The nominal (rated) shaft torque

$$T_n = \frac{P_{out}}{\Omega_n} = \frac{4000}{125.6} = 31.85\ \text{Nm}$$

The nominal electromagnetic torque

$$T_{elmn} = \frac{c_E \Phi}{2\pi} I_a = \frac{10.591}{2\pi} 20.6 = 34.72 \text{ Nm}$$

The torque corresponding to the rotational losses at no load

$$T_0 = T_{elmn} - T_n = 34.72 - 31.85 = 2.87 \text{ Nm}$$

The electromagnetic torque developed by the motor at $T = 20$ Nm

$$T_{elm20} = T_0 + T = 2.87 + 20.0 = 22.87 \text{ Nm}$$

The armature current at shaft torque $T = 20$ Nm

$$I_{a20} = \frac{T_{elm20}}{c_E \Phi} 2\pi = \frac{22.87}{10.591} 2\pi = 13.57 \text{ A}$$

The speed at $V = 110$ V and $T = 20$ Nm

$$n = \frac{1}{c_E \Phi}(0.5V - I_{a20} \sum R_a - \Delta V_{br})$$

$$= \frac{1}{10.591}(110.0 - 13.57 \times 0.3 - 2.0) = 9.81 \text{ rev/s} = 588 \text{ rpm}$$

The corresponding angular speed

$$\Omega = 2\pi n = 2\pi \times 9.81 = 61.64 \text{ rad/s}$$

(b) The speed at $V = 110$ V, $T = 20$ Nm and armature resistance $R_{rhe} = 5 \Omega$

$$n' = \frac{1}{c_E \Phi}[0.5V - I_{a20}(\sum R_a + R_{rhe}) - \Delta V_{br})]$$

$$= \frac{1}{10.591}[110.0 - 13.57(0.3 + 5.0) - 2.0] = 3.4 \text{ rev/s} = 204.4 \text{ rpm}$$

The corresponding angular speed

$$\Omega' = 2\pi n' = 2\pi \times 3.4 = 21.36 \text{ rad/s}$$

The no-load speed at $V = 220$ V

$$n_0 = \frac{1}{c_E \Phi}(V - \Delta V_{br}) = \frac{1}{10.591}(220 - 2) = 20.58 \text{ rev/s} = 1235 \text{ rpm}$$

The no-load speed at $V = 110$ V

$$n_0' = \frac{1}{c_E \Phi}(0.5V - \Delta V_{br}) = \frac{1}{10.591}(110 - 2) = 10.20 \text{ rev/s} = 611.8 \text{ rpm}$$

(c) The speed–torque curves for $\sum R_a = 0.3 \Omega$ and $\sum R_a + R_{rhe} = 5.3 \Omega$
The speed-torque characteristics $n = f(T)$ for $V = 220$ V ($R_{rhe} = 0$), $V = 110$ V ($R_{rhe} = 0$), and $V = 110$ V and with the armature circuit rheostat $R_{rhe} = 5.0 \Omega$ are plotted in Fig. 4.33.

4.11 Braking

In some duties that require rapid retardation, it is necessary to brake a motor during the operating period. A speed reduction from Ω_1 to Ω_2 during braking time $0 \leq t \leq t_b$ of a system of inertia J requires a braking power P_b (see Table 1.6). These quantities may be related by the equation

$$E_k = \int_0^{t_b} P_b dt = \frac{1}{2} J(\Omega_1^2 - \Omega_2^2) \tag{4.59}$$

In practice the braking time t_b is shortened by rotational losses and by the presence of any load torque.

The three electrical braking methods which are now considered in terms of shunt and series DC motors are

- Rheostatic braking;
- Counter current braking (plugging);
- Regenerative braking.

4.11.1 Braking a shunt DC motor

Rheostatic braking

The armature is disconnected from the supply and then connected across a resistor R_b. The machine then acts as a generator, driven by its stored kinetic energy and dissipating power in its armature circuit resistance. Since $V = 0$ and $E < 0$, the armature current is reversed, and

$$I_a = \frac{-E}{\sum R_a + R_b} < 0 \tag{4.60}$$

The brush voltage drop ΔV_{br} has been neglected.

Counter-current braking

The armature terminal connections are reversed so that the supply voltage, now augmented by the rotational EMF, imposes a large current and a strong braking torque. To intensify the braking, the resistance $\sum R_a + R_b$ is reduced by changing R_b. The braking current is very high (since $E < 0$), and is written as

$$I_a = \frac{-V - E}{\sum R_a + R_b} \tag{4.61}$$

Regenerative braking

If the rotational EMF is greater than the applied voltage (that is, if $V < E$), then the machine generates electrical energy and produces a braking torque by current reversal where

$$I_a = \frac{V - E}{\sum R_a + R_b} \tag{4.62}$$

This is maintained down to the speed at which the EMF and voltage balance. The speed of a shunt motor must be higher than the no-load speed n_o. A motor undergoes regenerative braking if forced by the driven machine to run at a speed exceeding n_o. Visualize, for example, a train riding downhill.

4.11.2 Braking a series DC motor

Rheostatic braking

The field and armature windings, being connected in series, are cut off from the power circuit and closed on a load resistance R_b.

Counter-current braking

This type of braking can be accomplished in two ways: (a) when the driven machine forces the motor to rotate in the opposite direction to the developed torque, and (b) by reversing both the armature current and the direction of rotation.

Regenerative braking

The field and armature windings must be connected in parallel or separately excited.

4.12 Permanent magnet DC commutator motors

4.12.1 Permanent magnet materials

A permanent magnet (PM) material is described by the *demagnetization curve* which is a portion of the full B—H hysteresis loop located in the second quadrant in the magnetic flux density B versus magnetic field intensity H coordinate system. The *coercive force* H_c corresponds to $B = 0$ and the *remanent magnetic flux density* B_r corresponds to $H = 0$.

There are three classes of PM materials that are used for electric motors. These are listed below, and their demagnetization curves are given in Fig. 4.34:

- Alnicos (Ni, Al, Fe, Co, Cu, Ti);
- Ceramics (ferrites), e.g., barium ferrite $BaO \times 6Fe_2O_3$;
- Rare-earth materials, e.g., samarium–cobalt SmCo and neodymium–iron–boron NdFeB.

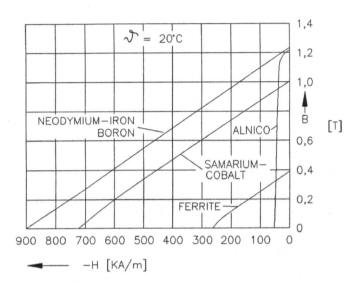

Fig. 4.34. Demagnetization curves for different PM materials.

The main advantages of *Alnico* are its high magnetic remanent flux density and low temperature coefficients for B_r and H_c. These advantages permit a high air gap flux density and high operating temperatures. Unfortunately, the demagnetization curve is extremely nonlinear and the coercive force is very low. Therefore, it is very easy not only to magnetize but also to demagnetize Alnico. Alnico is used in PM disk (axial flux) brush motors, where the air gap is relatively large. This results in a negligible armature reaction. Sometimes, the Alnico PMs are protected from the armature field, and consequently from demagnetization, by means of additional soft-iron pole shoes. Alnico magnets can still be used in DC brush motors having ratings in the range 500 W to 150 kW.

A *ferrite* has a higher coercive force than that of Alnico, but at the same time has a lower remanent magnetic flux density. The temperature coefficient of the remanent flux density is relatively high. The main advantages of ferrites are their low cost and their very high resistance which means that no eddy-current losses are suffered in the PM volume. Ferrite magnets are most economical when used in fractional horsepower motors and they tend to have an economic advantage over Alnico up to about 7.5 kW.

In the 1970s and the 1980s, great progress in improving the available energy density $(BH)_{max}$ was achieved with the development of the *rare-earth* PMs. The rare-earth elements are (in general) not rare at all, but their natural minerals are usually mixed compounds. To produce a given rare-earth metal, several others, for which no commercial application exists, may also have to be refined from the available ore, and this limits the availability of such metals. The first generation of these new alloys was based on the composition $SmCo_5$ (samarium-cobalt), and has been in commercial production since the early 1970s. Today it is a well-established magnetic material. SmCo has the advantage of having high remanent flux density, high coercive force, a high energy product, a linear demagnetization curve and a low temperature coefficient. It is very suitable for building motors that have low volume, and which consequently display high specific powers and low moments of inertia. The cost is the only drawback. Both Sm and Co are relatively expensive due to their restricted supply.

With the discovery in the early 1980s of a second generation of rare-earth magnets based on the inexpensive metals Fe and Nd (neodymium) — a much more abundant rare-earth element than Sm — remarkable progress in lowering of raw material costs has been achieved. The NdFeB magnets, which are now produced in increasing quantities, have better magnetic properties than those of SmCo, but unfortunately only at room temperature. Several of their properties, and particularly the coercive force, are strongly temperature dependent (Curie temperature $= 310^0 C$). However, the manufacturers of PMs promise NdFeB products that will have lower temperature coefficients and even lower cost in due course. Although NdFeB is unfortunately susceptible to corrosion, NdFeB magnets have a great potential for a considerably improved *performance–to–cost* ratio for many applications. For this reason they have a major impact on the development and application of PM equipment.

The rare-earth magnet materials are costly, but are the best economic choice in small and medium power motors. They might have a cost advantage in very large motors (ship propulsion) as well.

4.12.2 Construction of DC permanent magnet motors

PMs mounted on the stator (Fig. 4.35) produce a constant field flux Φ. Furthermore, the electromagnetic torque, T_d, and the armature EMF, E, are

$$T_{elm} = k_T I_a \qquad \text{and} \qquad E = k_E n \qquad (4.63)$$

where $k_T = c_T \Phi$ and $k_E = c_E \Phi$, assuming that the magnetic flux of PMs $\Phi = const$. From the above equations and eqn (4.7) it is possible to obtain the steady-state speed n as a function of T_d for a given V, to arrive at

$$n = \frac{1}{k_E}(V - \Delta V_{br}) - \frac{\sum R_a}{k_E k_T} T_{elm} \qquad (4.64)$$

(a)

(b)

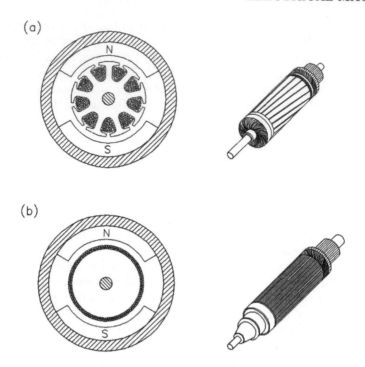

Fig. 4.35. Construction of DC PM commutator motors with laminated-core rotors: (a) slotted rotor, (b) slotless rotor.

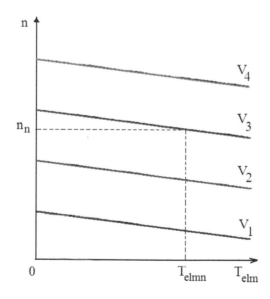

Fig. 4.36. Speed-torque curves at $V_1 < V_2 < V_3 < V_4$ for a DC PM commutator motor.

The family of curves $n = f(T_{elm})$ generated by eqn (4.64) are plotted in Fig. 4.36. As the torque is increased, the speed–torque characteristic at any given V is essentially horizontal, except for the drop due to the voltage drop $I_a \sum R_a$ across the armature-circuit resistance, to the brush voltage drop ΔV_{br}, and to the armature reaction. The speed–torque characteristics in Fig. 4.36 can be shifted vertically by controlling the applied terminal voltage V. Therefore, the speed of a load with an arbitrary speed–torque characteristic can be controlled by controlling V (with $\Phi = const$).

Magnetic circuit configurations of cylindrical-rotor PM motors for different types of PMs are shown in Figs 4.35, 4.37 and 4.38. The fundamnetal DC brush motors structures are

- The conventional slotted rotor (Fig.4.35a);
- The slotless (surface wound) rotor (Fig.4.35b);
- The moving-coil cylindrical rotor (Fig.4.37a, Fig.4.38);
- The moving-coil disk (pancake) rotor (Fig.4.37b,c).

The *slotted-rotor* and *slotless-rotor* PM brush motors have their armature winding fixed to the laminated core. Hence, the armature winding, the armature core, the commutator and the shaft comprise one integral part.

The *moving-coil DC motor* has its armature winding fixed to an insulating cylinder or disk which rotates between PMs or between PMs and a laminated core. Note that the moment of inertia of the rotor is very small, and that since in a moving-coil motor all iron cores are stationary (i.e., they do not move in the magnetic flux), no eddy currents or hysteresis losses are produced in them. The efficiency of a moving-coil motor is thus better than that of a slotted rotor motor.

The mechanical time constant

$$\tau_m = \frac{2\pi n_0 J}{T_{elmst}} = \frac{2\pi n_0 J}{c_T \Phi I_{ast}} = \frac{2\pi n_0 J}{k_T I_{ast}} \tag{4.65}$$

of moving-coil motors is much smaller than that of iron-core armature motors. In the above equation J is the moment of inertia of the rotor, n_0 is the no-load speed, T_{elmst} is the starting electromagnetic torque, c_T is the torque constant, Φ is the magnetic flux and I_{ast} is the starting armature current.

4.12.3 Slotted-rotor PM DC motors

The core of a slotted rotor (Fig. 4.35a) is made from laminated silicon steel sheet or carbon steel sheet, and the rotor windings are located in slots cut into the core. In the slotted rotor, the torque acts directly on the solid iron core, and not on the fragile coils alone. The slotted rotor is mechanically much more durable than a moving coil motor. A core having many slots is usually desirable because the greater the number of slots, the less the *cogging torque* and the less the electromagnetic noise. Cores having even numbers of slots are often

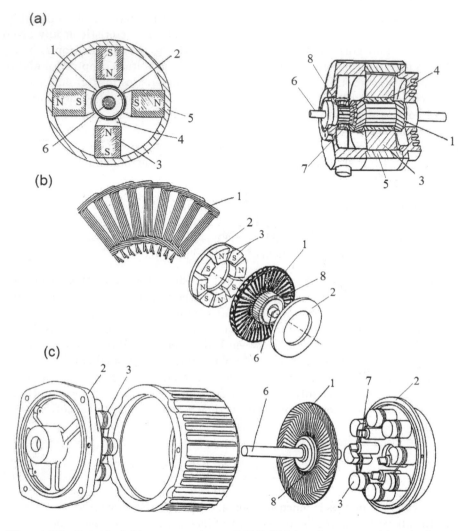

Fig. 4.37. Outside-field type moving coil DC PM brush motors: (a) cylindrical motor; (b) disk motor with wound rotor; (c) disc motor with printed rotor winding. 1 – moving coil armature winding, 2 – mild steel yoke, 3 – PM, 4 – pole shoe, 5 – mild steel frame, 6 – shaft, 7 – brush, 8 – commutator.

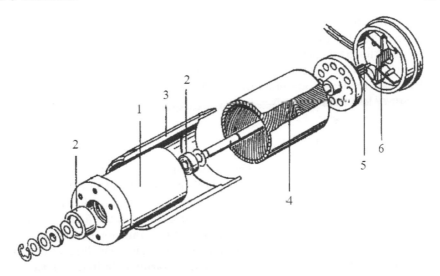

Fig. 4.38. Inside-field type cylindrical moving coil motor: 1 – PM, 2 – sleeve bearing, 3 – steel frame (magnetic circuit), 4 – armature winding, 5 – commutator, 6 – precious metal brushes.

used for the motors manufactured by an automated mass-production process because of their relative ease of production. However, cores with odd numbers of slots are often preferred because they exhibit lower cogging torque.

Twisting the rotor laminations (*skewing* as depicted in Fig. 4.35a) reduces the cogging torque that is produced by the interaction between the rotor teeth and the stator PMs.

4.12.4 Slotless rotor PM motors

Extremely low cogging torque can be produced by fixing the windings on a cylindrical iron core without using slots (as in Fig. 4.35b). In this case the torque is exerted directly on the conductors in accordance with Fleming's left-hand rule (Fig. 1.5a). However, it is likely that the flux will decrease since the gap between the rotor core and the pole shoes is large. Therefore, rare-earth magnets or large Alnico magnets must be used to ensure sufficient magnetic flux.

4.12.5 Moving-coil cylindrical motors

Cylindrical outside-field type

This type of motor (Fig. 4.37a) has the smallest electromechanical time constant. In order to obtain a small τ_m, the ratio Φ/J must be as large as possible.

One way to achieve this is to use an anisotropic Alnico magnet, which will produce more flux because it has a high remanence, B_r, and a low coercive force, H_c. Since Alnico magnets are easy to demagnetize, a long magnet, magnetized lengthwise, is used in order to avoid demagnetization. Some of the motors in which Alnico magnets are employed in this way can have a mechanical time constant of less than 1 ms.

Cylindrical inside-field type

Moving-coil motors of the inner-field type (Fig. 4.38), which are also known as coreless motors, are often used for applications of less than 10 W, but one also sometimes finds motors in this class with outputs near 30 W. This type of motor has a PM inside the moving-coil armature. Though the moment of inertia of this rotor is small, the mechanical time constant is not always low because little magnetic flux is obtained from the magnet which must be placed inside the armature and which is therefore necessarily of limited size. However, coreless motors are much used for driving the capstans of audio casette players and video tape recorders, the zoom lenses of cameras, etc. because: (a) they have very small size and high efficiency; and (b) they exhibit low cogging.

Fig. 4.39. Robotic vehicles for Mars Missions: (a) *Sojourner* (1996-97); (b) exploration rovers *Spirit* and *Opportunity* (2003 ongoing). Photo courtesy of NASA.

Inside-field type cylindrical moving coil motors[1] with NdFeB PMs and precious metal brushes have been used in robotic vehicles for Mars Missions [13]:

* *Sojourner* (Fig. 4.39a), built at NASA Jet Propulsion Laboratory, Los Angeles County, CA , used in the 1996-97 NASA Mars Pathfinder mission;
* *Spirit* and *Opportunity* (Fig. 4.39b), which landed on Mars in 2004.

---

[1] Manufactured by Maxon, Sachseln, Switzerland.

The most recent Mars lander called *Phoenix* (2007) was supported by Maxon inside-field type cylindrical moving coil motors too.

4.12.6 Disk-type motors

There are two main types of disk (pancake) brush motors: the wound-rotor motor and the printed armature winding motor.

In the *pancake wound-rotor motor* the winding is usually made of magnetic wires and moulded with resin (as in Fig. 4.37b). The type of commutator is identical to the conventional type. Motors of this type can often be found, for example, in radiator fans.

The *disk-type printed armature winding motor* is shown in Fig. 4.37c. The coils are stamped from pieces of sheet copper and then welded, forming a wave winding. When this motor was invented, the armature was made using a similar method to that by which printed circuit boards (PCBs) are manufactured. Hence this is called a printed winding motor. The magnetic flux of the printed motor can be produced using either Alnico or ferrite magnets.

Example 4.5

A two-pole, 380 W, 180 V, 1950 rpm, $\eta = 0.84$ DC brush motor with segmental PMs has $N = 920$ armature conductors and armature circuit resistance $\sum R_a = 5.84$ Ω. The effective length of the armature core $L_i = 0.064$ m, the armature diameter $D = 0.12$ m, the effective pole arc coefficient $\alpha_i = 0.75$, the number of pairs of armature parallel paths $a = 1$ and the brush voltage drop $\Delta V_{br} \approx 2$ V. Find: (a) the armature constant k_E and torque constant k_T, (b) the magnetic flux Φ and airgap magnetic flux density B_g, (c) the mechanical time constant T_m and (d) speed at 40% of the nominal voltage and nominal torque. The armature reaction is neglected.

Solution

(a) The armature constant and torque constant

The nominal (rated) armature current

$$I_a = \frac{P_{out}}{V\eta} = \frac{380}{180 \times 0.84} = 2.51 \text{ A}$$

The EMF

$$E = V - I_a \sum R_a - \Delta V_{br} = 180 - 2.51 \times 5.84 - 2 = 163.34 \text{ V}$$

The armature constant

$$k_E = \frac{E}{n} = \frac{163.34}{1950/60} = 5.026 \text{ Vs}$$

The torque constant

$$k_T = \frac{k_E}{2\pi} = \frac{5.026}{2\pi} = 0.8 \text{ Nm/A}$$

(b) The magnetic flux and air gap magnetic flux density

Since

$$k_T = \frac{N}{a}\frac{p}{2\pi}\Phi$$

the air gap magnetic flux density is

$$\Phi = \frac{2\pi a k_T}{Np} = \frac{2\pi \times 1 \times 0.8}{920 \times 1} = 0.00546 \text{ Wb}$$

The pole pitch

$$\tau = \frac{\pi D_a}{2p} = \frac{\pi \times 0.12}{2} = 0.1885 \text{ m}$$

The air gap magnetic flux density

$$B_g = \frac{\Phi}{\alpha_i \tau L_i} = \frac{0.00546}{0.75 \times 0.1885 \times 0.064} = 0.6 \text{ T}$$

(c) The mechanical time constant

The mass of the armature (rotor)

$$m_a = 1.1\rho\frac{\pi D^2}{4}L_i = 1.1 \times 8000\frac{\pi \times 0.12^2}{4} \times 0.064 \approx 6.37 \text{ kg}$$

It has been assumed that the average mass of the armature is $\rho = 8000$ kg/m^3 and 10% has been allowed for the commutator and shaft. The moment of inertia of the armature

$$J = m_a\frac{D^2}{8} = 6.37\frac{0.12^2}{8} = 0.01147 \text{ kgm}^2$$

The speed at no load

$$n_0 \approx \frac{V}{k_E} = \frac{180}{5.026} = 35.8 \text{ rev/s} = 2149 \text{ rpm}$$

The starting current $(E = 0)$

$$I_{ast} = \frac{V - \Delta V_{br}}{\sum R_a} = \frac{180 - 2}{5.84} = 30.48 \text{ A}$$

The mechanical time constant

$$\tau_m = \frac{2\pi n_0 J}{k_T I_{ast}} = \frac{2\pi 35.8 \times 0.01147}{0.8 \times 30.48} \approx 0.1 \text{ s}$$

(d) The speed at 40% nominal (rated) voltage and nominal torque

The speed–torque equation

$$n = \frac{1}{5.026}(V - \Delta V_{br}) - \frac{5.84}{0.8 \times 5.026} T_d = 0.199(V - \Delta V_{br}) - 1.452 T_d$$

For $V = 180$ V and $T_d = 2$ Nm the speed is $n = 35.422 - 2.904 = 32.5$ rev/s $= 1951$ rpm. For $V' = 0.4V = 0.4 \times 180 = 72$ V and $T_{elm} = 2$ Nm the speed is

$$n' = 0.199(72 - 2) - 1.452 \times 2 = 13.93 - 2.904 = 11.03 \text{ rev/s} \approx 662 \text{ rpm}$$

Summary

A *DC brush machine* converts the DC current energy into mechanical energy with controlled parameters, i.e., variable speed or variable torque (motor), or vice versa, i.e., mechanical energy into the DC energy (generator).

The *DC motor produces* electromagnetic torque as a result of interaction of the electric current in the rotor (armature) winding on the stator magnetic field. To obtain continuous rotation of the rotor, the current must be reversed when the coil with current is in neutral position with respect to the stator poles, i.e., the coils sides are in the space between the poles and the coil plane is perpendicular to the stator magnetic flux (Fig. 4.1). The so-called commutator is used to reverse the current, which is a *mechanical inverter*. The DC machine is *reversible machine* and can operate also as a DC generator. In a *DC generator*, the rotor (armature) is driven by a prime mover and as the coil rotates in the magnetic field excited by the stator pole, an EMF is induced in the coil. The commutator operates as a *mechanical rectifier*, because the armature EMF must always have the same polarity, independent of the rotor position. The DC EMF contains pulsations (Fig. 4.3). An increase in the number of commutator segments reduces the amplitude of pulsations and increases their frequency.

Fig. 4.4 shows the longitudinal and-cross sections of a typical DC brush motor with its electromagnetic excitation system and interpoles. The electromagnetic torque developed by this type of machine is created by two main windings: the armature winding, in which the EMF is induced, and the field winding, which produces the excitation magnetic flux. In typical designs, the armature winding is inserted into slots of a laminated rotor core (Fig. 4.5) while the field winding is located on salient stator poles (Fig. 4.6). The armature coils are interconnected through the commutator, which consists of a number of insulated copper segments (Fig. 4.7). The commutator is located on the same shaft as the armature (rotor) and rotates together with the armature winding. The armature (rotor) core must be laminated to reduce the core losses that may arise due to the AC current. The frequency of the armature current is $f = pn$, where p is the number of pole pairs and n is the speed [rev/s]. The electric contact between commutator and terminal board is maintained with the aid of brushes, often made of carbon or graphite (Fig. 4.8). The main pole consists of a pole core with a concentrated-parameter coil and a pole shoe (Fig. 4.11). There are smaller poles, which are called interpoles or commutating poles, between the main poles (Fig. 4.6). The interpole winding is connected in series with the armature winding. Interpoles produce an MMF in opposition to that of the armature winding in order to achieve commutation with reduced sparking at the brushes and to reduce the demagnetizing effect of the quadrature axis armature reaction MMF. In machines subject to heavy duties, the quadrature armature reaction is neutralized by means of a compensating winding embedded in the pole-shoe slots of the main poles (Fig. 4.12). Like the interpole winding, the compensating winding is connected in series with the armature winding.

There are two *basic armature windings* (Fig. 4.13) of DC brush machines: (a) *lap winding* and (b) *wave winding*. The armature windings of DC brush machines have neither beginning nor end and each coil is connected to the commutator segments. The brushes divide the armature winding into parallel paths (Fig. 4.15). The number of pairs of parallel paths a is equal to the number of pairs of brushes. If the armature current is I_a, the current of parallel path is $I_a = I_a/(2a)$. The voltage balance equation (4.7) for armature circuit of DC machines includes the terminal voltage V, EMF E, voltage drop across the armature circuit resistance $I_a R_a$ and brush voltage drop V_{br}. The armature circuit resistance $\sum R_a$ is the sum of the armature winding resistance R_a, resistance of the commutation winding located on interpoles R_{int}, resistance of compensating winding R_{comp} and resistance of series winding R_{se}. For small and medium-power shunt machines it is only the sum of R_a and R_{int}.

The EMF induced in the armature winding by the magnetic flux Φ of main poles is given by eqn (4.9), i.e.,

$$E = \frac{N}{a} pn\Phi = c_E n\Phi$$

where N is the number of armature conductors, p is the number of pole pairs, a is the number of pairs of armature current parallel paths and n is the rotational speed [rev/s]. The constant c_E is called the *EMF constant* or *armature constant*.

The electromagnetic torque developed by the DC machine is expressed by eqn (4.18), i.e.,

$$T_{elm} = c_T \Phi I_a \tag{4.66}$$

The MMF F_f per one pole pair can be calculated on the basis of Ampere's circuital law (4.12). This MMF balances the voltage drops across the air gap taken twice, magnetic voltage drop along the main pole core taken twice, magnetic voltage drop along the stator yoke, magnetic voltage drop along the rotor tooth taken twice and magnetic voltage drop along the rotor (armature) core.

The power *losses* in a DC brush machine (4.25) are the armature winding losses ΔP_a, the shunt-field winding losses ΔP_f, the series-field losses ΔP_{se}, the armature core losses ΔP_{Fe}, the brush drop losses ΔP_{br}, the rotation (mechanical) losses ΔP_{rot} and the stray load losses ΔP_{str}.

The *armature reaction* on the field excitation winding has the following effects:

- Distortion of the resultant field (shift of the neutral axis);
- Weakening of the magnetic flux when the magnetic circuit is saturated (Fig. 4.20).

The armature reaction can be compensated by:

- Magnetic flux of the interpoles (compoles);
- Magnetic flux of the commutation winding that is placed in slots of the main poles.

DC machines are *reversible machines* and can operate either as motors and generators. DC machines may be classified according to their armature and field winding connections as follows:

(a) *Separately-excited machines*, in which the field winding is fed from a source, which is separate than that of the armature winding or the machine is excited by a system of PMs;

(b) *Shunt machines*, in which the armature and field windings are connected in parallel;

(c) *Series machines*, in which the armature and field windings are connected in series;

(d) *Compound-wound machines* with two field windings, i.e., series and shunt (parallel).

In a *separately excited generator* the machine is excited from a separate DC power supply and the armature (rotor) winding is loaded with a resistive load (Fig. 4.21). The *external characteristic* $V = f(I_a)$ at constant speed n and constant excitation flux Φ is plotted in Fig. 4.22b.

In a DC *shunt generator* the armature winding feeds both the field excitation winding and the resistive load (Fig. 4.23). Conditions to be met for self-excitation are

- Residual magnetism;
- The magnetic excitation flux must be in the same direction as the residual magnetic flux;
- The resistance of the field excitation circuit R_f must be less than the critical resistance R_{fcr}, i.e., $R_f < R_{fcr}$.

The *build-up* of self-excitation of a DC shunt generator is illustrated in Fig. 4.24. The external characteristic $V = f(I_a)$ at constant speed n and constant resistance of the field winding R_f is shown in Fig. 4.25.

In the DC *shunt motor*, the *shunt field winding* is connected in parallel with the armature winding (Fig. 4.26). The total line current I is the sum of the armature current I_a and shunt-field current I_f. The mechanical characteristic, i.e., $n = f(T)$ is affected by the direct axis armature reaction (magnetizing or demagnetizing). If the brushes are in the *geometrical neutral line* or brushes are shifted ahead from the neutral line, the speed decreases as the torque increases (Fig. 4.27).

In a DC *series motor*, the armature and series field windings are connected in series (Fig. 4.28). The armature current and the field current are therefore the same, i.e., $I = I_a = I_f$. The speed-torque characteristic (Fig. 4.29b) has a hyperbolic shape. At low torque, the speed is very high. This is dangerous because the rotor and commutator can be mechanically due to high radial stresses imposed by high centrifugal forces. For this reason, a series motor should be operated so as to exclude the possibility of starting it without a load (by means of a permanent coupling, toothed gear, worm gear, etc.).

A DC *compound-wound* motor has two field windings: a *shunt field winding* and a *series field winding* (Fig. 4.30). When the field windings are *cumulatively compounded*, their magnetizing forces are added, increasing the resultant excitation flux. When the field windings are *differentially compounded*, the series field winding opposes the shunt winding. The speed is inversely proportional to the resultant excitation flux. Cumulative compound motors have speed-torque characteristics resembling those of a series motor (Fig. 4.31b).

To decrease the *inrush current* when starting a motor, a *starting rheostat* is inserted into the armature circuit (as in Fig. 4.32). At the first instant of starting, the EMF $E = 0$. The input terminal voltage is only balanced by the brush voltage drop and the armature current I_a is very high, several times higher than the nominal current. To reduce the armature current, a variable resistor called staring rheostat is connected in series with the armature circuit. As the speed increases, the EMF E increases too and the input voltage is somewhat compensated by the small amount of the EMF E. The resistance of the starting rheostat can be reduced. With further increase in the speed, the EMF E increases further and the resistance of the starting rheostat can be reduced further. At full speed, i.e., the nominal speed, the resistance of the starting rheostat can be reduced to zero because the magnetic flux excites the nominal EMF, which balances the input terminal voltage. The armature current I_a reaches its nominal value.

The *speed* of a DC motor *can be controlled* by changing:

- The supply mains voltage V;
- The armature-cirucit resistance $R_a + R_{rhe}$ where R_{rhe} is the speed control rheostat;
- The field flux Φ.

The speed control resistor is designed for higher power than the starting rheostat because the starting rheostat is used only for short-time duty.

The three *electrical braking methods*, which are now considered in terms of shunt and series DC motors, are

- Rheostatic braking;
- Counter current braking (plugging);
- Regenerative braking.

A *permanent magnet* (PM) material is described by the *demagnetization curve*, which is a portion of the full *B–H hysteresis loop* located in the second quadrant in the magnetic flux density B versus magnetic field intensity H coordinate system. The *coercive force* H_c or shortly *coercivity* corresponds to $B = 0$ and the *remanent magnetic flux density* B_r corresponds to $H = 0$. There are *three classes of PM materials* that are used for electric machines (Fig. 4.34):

- Alnicos (Ni, Al, Fe, Co, Cu, Ti);
- Ceramics (ferrites), e.g., barium ferrite $BaO \times 6Fe2O3$;
- Rare-earth materials, e.g., samarium-cobalt SmCo and neodymium iron-boron NdFeB.

The materials with the highest *magnetic energy density* are *rare-earth* materials. They are the best materials for magnetic field excitation in electrical machines.

Magnetic circuit configurations of cylindrical-rotor PM motors for different types of PMs are shown in Figs 4.35, 4.37 and 4.38. The four fundamental armature (rotor) and stator structures are

- The conventional slotted rotor (Fig. 4.35a);
- The slotless (surface wound) rotor (Fig. 4.35b);
- The moving-coil cylindrical rotor (Figs 4.37a, 4.38);
- The moving-coil disk (pancake) rotor (Fig. 4.37b,c).

In *axial flux* (disk-type) DC PM brush motors, the armature winding can be made of copper wires and moulded with resin (Fig. 4.37b), stamped from pieces of sheet copper and then welded or etched using a similar method to that by which PCBs are manufactured (Fig. 4.37c).

Problems

1. Sketch the diagram of wave winding with the following parameters: number of slots $S = 22$, number of poles $2p = 6$, number of coil sides in upper layer of a slot $u = 1$, number of turns per coil $N_c = 1$.

   Answer: $C = 22$, $y = 7$, $y_1 = 4$, $y_2 = 3$

2. A 15-kW, 230-V, 960-rpm DC shunt generator has the armature winding resistance $R_a = 0.15\ \Omega$ and interpole winding resistance $R_{int} = 0.06\ \Omega$. The brush voltage drop $\Delta V_{br} = 2.0$ V. The open circuit characteristic $E = f(I_f)$ and characteristic $E' = f(I_f)$ under load at 960 rpm are given below:

I_f	0.2	0.4	0.6	0.8	1.0	1.2	1.4	1.6	A
E	68	135	200	238	257	272	282	290	V
E'	60	120	180	220	245	260	275	287	V

   Calculate:
   (a) Field excitation current corresponding to nominal operating conditions and total resistance of the shunt winding;
   (b) Voltage at the output terminals when the generator is unloaded;
   (c) No-load speed and speed under nominal load when the machine operates as a motor, the input terminal voltage is 230 V and resistance of the shunt winding is the same as for nominal operating conditions.

   Answer: (a) $I_f = 1.02$ A, $R_f = 226.0\ \Omega$; (b) $V = 257.7$ V; (c) $n_0 = 887$ rpm, $n_n = 835$ rpm

3. On no-load, the speed of a DC motor is $n_0 = 950$ rpm. Calculate the speed when the load is such that the armature current is $I_a = 25$ A. The terminal voltage is $V = 400$ V $= const$, the armature circuit resistance is $\sum R_a = 0.15\ \Omega$, the brush voltage drop is $\Delta V_{br} = 2$ V, and the armature reaction is to be neglected.

   Answer: $n = 941$ rpm

4. A 25-kW DC shunt motor has the following nominal parameters: $V_n = 220$ V, $I_n = 127$ A, $n_n = 1800$ rpm, $I_{fn} = 2.1$ A. The no-load speed is $n_o = 1850$ rpm. Find: (a) the resistance R_{st} of a starting rheostat to obtain the starting current $I_{st} = 2I_n$ and (b) the starting torque corresponding to R_{st} if the voltage across the shunt-field winding terminals is $V_f = V_n = 220$ V.
   Assumptions: (i) the open circuit characteristic (magnetization curve) $E = f(I_f)$ is linear; (ii) the armature reaction is neglected; (c) the brush contact voltage drop is neglected.

   Answer: (a) $R_{st} = 0.827\ \Omega$, (b) $T_{st} = 2.01T_n$

5. A 5.5-kW DC separately excited motor has a terminal voltage $V_n = 220$ V at nominal (rated) current $I_{an} = 30.5$ A and nominal speed $n_n = 750$ rpm. The resistance of the armature circuit is $\sum R_a = 0.4\ \Omega$. Find: (a) the armature current at the first instant when the terminal voltage drops from $V_n = 220$ V down to $V' = 200$ V and (b) the steady-state speed n' at $V' = 200$ V.
The armature reaction, brush contact voltage drop and armature winding inductance are neglected.

Answer: (a) $I_a(t = 0) = -19.5$ A $(E > V)$, (b) $n' = 678$ rpm

6. A 13-kW, 220-V, 71-A, 600-rpm DC shunt motor has armature circuit resistance $\sum R_a = 0.25\ \Omega$ and shunt field resistance $R_f = 110\ \Omega$. The motor is loaded with the rated torque at rated voltage. Find: (a) the resistance of an armature rheostat needed to obtain the speed $n' = 200$ rpm and (b) the output power P'_{out} and input power P'_{in} required to obtain the speed $n' = 200$ rpm.
The armature reaction and brush contact voltage drop are neglected.

Answer: (a) $R_{rhe} = 1.96\ \Omega$, (b) $P'_{out} = 4.33$ kW, $P'_{in} = 15.6$ kW

7. A 22-kW, 220-V, 111-A, 1200-rpm DC separately excited motor has armature circuit resistance $\sum R_a = 0.1\ \Omega$. The nominal field excitation current is $I_{fn} = 2$ A and the nominal voltage across the excitation winding is $V_{fn} = 220$ V. Find: (a) the input voltage V' at the rated field current, rated torque and speed $n' = 0.5 n_n$ and (b) the input and output power under the conditions given above.
The armature reaction and brush contact voltage drop are neglected.

Answer: (a) $V' = 115.45$ V, (b) $P'_{out} = 11$ kW, $P'_{in} = 13.1$ kW

8. A 36-kW, 600-V, 860-rpm, 73-A DC series motor has armature circuit resistance $\sum R_a = 0.35\ \Omega$ (including the series-field resistance R_{se}). The motor has been cut off from the power supply and closed on a load resistance $R_{rhe} = 8\ \Omega$ (this is dynamic braking). The open circuit characteristic (also called the magnetization curve) in relative units is given below:

E/E_r	0.058	0.87	1.16	1.28	1.40
I_a/I_{ar}	0	0.50	1.0	1.25	1.60

Find the armature current I_{ab} and braking torque T_b if the rotor speed is $n_b = 1000$ rpm.
Assumptions: (i) the nonlinearity of the magnetization curve is to be included; (b) the armature reaction is neglected; (c) the brush contact voltage drop is neglected.

Answer: (a) $I_{ab} = 86.0$ A; (b) $T_b = 555.6$ Nm

9. A 25-kW, 220-V, 1000-rpm, $\eta = 87\%$ DC series motor operates under nominal load torque and nominal input voltage. Resistances of winding at steady-temperature (corresponding to nominal operation) are: armature winding resistance $R_a = 0.085$ Ω, resistance of series winding $R_{se} = 0.045$ Ω, resistance of interpole winding $R_{int} = 0.03$ Ω. The brush voltage drop is $\Delta V_{br} = 1.8$ V. Calculate the speed, shaft power and efficiency when at the nominal load torque the input voltage

   (a) Increases 10%;
   (b) Decreases 10%.

   The variation of rotational losses with speed is neglected.

   Answer: (a) $n = 888.5$ rpm, $P = 22.2$ kW, $\eta = 85.9\%$; (b) $n = 1111.5$ rpm, $P = 27.8$ kW, $\eta = 87.9\%$

10. A 30-kW, 220-V, 153-A, 3000-rpm DC shunt motor has the armature circuit resistance $\sum R_a = 0.07$ Ω and the field excitation current $I_f = 2$ A. The motor had worked under nominal (rated) conditions. At an instant $t = 0$ the armature terminal connections were reversed and a resistance R_b was inserted in the armature circuit (counter-current braking). Find: (a) the resistance R_b if the braking torque at $t = 0$ is $T_b = 2T_n$, (b) the input power and power dissipated in the resistance R_b at $t = 0$ and (c) the energy dissipated in the resistance R_b during braking, if the speed changes with time according to equation $n_b = n(1 - t/\tau)$ for $0 \leq t \leq \tau$ and $\tau = 10$ s. At $t = \tau$ the machine was switched off.

    Assumptions: (i) the characteristics $E = f(I_f)$ of the magnetic circuit is linear, (ii) the armature reaction is neglected, (c) the inductance of the armature winding is negligible, (d) the brush contact voltage drop is neglected.

    Answer: (a) $R_b = 1.63$ Ω, (b) $P_{inb}(t = 0) = 66.7$ kW, $P_b(t = 0) = 12.3$ kW, (c) 0.203 kWh

11. A two-pole, 750-W, 220-V, 1200-rpm, $\eta = 0.85$ DC brush motor with segmental PMs has the armature circuit resistance $\sum R_a = 4.06$ Ω. The air gap magnetic flux density is $B_g = 0.63$ T, the effective length of the armature core $L_i = 0.084$ m, the armature diameter $D = 0.12$ m, the effective pole arc coefficient $\alpha_i = 0.8$, the number of pairs of armature parallel paths $a = 1$ and the brush voltage drop $\Delta V_{br} \approx 2$ V. Find: (a) the armature constant k_E and torque constant k_T, (b) the electromagnetic torque T_{elm} and shaft torque T, (c) the number of armature conductors N, (d) the armature electric loading A and (e) the speed at 60% of the nominal voltage and shaft torque $T = 2.5$ Nm. The armature reaction is neglected.

    Answer: (a) $k_E = 10.09$ Vs, $k_T = 1.6$ Nm/A, (b) $T_{elm} = 6.4$ Nm, $T = 5.97$ Nm, (c) $N = 1260$, (d) $A = 6684.5$ A/m, (e) $n' \approx 729$ rpm.

WINDINGS OF AC MACHINES

5.1 Construction or windings

The armature windings of AC electrical machines are made of coils, which create coil groups. Each phase winding has an entry and exit terminals. Fig. 5.1 shows a portion of the stator (armature) of an AC electrical machine.

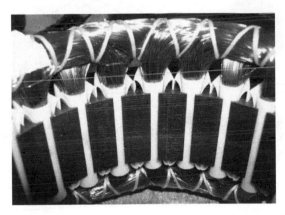

Fig. 5.1. Stator of an AC machine with winding distributed in slots.

A single-phase winding is a system of groups of coils connected in series or in parallel. In one phase of a three-phase winding consisting of $q_1 =$ integer coils there is at least p coil groups (p is the number of pole pairs) in which the EMFs have the same amplitude and phase angle. Basic parts of a winding with distributed parameters are

- Turn;
- Coil (Fig. 5.2);
- Coil group (section) consisting of several coils (Fig. 5.3).

When a coil has one turn only, it is called a single turn coil. When there are more than one turns per coil, then it is called a multiturn coil.

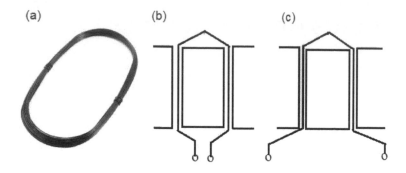

Fig. 5.2. Coils: (a) coil wound with round wire; (b) lap coil; (c) wave coil.

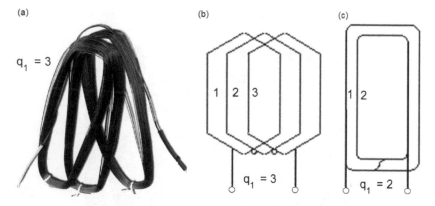

Fig. 5.3. Coil groups: (a) coil group wound with round wire $q_1 = 3$; (b) coil group of a lap winding $q_1 = 3$; (c) coil group consiting of concentric coils $q_1 = 2$.

The number of slots per pole is

$$Q_1 = \frac{s_1}{2p} \tag{5.1}$$

where s_1 is the number of stator slots, $2p$ is the number of poles and p is the number of pole pairs. The "1" subscript is for the stator and the "2" subscript is for the rotor. The number of slots per pole per phase

$$q_1 = \frac{s_1}{2pm_1} \tag{5.2}$$

where m_1 is the number of stator phases. The pole pitch expressed as a segment of a circle and in units of length

$$\tau = \frac{\pi D}{2p} \qquad (5.3)$$

where D is the stator inner diameter. The pole pitch expressed in the number of slots

$$\tau = \frac{s_1}{2p} = Q_1 \qquad (5.4)$$

The distance from one coil side to the second coil side is called coil span. The coil span expressed in the units of length

$$w_c \leq \tau \qquad \Rightarrow \qquad \frac{w_c}{\tau} \leq 1 \qquad (5.5)$$

Usually the coil span is made nearly equal to pole pitch. However, for special reasons, coils are chorded ($w_c \leq \tau$).

The coil span can also be measured in the number of slots from the left side to the right side of the coil, i.e.,

$$y \leq Q \qquad \Rightarrow \qquad \frac{y}{Q_1} \leq 1 \qquad (5.6)$$

With respect to arrangement of coil sides in slots, the windings are divided into single layer and double layer windings. The single-layer winding is when every coil side occupies the whole slot (Fig. 5.4a). The double-layer winding is when the coil side occupies only half of the slot and the other half is occupied by another coil side (Fig. 5.4b).

With respect to shape of end turns, the windings are divided into (a) double-tier, (b) three-tier and (c) made of former coils (Fig. 5.5).

Dependent on the design or mode of operation, a three-phase winding can be Y-connected or Δ-connected (Fig. 5.6).

Double-layer windings can be divided into lap windings and wave windings (Fig. 5.2). Concentric windings are single layer windings made of concentric type coils (Fig. 5.3c) with different coil span.

With respect to the number of slots per pole per phase $q_1 = s_1/(2pm_1)$ the windings can be classified into windings with q_1 an integer and windings with fractional q_1 (fractional slot winding).

With respect to coil span w_c the windings are divided into full pitch windings with $w_c = \tau$ and chorded (short-pitch) windings with $w_c < \tau$ (Fig. 5.7).

With respect to fabrication of coils the windings are divided into windings inserted in open slots (Fig. 5.8a), dropped in slots wire by wire (Fig. 5.8b), sewed up (Fig. 5.8c), and pushed through semi-closed slots (Fig. 5.8d). The winding shown in Fig. 5.8d is made of "hairpin" shape coils.

The winding made of elastic coils is shown in Fig. 5.9 and the winding made of stiff diamond-shaped coils is shown in Fig. 5.10.

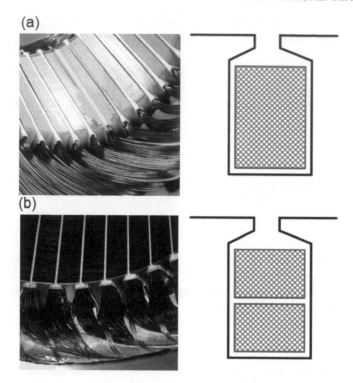

Fig. 5.4. Arrangement of coil sides in slots: (a) single-layer winding; (b) double-layer winding.

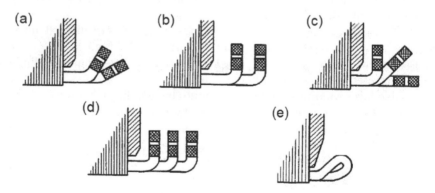

Fig. 5.5. End turns of AC windings: (a), (b) double-tier; (c), (d) three-tier; (e) made of former coils.

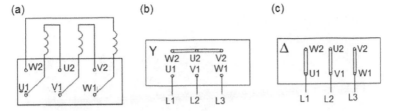

Fig. 5.6. Terminal board of three-phase AC windings: (a) connection of phase windings to terminals; (b) Y-connection; (c) Δ-connection.

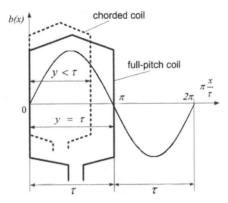

Fig. 5.7. Full-pitch and corded coil.

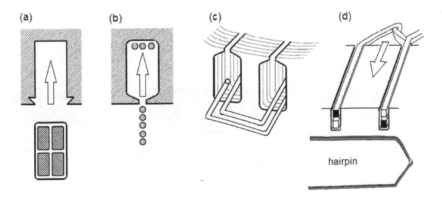

Fig. 5.8. Coils in slots: (a)inserted in open slots, (b) dropped in semi-closed slots wire by wire (mush winding), (c) sewed up, (d) pushed through a pair of slots.

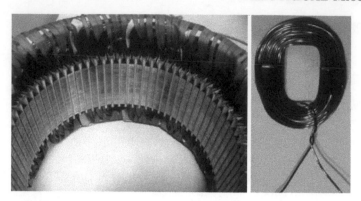

Fig. 5.9. Winding made of elastic coils inserted in slots or sewed up: (a) stator; (b) coil.

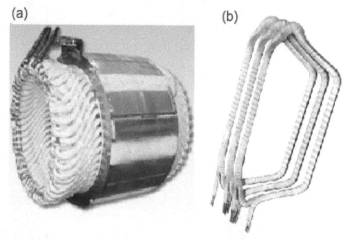

Fig. 5.10. Winding made of stiff former coils: (a) stator; (b) diamond coil group.

Fig. 5.11. Different techniques of stator coil winding: (a) concentric coils wound of round wire; (b) lap winding wound of round wire; (c) lap winding made of stiff diamond-type coils.

Fig. 5.11 shows different techniques of stator coil winding: (a) concentric coils wound of round wire; (b) lap winding wound of round wire; (c) lap winding made of stiff diamond-type coils.

Assuming symmetrical distribution of conductors in slots around the stator core periphery, the voltage induced (EMF) by the magnetic rotating field in all conductors have the same amplitude and frequency. It is possible to represent the induced voltages with the aid of phasors of the same module and uniformly shifted in phase (time). Such a diagram is called *star of slot voltages* of a given winding.

Fig. 5.12 also shows the mechanical angle between slots (conductors) and electrical angle between phasors of slot voltages. In general, the following relationship exists between the electrical and mechanical angle:

$$\alpha_{el} = p\alpha_m \tag{5.7}$$

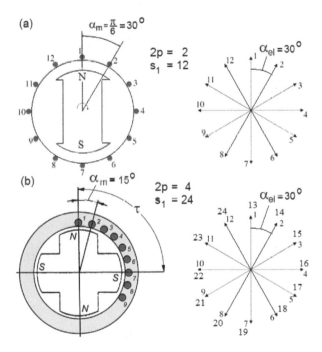

Fig. 5.12. Explanation of electrical and mechanical angle using a sketch of the stator and phasor diagram of slot voltages: (a) two-pole stator $2p = 2$, $s_1 = 12$; (b) four-pole stator $2p = 4$, $s_1 = 24$.

The mechanical angle corresponding to pole pitch

$$\alpha_m = \frac{360°}{2p} = \frac{180°}{p} \equiv \frac{\pi}{p} \tag{5.8}$$

The electrical angle corresponding to pole pitch

$$\alpha_{el\tau} = p\alpha_m = p\frac{360°}{2p} = 180° \equiv \pi \qquad (5.9)$$

The electrical angle between neighboring slots (Figs 5.12 and 5.13)

$$\alpha_{el} = p\frac{360°}{s_1} \equiv p\frac{2\pi}{s_1} \qquad (5.10)$$

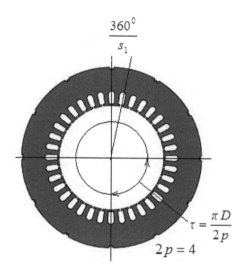

Fig. 5.13. Pole pitch and angle between neighboring slots.

If the current in the phase winding or voltage across the phase windings are too high, the winding can be divided into *parallel current paths*. For single layer windings, the maximum possible number of parallel paths of one phase is

$$a = p \qquad (5.11)$$

For double layer windings, the maximum possible number of parallel paths of one phase is

$$a = 2p \qquad (5.12)$$

5.2 Winding diagrams

It is easy to draw a winding diagram for a simple stator of a three-phase ($m_1 = 3$) machine with $2p = 2$ and $s_1 = 6$ (Fig. 5.14). The winding is distributed in $s_1 = 6$ slots, $Q_1 = s_1/(2p) = 6/2 = 3$ and the electrical angle between neighboring slots is $\alpha_{el} = 1 \times 360°/6 = 60°$. The winding shown in Fig. 5.14b consists of one coil group $q_1 = 6/(2 \times 3) = 1$ per phase. In the case of more coil groups than one, the coils group can be series or parallel connected to create a phase winding. The phase windings of a three-phase stator are shifted in space one to each other by $120°$ electrical degrees.

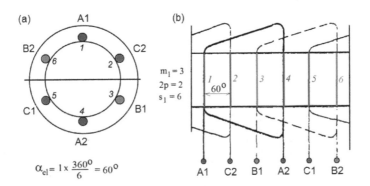

Fig. 5.14. A simple three-phase winding with $m_1 = 3$, $2p = 2$, $s_1 = 6$ and $q_1 = 1$: (a) model of the stator; (b) flat winding diagram.

The phasor diagram of the stator (armature) EMFs or star of slot voltages is the most precise tool for the winding design and evaluation if all coil connections are correct. When designing a symmetrical well-utilized three-phase winding, it is necessary to take into account the following rules:

- EMFs of three phases should be equal and shifted symmetrically in phase by $2\pi/3 = 120°$;
- A coil should be designed in such a way as to obtain EMFs in the left and right coil side shifted in phase by an angle $\pi = 180°$ (full-pitch coil) or slightly less (chorded coil);
- It is necessary to create a phase winding by connecting conductors (coil sides), in which EMFs are least shifted in phase.

Fig. 5.15 shows a three-phase, single-layer winding with $2p = 4$, $s_1 = 24$, $q_1 = 2$, the star of slot EMFs and polygon of EMFs for phase A.

Figs 5.16 to 5.19 show diagrams for double-layer stator winding with $m_1 = 3$, $2p = 2$, $s_1 = 24$. For this winding $Q_1 = 24/2 = 12$. This is also the coil span of a full pitch coil. In this case chorded coils with $y = 8$ slots have been designed. The chorded coil pitch related to the full coil pitch is $8/12 = 2/3$. The electrical angle between neighboring slots is $\alpha_{el} = 1 \times 360/24 = 15°$ electrical and the number of slots per pole per phase $q_1 = 24/(2 \times 3) = 4$.

Fig. 5.16 shows a flat full winding diagram when two coil groups of the same phase are series connected and the number of parallel paths $a = 1$.

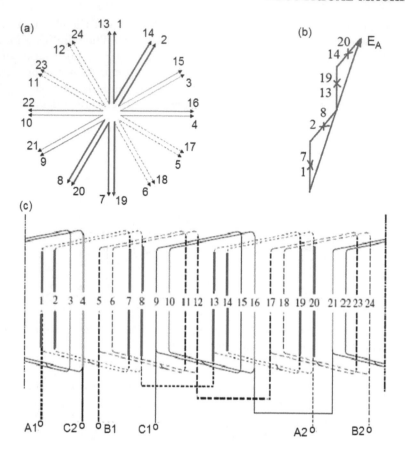

Fig. 5.15. A three-phase winding with $m_1 = 3$, $2p = 4$, $s_1 = 24$ and $q_1 = 2$: (a) star of induced slot voltages (EMFs); (b) polygon of slot EMFs for the phase A;(c) flat winding diagram.

Fig. 5.17 shows each of the phase windings drawn separately, the coil group connections are well visible. Fig. 5.18 shows coil connection to create one parallel path $a = 1$ (Fig. 5.18a) and two parallel paths $a = 2$ (Fig. 5.18b). Fig. 5.19 shows a flat full winding diagram when two coil groups of the same phase are parallel connected and the number of parallel paths $a = 2$.

5.3 Electromotive force induced in a winding by rotating magnetic field

In Fig. 5.20 a single-turn coil placed in two slots of the stator core has been shown. The magnetic field coupled with the coil is sinusoidal and rotates with rotational speed n_s (linear speed v_s) with respect to the coil. The sinusoidal

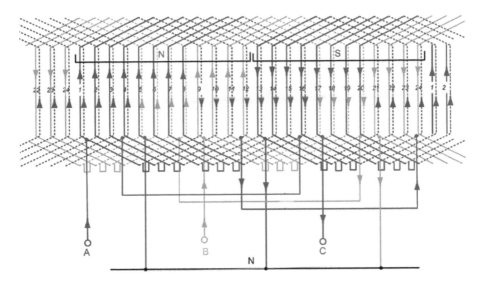

Fig. 5.16. Full diagram of a three-phase winding with $m_1 = 3$, $2p = 2$, $s_1 = 24$, $y = 8$ slots, $a = 1$ and $q_1 = 4$.

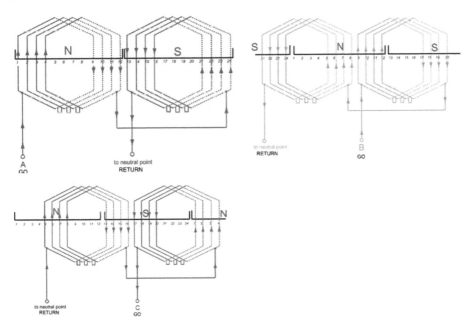

Fig. 5.17. Three-phase winding with $m_1 = 3$, $2p = 2$, $s_1 = 24$, $y = 8$ slots, $a = 1$ and $q_1 = 4$: (a) phase A, (b) phase B, (c) phase C.

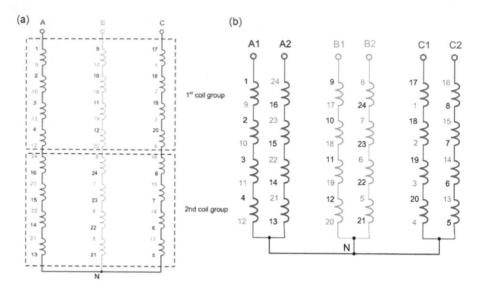

Fig. 5.18. Coil connection for three-phase winding with $m_1 = 3$, $2p = 2$, $s_1 = 24$, and $q_1 = 4$: (a) $a - 1$; (b) $a = 2$.

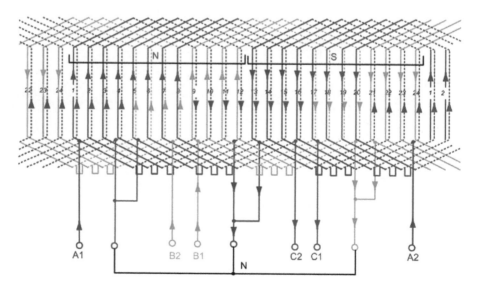

Fig. 5.19. Full diagram of a three-phase winding with $m_1 = 3$, $2p = 2$, $s_1 = 24$, $y = 8$ slots, $a = 2$ and $q_1 = 4$.

wave of magnetic flux density induces EMFs in two side of the coil. The EMF induced in a single conductor (one coil side) is

$$e_s = BLv_s \tag{5.13}$$

The EMF induced in the coil (two coil sides) consisting of one turn

$$e_t = 2BLv_s \tag{5.14}$$

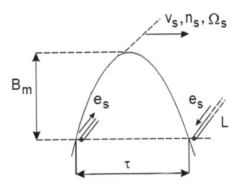

Fig. 5.20. A single-turn coil placed in two slots in rotating magnetic field.

The magnetic flux density as a function of time

$$B(t) = B_m \sin(\omega t) = B_m \sin(2\pi f t) \tag{5.15}$$

The pole pitch is equal to one half of the wave of the magnetic flux density, i.e.,

$$\tau = v_s \frac{T}{2} = \frac{v_s}{2f} \tag{5.16}$$

where the mechanical angular speed $\Omega_s = 2\pi n_s$ and linear speed $v_s = \pi D n_s = 0.5\Omega_s D$. Thus, the frequency of current in the coil

$$f = \frac{v_s}{2\tau} = \frac{0.5\Omega_s D}{2\pi D/(2p)} = \frac{p\Omega_s}{2\pi} \tag{5.17}$$

In the case of N_1 turns of a phase winding

$$e_{Nt} = 2NL(2\tau f)B_m \sin(\omega t) = E_m \sin(\omega t) \tag{5.18}$$

where $E_m = 4NfB_mL\tau$. Thus, the *rms* EMF

$$E_1 = \frac{4}{\sqrt{2}}N_1 f\tau LB_m = 2\sqrt{2}N_1 f\tau LB_m \tag{5.19}$$

On the basis of Fig. 5.20 the mean magnetic flux in the air gap can be found

$$\Phi = L \int_0^\tau B_m \sin\left(\frac{\pi}{\tau}x\right) dx = \frac{2}{\pi}B_m\tau L \tag{5.20}$$

Thus,

$$B_m = \frac{\pi\Phi_m}{2\tau L} \tag{5.21}$$

Putting eqn (5.21) to eqn (5.19) the *rms* EMF is

$$E_1 = 2\sqrt{2}N_1 f\tau L\frac{\pi\Phi}{2\tau L} = \pi\sqrt{2}fN_1\Phi \tag{5.22}$$

To take into accout the distribution of the winding in slots and coil span of chorded coils ($w_c \leq \tau$), the *winding factor* k_{w1} for the fundamental space harmonic of the air gap magnetic flux density must be introduced, i.e.,

$$k_{w1} = k_{d1}k_{p1} \leq 1 \tag{5.23}$$

where k_{d1} is the *distribution factor* for the fundamental space harmonic and k_{p1} is the *pitch factor* for the fundamental space harmonic. Thus, the *rms* value of the first harmonic of the EMF per phase including the winding factor is

$$E_1 = \pi\sqrt{2}fN_1k_{w1}\Phi \approx 4.44fN_1k_{w1}\Phi \tag{5.24}$$

The above eqn (5.24) is for sinusoidal variation of the magnetic and electric quantities, i.e., for the first harmonic of the magnetic flux density produced by the stator winding. This is one of the most important equations in the theory of AC electrical machines.

5.4 Distribution factor and pitch factor

The distribution factor is defined as the phasor sum–to the arithmetic sum ratio of EMFs induced in coils belonging to the same coil group, i.e.,

$$k_{d1} = \frac{\text{phasor sum of EMFs induced in coils}}{\text{arithmetic sum of EMFs induced in coils}} \tag{5.25}$$

From Fig. 5.21 for $q_1 = $ integer the EMF induced in one coil side (one slot)

$$e_1 = 2R\sin\left(\frac{\alpha_{el}}{2}\right) \tag{5.26}$$

The EMF of a group of coils consisting of q_1 coils

$$e = 2R\sin\left(\frac{q_1\alpha_{el}}{2}\right) \tag{5.27}$$

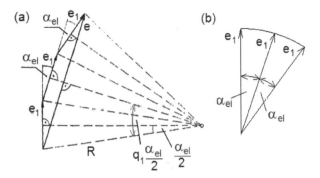

Fig. 5.21. SEMs of a coil group: (a) star of slot induced voltages; (b) polygon of slot induced voltages.

Thus, the distribution factor

$$k_{d1} = \frac{e}{q_1 e_1} = \frac{2R\sin\left(q_1 \frac{\alpha_{el}}{2}\right)}{q_1 2R\sin\left(\frac{\alpha_{el}}{2}\right)} = \frac{\sin\left(q_1 \frac{\alpha_{el}}{2}\right)}{q_1 \sin\left(\frac{\alpha_{el}}{2}\right)} \qquad (5.28)$$

With the aid of eqns (5.2), (5.7) and (5.8) the distribution factor can be expressed as

$$k_{d1} = \frac{\sin\left(q_1 \frac{\alpha_{el}}{2}\right)}{q_1 \sin\left(\frac{\alpha_{el}}{2}\right)} = \frac{\sin\left(\frac{\pi}{2m_1}\right)}{q_1 \sin\left(\frac{\pi}{2m_1 q_1}\right)} \qquad (5.29)$$

where s_1 is the number of stator slots, m_1 is the number of phases and q_1 is the number of slots per pole per phase.

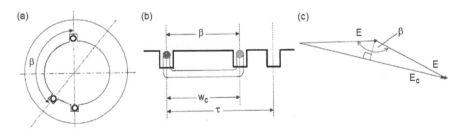

Fig. 5.22. SEMs of a coil: (a) full pitch and chorded coil; (b) distribution of corded coil in slots; (c) triangle of EMFs.

The pitch factor of the stator winding (Fig. 5.22) is defined as the ratio of the phasor sum–to–arithmetic sum of EMFs induced in two sides of a coil, i.e.,

$$k_{p1} = \frac{\text{phasor sum of EMFs induced in coil sides}}{\text{arithmetic sum of EMFs induced in coil sides}} \tag{5.30}$$

according to Fig. 5.22c, the pitch factor is

$$k_{p1} = \frac{E_c}{2E} = \frac{2E \sin\left(\frac{\beta}{2}\right)}{2E} = \sin\left(\frac{\beta}{2}\right) \tag{5.31}$$

On the basis of Fig. 5.22a and 5.22b

$$\frac{\beta}{\pi} = \frac{w_c}{\tau} \quad \Rightarrow \quad \beta = \pi\frac{w_c}{\tau}$$

where $\beta \le \pi$ is the coil span measured in radians or electrical degrees and $w_c \le \tau$ is the coil span measured in meters. Thus

$$k_{p1} = \sin\left(\frac{w_c}{\tau}\frac{\pi}{2}\right) \tag{5.32}$$

where w_c/τ is the *relative winding pitch*. The resultant winding factor k_{w1} is given by eqn (5.23).

Example 5.1

The stator of a 3-phase, 6-pole, 50-Hz AC electrical machine has the inner diameter $D = 0.1$ m and the stack length $L = 0.11$ m. The number of stator slots is $s_1 = 54$ and peak value of the magnetic flux density in the air gap $B_m = 0.7$ T. The number of series turns per phase is $N_1 = 360$ and the coil span measured in slots is $w_c = 8$. Find the EMF E_1 per phase.

Solution

Pole pitch

$$\tau = \frac{\pi D}{2p} = \frac{\pi \times 0.1}{6} = 0.052 \text{ m}$$

Electrical angle between slots

$$\alpha_{el} = p\frac{360°}{s_1} = 3\frac{360°}{54} = 20°$$

Number of slots per pole per phase (pole pitch measured in number of slots)

$$Q_1 = \frac{s_1}{2p} = \frac{54}{6} = 9$$

Number of slots per pole per phase

$$q_1 = \frac{s_1}{2pm_1} = \frac{54}{6 \times 3} = 3$$

Distribution factor

$$k_{d1} = \frac{\sin\left(\frac{\pi}{2m_1}\right)}{q_1 \sin\left(\frac{\pi}{2m_1 q_1}\right)} = \frac{\sin\left(\frac{\pi}{2\times 3}\right)}{3\sin\left(\frac{\pi}{2\times 3\times 3}\right)} = 0.96$$

Pitch factor

$$k_{p1} = \sin\left(\frac{w_c}{Q_1}\frac{\pi}{2}\right) = \sin\left(\frac{8}{9}\frac{\pi}{2}\right) = 0.985$$

Since the coils span $w_c = 8$ is expressed in the number of slots, the pole pitch must also be expressed in the number of slots, i.e., $Q_1 = 9$. The winding factor for fundamental

$$k_{w1} = k_{d1}k_{p1} = 0.96 \times 0.985 = 0.945$$

Magnetic flux

$$\Phi = \frac{2}{\pi}B_m\tau L = \frac{2}{\pi}0.75 \times 0.052 \times 0.11 = 0.002567 \text{ Wb}$$

EMF per phase

$$E_1 = \pi\sqrt{2}fN_1 k_{w1}\Phi = \pi\sqrt{2} \times 50 \times 360 \times 0.945 \times 0.002567 = 194.0 \text{ V}$$

5.5 Higher harmonics of EMF

In a practical machine the distribution of the normal component of the magnetic flux density in the air gap around the stator periphery is not sinusoidal and contains higher harmonics (Fig. 5.23). For higher space harmonics

- frequency

$$f_\nu = \nu f \tag{5.33}$$

- pole pitch

$$\tau_\nu = \frac{\tau}{\nu} \tag{5.34}$$

- electrical angle

$$\alpha_{el\nu} = \nu\alpha_{el} \tag{5.35}$$

- winding factor

$$k_{w1\nu} = k_{d1\nu}k_{p1\nu} \tag{5.36}$$

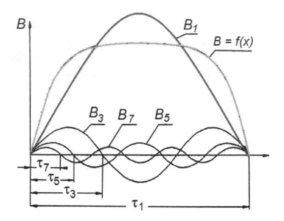

Fig. 5.23. Nonsinusoidal distribution of normal component of the magnetic flux density in the air gap $B = f(x)$, fundamental harmonic B_1 and higher odd harmonics B_3, B_5, B_7.

- distribution factor

$$k_{d1\nu} = \frac{\sin\left(q_1\nu\frac{\alpha_{el}}{2}\right)}{q_1\sin\left(\nu\frac{\alpha_{el}}{2}\right)} \tag{5.37}$$

- pitch factor

$$k_{p1\nu} = \sin\left(\frac{w_c}{\tau_\nu}\frac{\pi}{2}\right) = \sin\left(\nu\frac{w_c}{\tau}\frac{\pi}{2}\right) \tag{5.38}$$

Peak value of the magnetic flux density

$$B_{m\nu} = \frac{2}{\pi}\int_0^\tau B\sin\left(\nu\frac{\pi}{\tau}\right)dx = \frac{4}{\nu\pi}B \tag{5.39}$$

Mean magnetic flux for the νth space harmonic

$$\Phi_\nu = \frac{2}{\pi}B_{m\nu}\tau_\nu L = \frac{2}{\pi}B_{m\nu}\frac{\tau}{\nu}L \tag{5.40}$$

The rms value of the νth harmonic of EMF

$$E_{1\nu} = \pi\sqrt{2}f_\nu k_{w1\nu}N_1\Phi_\nu \approx 4.44 f_\nu k_{w1\nu}N_1\Phi_\nu \tag{5.41}$$

The *rms* value of distorted EMF

$$E_1 = \sqrt{\sum_{\nu=1}^\infty E_{1\nu}^2} = \sqrt{E_1^2 + E_3^2 + E_5^2 + E_7^2 + \ldots} \tag{5.42}$$

In a three-phase AC machine with symmetrical phase winding fed with balanced three-phase voltage system, the third harmonic $\nu = 3$ and its multiples theoretically do not exist, i.e.,

$$\nu = 3k = 0 \qquad \text{where} \qquad k = 1, 2, 3, \ldots \qquad (5.43)$$

The harmonics $\nu = 6k + 1 - 7, 13, 19, 25, \ldots$ rotate in the direction of the fundamental harmonic $\nu = 1$. The harmonics $\nu = 6k - 1 = 5, 11, 17, 23, \ldots$ rotate in the opposite direction. This applies both to the magnetic flux density and MMF.

Example 5.2

For the stator of Example 5.1 find the pole pitch and winding factors for higher space harmonics $\nu = 5$ and $\nu = 7$.

Solution

Pole pitches for higher space harmonics

$$\tau_5 = \frac{\tau}{5} = \frac{0.052}{5} = 0.0105 \text{ m} \qquad \tau_7 = \frac{\tau}{7} = \frac{0.052}{7} = 0.0075 \text{ m}$$

Electrical angles

$$\alpha_{el5} = 5\alpha_{el} - 5 \times 20 = 100° \qquad \alpha_{el7} = 7\alpha_{el} = 7 \times 20 = 140°$$

Distribution factors

$$k_{d15} = \frac{\sin(0.5q_1 \times 5\alpha_{el})}{q_1 \sin(0.5 \times 5\alpha_{el})} = \frac{\sin(0.5 \times 3 \times 5 \times 20)}{3 \sin(0.5 \times 5 \times 20)} = 0.218$$

$$k_{d17} = \frac{\sin(0.5q_1 \times 7\alpha_{el})}{q_1 \sin(0.5 \times 7\alpha_{el})} = \frac{\sin(0.5 \times 3 \times 7 \times 20)}{3 \sin(0.5 \times 7 \times 20)} = -0.177$$

Pitch factors

$$k_{p15} = \sin\left(5\frac{w_c}{Q_1}\frac{\pi}{2}\right) = \sin\left(5\frac{8}{9}\frac{\pi}{2}\right) = 0.643$$

$$k_{p17} = \sin\left(7\frac{w_c}{Q_1}\frac{\pi}{2}\right) = \sin\left(7\frac{8}{9}\frac{\pi}{2}\right) = -0.342$$

Winding factors

$$k_{w15} = k_{d15}k_{p15} = 0.218 \times 0.643 = 0.14$$

$$k_{w17} = k_{d17}k_{p17} = (-0.177) \times (-0.342) = 0.061$$

5.6 Magnetic field produced by a single coil

Fig. 5.24 shows a model of an electrical machine with a single coil placed in the stator core and fed with sinusoidal current

$$i(t) = I_m \sin(\omega t) = \sqrt{2}I \sin(\omega t) \qquad (5.44)$$

The magnetic flux produced by a coil with $N_c = 1$ turn

$$\Phi(t) = \frac{i(t)}{R_\mu} \qquad (5.45)$$

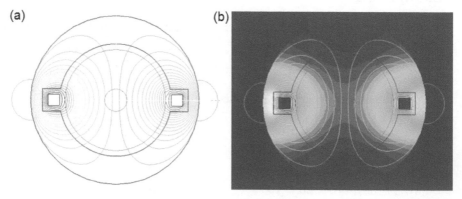

Fig. 5.24. Magnetic field of a single full pitch coil placed in two slots of the stator core: (a) magnetic flux lines; (b) magnetic flux density distribution.

where the reluctance of the magnetic flux path

$$R_\mu = 2R_{\mu g} + R_{\mu Fe} = 2\frac{k_C g}{\mu_0 S_g} + \sum_i \frac{l_{Fei}}{\mu_0 \mu_{ri} S_{Fei}} = 2\frac{k_{sat} k_C g}{\mu_0 S_g} \qquad (5.46)$$

In the above equation S_g is the cross-section of the air gap per pole, k_{sat} is the saturation factor of the magnetic circuit and k_C is Carter's coefficient of the air gap. The reluctance of the air gap g

$$R_{\mu g} = \frac{k_C g}{\mu_0 S_g} \qquad (5.47)$$

The reluctance $R_{\mu Fei}$ of the ith part of the ferromagnetic core with its length l_{Fei}, cross section S_{Fei} and relative magnetic permeability μ_{ri}

$$R_{\mu Fei} = \frac{l_{Fei}}{\mu_0 \mu_{ri} S_{Fei}} \qquad (5.48)$$

Fig. 5.25. Stator and rotor slots of an induction machine.

The saturation factor of the magnetic circuit

$$k_{sat} = \frac{2R_{\mu g} + R_{\mu Fe}}{2R_{\mu g}} = 1 + \frac{R_{\mu Fe}}{2R_{\mu g}}$$

$$= 1 + \frac{\sum \frac{L_{Fei}}{\mu_0 \mu_{ri} S_{Fei}}}{2 \frac{k_C g}{\mu_0 S_g}} \geq 1 \qquad (5.49)$$

$$R_{\mu Fe} = \sum_{i=1}^{n} \frac{l_{Fei}}{\mu_0 \mu_{ri} S_{Fei}} \qquad (5.50)$$

In general, the saturation factor of the magnetic circuit is defined as

$$k_{sat} = \frac{\text{total sum of reluctances}}{\text{reluctance of air gaps}} \qquad (5.51)$$

Thus

$$2R_{\mu g} + R_{\mu Fe} = 2R_{\mu g} k_{sat} \qquad (5.52)$$

The magnetic flux and magnetic flux density with reluctances of the stator and rotor cores being included

$$\Phi(t) = \frac{F(t)}{2R_{\mu g} + R_{\mu Fe}} = \frac{F(t)}{2R_{\mu g} k_{sat}} \qquad (5.53)$$

$$B(t) = \frac{\Phi(t)}{S_g} = \frac{\frac{F(t)}{2R_{\mu g} k_{sat}}}{S_g} = \frac{\mu_0}{2k_{sat} k_C g} F(t) \qquad (5.54)$$

where, in general case, for $N_c > 1$

$$F(t) = i(t) N_c \qquad (5.55)$$

In most cases, the stator core and rotor core too, e.g., induction and synchronous machines, have slots (Fig. 5.25). The stator core and rotor core slot openings are included in eqns (5.46), (5.47), (5.49), (5.54) with the aid of Carter's coefficient k_C. Carter's coefficient is defined as

$$k_C = \frac{\text{equivalent thickness of air gap with slots}}{\text{thickness of slotless air gap}} \tag{5.56}$$

The equivalent air gap with slot openings taken into account

$$g' = k_C g \qquad \text{where} \qquad k_C \geq 1 \tag{5.57}$$

The reluctance of a single air gap with slot openings and magnetic saturation being included

$$R_{\mu g} = \frac{g' k_{sat}}{\mu_0 S_g} = \frac{g k_C k_{sat}}{\mu_0 S_g} \tag{5.58}$$

For a single coil consisting of $N_c > 1$ turns the magnetic flux is

$$\Phi(t) = \mu_0 \frac{i(t) N_c}{2 g k_C k_{sat}} S_g \tag{5.59}$$

and the magnetic flux density

$$B(t) = \frac{\Phi(t)}{S_g} = \mu_0 \frac{i(t) N_c}{2 g k_C k_{sat}} = \frac{\sqrt{2}}{2} I N_c \sin(\omega t) \frac{\mu_0}{g k_C k_{sat}} = B_m \sin(\omega t) \tag{5.60}$$

where $i(t) = \sqrt{2} I \sin(\omega t)$. For sinusoidal variation of $i(t)$ with time, the magnetic flux density also changes sinusoidally with time. The peak value of the magnetic flux density excited by a single coil

$$B_m = \frac{\sqrt{2} I N_c}{2} \frac{\mu_0}{g k_C k_{sat}} = F_{mc} \lambda_g \tag{5.61}$$

where the peak value of the MMF of a single coil that is necessary to excite the peak value B_m of the magnetic flux density in the air gap g

$$F_{mc} = \frac{\sqrt{2} I N_c}{2} \tag{5.62}$$

and the unit permeance of the air gap including the magnetic saturation and slots

$$\lambda_g = \frac{\mu_0}{g k_C k_{sat}} \ [\text{H/m}^2] \tag{5.63}$$

The following relationship exists between the permeance Λ_g and unit permeance λ_g of the air gap

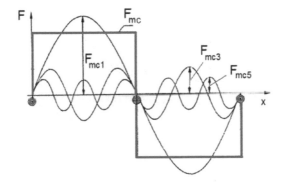

Fig. 5.26. MMF F_m of a single coil, fundamental harmonic F_{mc1} and higher space harmonics F_{mc3} and F_{mc5}.

$$\Lambda_g = \lambda_g S_g = \frac{\mu_0 S_g}{g k_C k_{sat}} \ [\text{H}] \tag{5.64}$$

The peak value of the MMF according to eqn (5.62) is constant along the whole periphery of the machine and only changes its sign in these points where the coil sides are located (rectangular waveform). The magnitude of the fundamental space harmonic is

$$F_{mc1} = \frac{4}{\pi} \frac{\sqrt{2} I N_c}{2} = \frac{2\sqrt{2}}{\pi} I N_c \tag{5.65}$$

The magnitudes of higher space harmonics $\nu = 3, 5, 7, \ldots$ of the MMF of a single coil

$$F_{mc\nu} = \frac{4}{\pi} \frac{1}{\nu} \frac{\sqrt{2} I N_c}{2} = \frac{2\sqrt{2}}{\nu \pi} I N_c \tag{5.66}$$

Example 5.3

The inner diameter of stator of a 4-pole electrical machine is $D = 0.092$ m, the stack length $L = 0.12$ m and the air gap is $g = 0.8$ mm. If the saturation factor $k_{sat} = 1.2$ and Carter's coefficient $k_C = 1.15$, find the reluctance of the ferromagnetic core. What will be the air gap magnetic flux density, MMF, and air gap permeance, if in the stator core a single coil consisting of $N_c = 210$ turns with *rms* current $I = 2.1$ A is placed?

Solution

Pole pitch

$$\tau = \frac{\pi \times 0.092}{4} = 0.0723 \text{ m}$$

Area of the air gap per pole

$$S_g = \tau L = 0.0723 \times 0.12 = 8.671 \times 10^{-3} \text{ m}^2$$

Reluctance of air gap on the basis of eqn (5.47)

$$R_{\mu g} = \frac{k_C g}{\mu_0 S_g} = \frac{1.15 \times 0.0008}{04\pi \times 10^{-6} \times 8.671 \times 10^{-3}} = 84,430 \ \frac{1}{\text{H}}$$

Reluctance of ferromagnetic core on the basis of eqn (5.52)

$$R_{\mu Fe} = 2R_{\mu g}(k_{sat} - 1.0) = 2 \times 84,430 \times (1.2 - 1.0) = 33,770 \ \frac{1}{\text{H}}$$

The reluctance of the ferromagnetic core for magnetic flux is much lower than that of the air gap. MMF of a single coil — eqn (5.62)

$$F_{mc} = \frac{\sqrt{2}IN_c}{2} = \frac{\sqrt{2} \times 2.1 \times 210}{2} = 311.8 \text{ A}$$

Unit permeance of the air gap

$$\lambda_g = \frac{\mu_0}{gk_C k_{sat}} = \frac{04\pi \times 10^{-6}}{0.0008 \times 1.15 \times 1.2} = 1.138 \times 10^{-3} \ \frac{\text{H}}{\text{m}^2}$$

Magnetic flux density — eqn (5.61)

$$B_m = \frac{\sqrt{2}IN_c}{2}\lambda_g = \frac{\sqrt{2} \times 2.1 \times 210}{2} \times 1.138 \times 10^{-3} = 0.355 \text{ T}$$

Permeance of the air gap — eqn (5.64)

$$\Lambda_g = \lambda_g S_g = 1.138 \times 10^{-3} \times 8.671 \times 10^{-3} = 9.87 \times 10^{-6} \text{ H}$$

First harmonic of the MMF — eqn (5.65)

$$F_{mc1} = \frac{2\sqrt{2}}{\pi}IN_c = \frac{2\sqrt{2}}{\pi}2.1 \times 219 = 397.0 \text{ A}$$

Higher space harmonics of MMF are calcuated on the basis of eqn (5.66), i.e., $F_{mc3} = F_{mc1}/3 = 397.0/3 = 132.3$ A, $F_{mc5} = F_{mc1}/5 = 397.0/5 = 79.4$ A, $F_{mc7} = F_{mc1}/7 = 397.0/7 = 56.7$ A, etc.

5.7 Magnetic field of a phase winding

For a winding that consists of q_1 coils fed with an *rms* phase current I_1 the resultant MMF is the algebraic sum of the MMF of each coil $F_{mc\nu}q_1$ multiplied by the winding factor, i.e.,

- for the fundamental harmonic $\nu = 1$

$$F_{m1} = \frac{2\sqrt{2}}{\pi} I_1 N_c q_1 k_{w1} \tag{5.67}$$

- for higher space harmonic $\nu > 1$

$$F_{m\nu} = \frac{2\sqrt{2}}{\nu\pi} I_1 N_c q_1 k_{w1\nu} \tag{5.68}$$

where the number of slots per pole per phase q_1 is according to eqn (5.2) and the winding factors k_{w1}, $k_{w1\nu}$ are accordning to eqns (5.23), (5.36), respectively. Since the number of turns in series per phase is

$$N_1 = N_c q_1 p \qquad \Rightarrow \qquad N_c q_1 = \frac{N_1}{p} \tag{5.69}$$

Thus, the MMF of one phase is

- For the fundamental harmonic $\nu = 1$

$$F_{m1} = \frac{2\sqrt{2}}{\pi} I_1 \frac{N_1 k_{w1}}{p} \approx 0.9 I_1 \frac{N_1 k_{w1}}{p} \tag{5.70}$$

- For higher space harmonic $\nu > 1$

$$F_{m\nu} = \frac{2\sqrt{2}}{\pi} \frac{1}{\nu} I_1 \frac{N_1 k_{w1\nu}}{p} \approx 0.9 \frac{1}{\nu} I_1 \frac{N_1 k_{w1\nu}}{p} \tag{5.71}$$

According to Fig. 5.26 and eqn (5.44) the spacial-time varying MMF of one phase including higher space harmonics

$$F_{m\nu} = \frac{2\sqrt{2}}{\pi} \frac{1}{\nu} I_1 \frac{N_1 k_{w1\nu}}{p} \times \cos\left(\nu \frac{\pi}{\tau} x\right) \times \sin(\omega t) \tag{5.72}$$

The current I_1 is the *rms* phase current in the stator (armature) winding.

5.8 Magnetic field of a three-phase winding

Fig. 5.27 shows the simplified distribution of the stator three-phase winding. Each phase winding in Fig. 5.27a has been replaced by a concentrated-parameter coil. Each phase winding in Fig. 5.27b consists of one full-pitch coil. For instantaneous values of fluxes and currents the following equations can be written:

$$\phi_A + \phi_B + \phi_C = 0 \qquad\qquad i_A + i_B + i_C = 0 \tag{5.73}$$

$$i_A = I_m \sin(\omega t) = \sqrt{2} I_1 \sin(\omega t)$$

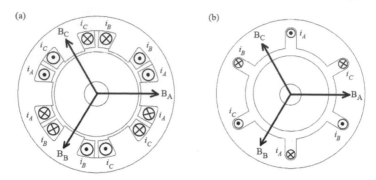

Fig. 5.27. Stator of a three-phase machine: (a) three-phase winding with concentrated parameters; (b) three-phase winding with one full pitch coil per phase.

$$i_B = I_m \sin\left(\omega t - \frac{2}{3}\pi\right) = \sqrt{2}I_1 \sin\left(\omega t - \frac{2}{3}\pi\right) \qquad (5.74)$$

$$i_C = I_m \sin\left(\omega t - \frac{4}{3}\pi\right) = \sqrt{2}I_1 \sin\left(\omega t - \frac{4}{3}\pi\right)$$

The *rms* current I_1 in each phase is the same (balanced current system). The angular frequency

$$\omega = 2\pi f \qquad (5.75)$$

The spacial distribution of the wave of the magnetic flux density has the form

$$B(x,t) = B_m \sin\left(\omega t - \frac{\pi}{\tau}x\right) \qquad (5.76)$$

Figs 5.28, 5.29, 5.30 and 5.31 explain graphically the production of rotating magnetic field by a three-phase symmetrical winding excited with a three-phase balanced current system (5.74).

For an observer that moves with the velocity v_s of the magnetic flux density wave, the rotating magnetic field is motionless. Thus,

$$B_m \sin\left(\omega t - \frac{\pi}{\tau}x\right) = 0 \quad \Rightarrow \quad \sin\left(\omega t - \frac{\pi}{\tau}x\right) = 0 \quad \Rightarrow \quad \omega t - \frac{\pi}{\tau}x = 0$$
$$(5.77)$$

From the above equation (5.77)

$$\omega dt - \frac{\pi}{\tau}dx = 0 \quad \Rightarrow \quad \frac{dx}{dt} = v_s = \omega\frac{\tau}{\pi} = 2f\tau \qquad (5.78)$$

The angular velocity of the magnetic field

$$\Omega_s = \frac{v_s}{0.5D} = 2\pi\frac{f}{p} = \frac{\omega}{p} \qquad (5.79)$$

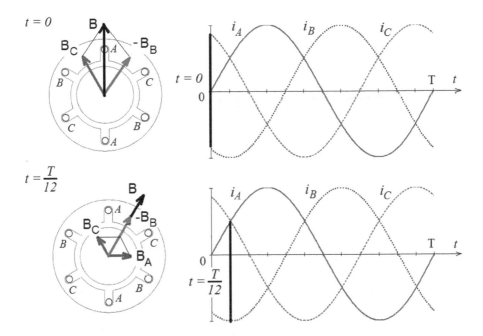

Fig. 5.28. Magnetic field of a three-phase stator winding at the time instants $t = 0$ and $t = T/12$.

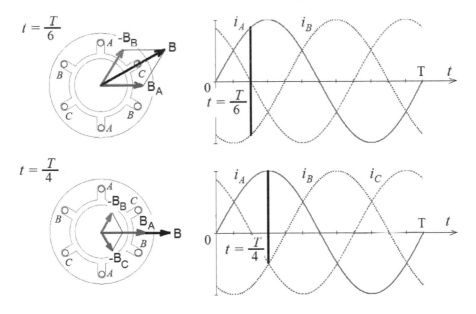

Fig. 5.29. Magnetic field of a three-phase stator winding at the time instants $t = T/6$ and $t = T/4$.

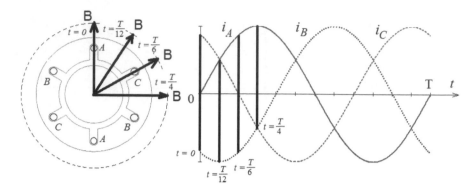

Fig. 5.30. Magnetic field of a three-phase stator winding: superposition for time instants $t = 0$, $t = T/12$, $t = T/6$ and $t = T/4$.

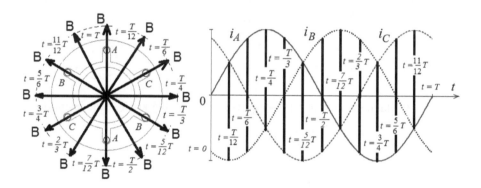

Fig. 5.31. Superposition of phasors of the magnetic flux density for one full period $0 \leq t \leq T$.

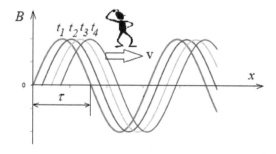

Fig. 5.32. Propagation of magnetic flux density wave along the pole pitch.

The rotational speed of the magnetic field, i.e., the so-called *synchronous speed*

$$n_s = \frac{f}{p} \qquad (5.80)$$

The synchronous speed n_s depends on the frequency f and number of pole pairs p. Table 5.1 specifies synchronous speeds as functions of the number of pole pairs for power frequencies 50 Hz and 60 Hz.

Table 5.1. Synchronous speeds for different numbers of pole pairs.

p	$f = 50$ Hz		$f = 60$ Hz	
	n_s [rev/s]	n_s [rpm]	n_s [rev/s]	n_s [rpm]
1	50	3000	60	3600
2	25	1500	30	1800
3	$16\frac{2}{3}$	1000	20	1200
4	12.5	750	15	900
5	10	600	12	720

The instantaneous values of phase currents are expressded by eqn (5.74). Similar equations can be written for the first space harmonics of phase MMFs assuming that the space distribution of the MMF is according to $\cos(\pi x/\tau)$, i.e.,

$$i_A(t) = I_m \sin(\omega t) \qquad F_{1A}(x,t) = F_{m1A} \cos\left(\frac{\pi}{\tau}x\right) \sin(\omega t) \qquad (5.81)$$

$$i_B(t) = I_m \sin\left(\omega t - \frac{2}{3}\pi\right) \quad F_{1B}(x,t) = F_{m1B} \cos\left(\frac{\pi}{\tau}x - \frac{2}{3}\pi\right) \sin\left(\omega t - \frac{2}{3}\pi\right)$$

$$i_C = I_m \sin\left(\omega t - \frac{4}{3}\pi\right) \quad F_{1C}(x,t) = F_{m1C} \cos\left(\frac{\pi}{\tau}x - \frac{4}{3}\pi\right) \sin\left(\omega t - \frac{4}{3}\pi\right)$$

Using the following trigonometric identity

$$\sin x \cos y = \frac{1}{2}\left[\sin(x+y) + \sin(x-y)\right] \qquad (5.82)$$

and then adding the MMFs $F_{1A}(x,t)$, $F_{1B}(x,t)$, $F_{1C}(x,t)$ and assuming that $F_{1mA} = F_{1mB} = F_{1mC} = F_{1m}$, the resultant MMF is

$$F_1(x,t) = F_{1A}(x,t) + F_{1B}(x,t) + F_{1C}(x,t) = \frac{3}{2}F_{m1} \sin\left(\omega t - \frac{\pi}{\tau}x\right) \qquad (5.83)$$

where the amplitude of the MMF produced by three phases

$$F_m = \frac{3}{2}F_{m1} \qquad (5.84)$$

Putting F_{m1} given by eqn (5.70) the resultant MMF of a three-phase stator winding is

$$F_1(x,t) = F_m \sin\left(\omega t - \frac{\pi}{\tau}x\right)$$

$$= \frac{3\sqrt{2}}{\pi}I_1\frac{N_1 k_{w1}}{p}\sin\left(\omega t - \frac{\pi}{\tau}x\right) \approx 1.35 I_1 \frac{N_1 k_{w1}}{p}\sin\left(\omega t - \frac{\pi}{\tau}x\right) \qquad (5.85)$$

In the above eqn (5.85)

$$F_m = \frac{3}{2}F_{m1} = \frac{3\sqrt{2}}{\pi}I_1\frac{N_1 k_{w1}}{p} \qquad (5.86)$$

is the amplitude of the three-phase MMF. Eqns (5.83) and (5.85) describe the fundamental harmonic of the resultant MMF under assumption that the three-phase windings are symmetrical and the three-phase current system is balanced. The symmetry of windings means that for each phase winding the number of turns in series N_1 and winding factors k_{w1} are the same.

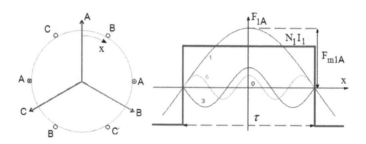

Fig. 5.33. MMF of one phase.

The rectangular distribution of the MMF excited by one phase and its harmonics are shown in Fig. 5.33. The resultant MMFs for all higher space harmonics can be found in a similar way as for the fundamental harmonics, i.e.,

$$F_{\nu A}(x,t) = \frac{2\sqrt{2}}{\pi}I_1\frac{1}{\nu}\frac{N_1 k_{w\nu}}{p}\cos\left(\nu\frac{\pi}{\tau}x\right)\sin(\omega t)$$

$$F_{\nu B}(x,t) = \frac{2\sqrt{2}}{\pi}I_1\frac{1}{\nu}\frac{N_1 k_{w\nu}}{p}\cos\left[\nu\left(\frac{\pi}{\tau}x - \frac{2}{3}\pi\right)\right]\sin\left(\omega t - \frac{2}{3}\pi\right) \qquad (5.87)$$

$$F_{\nu C}(x,t) = \frac{2\sqrt{2}}{\pi}I_1\frac{1}{\nu}\frac{N_1k_{w\nu}}{p}\cos\left[\nu\left(\frac{\pi}{\tau}x - \frac{4}{3}\pi\right)\right]\sin\left(\omega t - \frac{4}{3}\pi\right)$$

After adding the MMFs $F_{\nu A}(x,t)$, $F_{\nu B}(x,t)$, $F_{\nu C}(x,t)$

$$F_3(x,t) = 0 \tag{5.88}$$

$$F_5(x,t) = \frac{3}{2}I_1\frac{1}{5}\frac{2\sqrt{2}}{\pi}\frac{N_1k_{w5}}{p}\sin\left(\omega t + 5\frac{\pi}{\tau}x\right) \tag{5.89}$$

$$F_7(x,t) = \frac{3}{2}I_1\frac{1}{5}\frac{2\sqrt{2}}{\pi}\frac{N_1k_{w5}}{p}\sin\left(\omega t - 7\frac{\pi}{\tau}x\right) \tag{5.90}$$

$$F_9(x,t) = 0 \tag{5.91}$$

$$\dotsi\dotsi\dotsi\dotsi\dotsi$$

For the 3rd harmonics the simplified notation of phase MMFs has the form:

$$F_{3A}(x,t) = -\frac{1}{3}F_{m1}\frac{k_{w3}}{k_{w1}}\cos\left(3\frac{\pi}{\tau}x\right)\sin(\omega t)$$

$$F_{3B}(x,t) = -\frac{1}{3}F_{m1}\frac{k_{w3}}{k_{w1}}\cos\left[3\left(\frac{\pi}{\tau}x - \frac{2}{3}\pi\right)\right]\sin\left(\omega t - \frac{2}{3}\pi\right) \tag{5.92}$$

$$F_{3C}(x,t) = -\frac{1}{3}F_{m1}\frac{k_{w3}}{k_{w1}}\cos\left[3\left(\frac{\pi}{\tau}x - \frac{4}{3}\pi\right)\right]\sin\left(\omega t - \frac{4}{3}\pi\right)$$

where F_{m1} is according to eqn (5.70). In a three-phase AC machine the 3rd harmonics and their multiples $\nu = 3, 9, 15, 21, \ldots$ do not excite the resultant magnetic field. The harmonics $\nu = 1, 5, 11, 17, 23, \ldots$ rotate in the direction opposite from the direction of rotation of the fundamental harmonic $\nu = 1$. The harmonics $\nu = 7, 13, 19, 25, \ldots$ rotate in the same direction as the fundamental harmonic $\nu = 1$. In conclusion:

- In the spatial distribution of the magnetic field excited by a three-phase winding there are only fundamental harmonic $\nu = 1$ and higher space harmonics $\nu = 5, 7, 11, 13, 17, 19, 23, 25, \ldots$, in general $\nu = 6k \pm 1$, where $k = 0, 1, 2, 3, \ldots$. For an m_1-phase winding

$$\nu = 2m_1k \pm 1 \qquad \text{where} \qquad k = 0, 1, 2, 3, \ldots \tag{5.93}$$

- Every harmonic rotates around the periphery of the machine with the speed

$$\frac{n_s}{\nu} \qquad \Rightarrow \qquad \frac{n_s}{1}, -\frac{n_s}{5}, \frac{n_s}{7}, -\frac{n_s}{11}, \frac{n_s}{13}, -\frac{n_s}{17}, \frac{n_s}{19}, \ldots \tag{5.94}$$

- The magnitudes of higher space harmonics of the magnetic field are inversely proportional to the harmonic number ν.

The rotating magnetic field created by a symmetrical three-phase winding fed with balanced three-phase current system is called "circular magnetic field" because the locus of the resultant rotating magnetic flux density phasor is a circle. If the winding is asymmetrical and/or the three-phase current system is unbalanced, the rotating magnetic field is called "elliptical magnetic field" because the locus of the resultant magnetic flux density is an ellipse.

Example 5.4

If the four-pole stator of Example 5.3 has $s_1 = 24$ slots that accommodates three-phase winding consisting of full pitch coils, find the first harmonic of the MMF of a single phase and all three phases. The phase rms current is $I_1 = 2.1$ A.

Solution

Number of slots per pole

$$Q_1 = \frac{24}{4} = 6$$

This is the coil span or a full pitch coil measured in the number of slots. Number of slots per pole per phase $q_1 = 24/(4 \times 3) = 2$. Number of turns in series per phase — eqn (5.69)

$$N_1 = N_c q_1 p = 210 \times 2 \times 2 = 840$$

Distribution factor

$$k_{d1} = \frac{\sin\left(\frac{\pi}{2 \times 3}\right)}{2\sin\left(\frac{\pi}{2 \times 3 \times 2}\right)} = 0.966$$

Pitch factor $k_{p1} = 1.0$ because there are full pitch coils. Winding factor

$$k_{w1} = 0.966 \times 1.0 = 0.966$$

First harmonic of the MMF of one phase — eqn (5.70)

$$F_{m1} = \frac{2\sqrt{2}}{\pi} \times 2.1 \frac{840 \times 0.966}{2} = 767.0 \text{ A}$$

First harmonic of MMF of three-phase winding — eqn (5.84)

$$F_m = \frac{3}{2} F_{m1} = \frac{3}{2} \times 767.0 = 1150.5 \text{ A}$$

or, using eqn (5.86)

$$F_m = \frac{3\sqrt{2}}{\pi} \times 2.1 \times \frac{840 \times 0.966}{2} = 1150.5 \text{ A}$$

5.9 Influence of magnetic saturation

The magnetic flux can be obtained by dividing the MMF by the reluctance of two air gaps (Fig. 5.24), i.e.,

$$\Phi(x,t) = \frac{F(x,t)}{2R_{\mu g}} \tag{5.95}$$

where

$$2R_{\mu g} = 2\frac{g}{\mu_0 S_g} \tag{5.96}$$

Dividing the magnetic flux $\Phi(x,t)$ by the area of the air gap, the magnetic flux density is obtained

$$B(x,t) = \frac{\Phi(x,t)}{S_g} = \frac{\frac{F(x,t)}{2R_{\mu g}}}{S_g} = \frac{\mu_0}{2g}F(x,t) \tag{5.97}$$

Eqns (5.95), (5.96) and (5.97) do not take into account magnetic voltage drops in ferromagnetic cores of the stator and rotor with the total reluctance $R_{\mu Fe}$. Including the reluctance of the stator and rotor cores with slots with the aid of Carter's coefficient k_C defined by eqn (5.56), the above eqns (5.95) and (5.97) take the forms, respectively,

$$\Phi(x,t) = \frac{F(x,t)}{2R_{\mu g} + R_{\mu Fe}} = \frac{F(x,t)}{2R_{\mu g}k_{sat}k_C} \tag{5.98}$$

$$B(x,t) = \frac{\Phi(x,t)}{S_g} = \frac{\frac{F(x,t)}{2R_{\mu g}k_{sat}k_C}}{S_g} = \frac{\mu_0}{2k_{sat}k_C g}F(x,t) \tag{5.99}$$

The saturation factor of the magnetic circuit defined with the aid magnetic voltage drops $V_\mu = R_\mu \Phi$ is

$$k_{sat} = \frac{\text{total sum of magnetic voltage drops}}{\text{magnetic voltage drops of air gaps}} \tag{5.100}$$

Thus,

$$k_{sat} = \frac{2R_{\mu g} + R_{\mu Fe}}{2R_{\mu g}}$$

$$= \frac{2R_{\mu g}\Phi + R_{\mu Fe}\Phi}{2R_{\mu g}\Phi} = \frac{2V_{\mu g} + V_{\mu Fe}}{2V_{\mu g}} = 1 + \frac{V_{\mu Fe}}{2V_{\mu g}} \geq 1 \tag{5.101}$$

The magnetic voltage balance equation results from eqn (5.101), i.e.,

$$2V_{\mu g} + V_{\mu Fe} = 2V_{\mu g}k_{sat} \tag{5.102}$$

5.10 MMF of two-phase winding

A symmetrical two-phase stator winding (Fig. 5.34) fed with balanced system of two-phase currents also excites a circular magnetic rotating field. The windings on the periphery of the stator are shifted by an electrical angle $\beta = 90°$ ($\pi/2$) and the currents are shifted in phase also by the same angle. Thus, the fundamental harmonics of MMFs are described by the following equations:

$$F_A(x,t) = \frac{2\sqrt{2}}{\pi} I_1 \frac{N_1 k_{w1}}{p} \sin(\omega t) \cos\left(\frac{\pi}{\tau}x\right)$$

$$(5.103)$$

$$F_B(x,t) = \frac{2\sqrt{2}}{\pi} I_1 \frac{N_1 k_{w1}}{p} \sin\left(\omega t - \frac{\pi}{2}\right) \cos\left(\frac{\pi}{\tau}x - \frac{\pi}{2}\right)$$

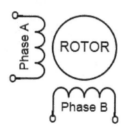

Fig. 5.34. Two-phase stator winding.

The sum of these MMFs is

$$F(x,t) = F_A(x,t) + F_B(x,t) = \frac{2\sqrt{2}}{\pi} I_1 \frac{N_1 k_{w1}}{p} \sin\left(\omega t - \frac{\pi}{\tau}x\right)$$

$$\approx 0.9 I_1 \frac{N_1 k_{w1}}{p} \sin\left(\omega t - \frac{\pi}{\tau}x\right) \qquad (5.104)$$

The amplitude of the rotating MMF is

$$F_{m1} = \frac{2\sqrt{2}}{\pi} I_1 \frac{N_1 k_{w1}}{p} \approx 0.9 I_1 \frac{N_1 k_{w1}}{p} \qquad (5.105)$$

In two-phase machines, in general, the windings are asymmetrical because the windings may have different numbers of turns N_{1A}, N_{1B}, different numbers of slots per pole Q_{1A}, Q_{1B}, different winding factors k_{w1A}, k_{w1B} or different diameters of wires d_{1A}, d_{1B}, i.e.,

$$N_{1A} \neq N_{1B}, \qquad Q_{1A} \neq Q_{1B}, \qquad k_{w1A} \neq k_{w1B} \qquad d_{1A} \neq d_{1B} \quad (5.106)$$

Two-phase machines can also be fed with current with different *rms* values and different phase shift $\beta \leq \pi/2$, i.e.,

$$I_{1A} \neq I_{1B}, \qquad\qquad \beta \neq \frac{\pi}{2} \qquad\qquad (5.107)$$

Thus

$$F_A(x,t) = \frac{2\sqrt{2}}{\pi} I_{1A} \frac{N_{1A} k_{w1A}}{p} \sin(\omega t) \cos\left(\frac{\pi}{\tau} x\right)$$

$$(5.108)$$

$$F_B(x,t) = \frac{2\sqrt{2}}{\pi} I_{1A} \frac{N_{1A} k_{w1A}}{p} \sin\left(\omega t - \beta\right) \cos\left(\frac{\pi}{\tau} x - \frac{\pi}{2}\right)$$

Using trigonometric identity (5.82) and denoting the amplitudes of MMFs of phases A and B as

$$F_{m1A} = \frac{2\sqrt{2}}{\pi} I_{1A} \frac{N_{1A} k_{w1A}}{p} \approx 0.9 I_{1A} \frac{N_{1A} k_{w1A}}{p}$$

$$(5.109)$$

$$F_{m1B} = \frac{2\sqrt{2}}{\pi} I_{1B} \frac{N_{1B} k_{w1B}}{p} \approx 0.9 I_{1B} \frac{N_{1B} k_{w1B}}{p}$$

each MMF per phase can be described as a sum of two MMFs with the same amplitude rotating in opposite directions, i.e.,

$$F_A(x,t) = \frac{1}{2} F_{m1A} \sin\left(\omega t - \frac{\pi}{\tau} x\right) + \frac{1}{2} F_{m1A} \sin\left(\omega t + \frac{\pi}{\tau} x\right)$$

$$(5.110)$$

$$F_B(x,t) = \frac{1}{2} F_{m1B} \sin\left(\omega t - \frac{\pi}{\tau} x - \beta + \frac{\pi}{2}\right) + \frac{1}{2} F_{m1B} \sin\left(\omega t + \frac{\pi}{\tau} x - \beta - \frac{\pi}{2}\right)$$

The constant value of each rotating MMF is equal to half of the amplitude of the respective MMF per phase. On the basis of Fig. 5.35

$$F_1(x,t) = \frac{1}{2} F_{m1A} \sin\left(\omega t - \frac{\pi}{\tau} x\right) + \frac{1}{2} F_{m1B} \sin\left(\omega t - \frac{\pi}{\tau} x - \beta - \frac{\pi}{2}\right)$$

$$= F_{m1} \sin\left(\omega - \frac{\pi}{\tau} x + \gamma\right) \qquad (5.111)$$

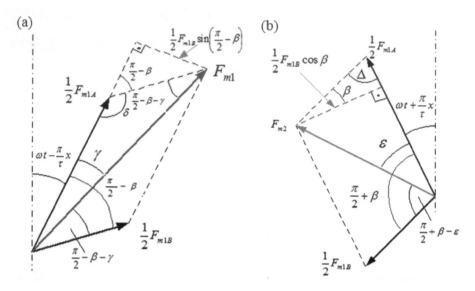

Fig. 5.35. MMFs of two-phase winding: (a) excitation of forward-rotating field — eqn (5.111); (b) excitation of backward-rotating field — eqn (5.112).

$$F_2(x,t) = \frac{1}{2}F_{m1A}\sin\left(\omega t + \frac{\pi}{\tau}x\right) + \frac{1}{2}F_{m1B}\sin\left(\omega t + \frac{\pi}{\tau}x - \beta - \frac{\pi}{2}\right)$$

$$= F_{m2}\sin\left(\omega + \frac{\pi}{\tau}x + \epsilon\right) \tag{5.112}$$

Excitation of the *forward-rotating field* is visualized in Fig. 5.35a, while excitation of the *backward-rotating field* is visualized in Fig. 5.35b. The angles δ and γ have been found on the basis of the theorem that on a Euclidean plane the sum of inner angles of a triangle is equal to $180°$.

Thus, the resultant MMF is

$$F(x,t) = F_{m1}\sin\left(\omega - \frac{\pi}{\tau}x + \gamma\right) + F_{m2}\sin\left(\omega + \frac{\pi}{\tau}x + \epsilon\right) \tag{5.113}$$

Using the cosine theorem

$$c^2 = a^2 + b^2 - 2ab\cos(\alpha) \tag{5.114}$$

the amplitudes of MMFs, $\sin(\gamma)$ and $\sin(\epsilon)$ are

$$F_{m1} = \frac{1}{2}\sqrt{F_{m1A}^2 + F_{m1B}^2 + 2F_{m1A}F_{m1B}\sin(\beta)} \tag{5.115}$$

$$F_{m2} = \frac{1}{2}\sqrt{F_{m1A}^2 + F_{m1B}^2 - 2F_{m1A}F_{m1B}\sin(\beta)} \qquad (5.116)$$

$$\sin(\gamma) = \frac{1}{2}\frac{F_{m1B}}{F_1}\sin\left(\frac{\pi}{2} - \beta\right) = \frac{1}{2}\frac{F_{m1B}}{F_{m1}}\cos(\beta) \qquad (5.117)$$

$$\sin(\epsilon) - \frac{1}{2}\frac{F_{m1B}}{F_{m2}}\cos(\beta) \qquad (5.118)$$

since $\sin(\pi/2 \pm \beta) = \cos(\beta)$.

The MMFs $F_1(x,t)$ and $F_2(x,t)$ given by eqns (5.111) and (5.112) are called *forward-rotating MMF* and *backward-rotating MMF*. The magnetic field is pure *circular* if either the *backward-rotating field* or *forward-rotating field* is equal to zero, e.g., for $F_{m2} = 0$ it must be

$$F_{m1A} = F_{m1B} \qquad\qquad \beta = \frac{\pi}{2} \qquad (5.119)$$

$$F_{m2} = \frac{1}{2}\sqrt{F_{m1A}^2 + F_{m1B}^2 - 2F_{m1A}F_{m1B}\sin(\beta)}$$

$$= \frac{1}{2}\sqrt{2F_{m1A}^2 - 2F_{m1A}^2\sin(90°)} = 0 \qquad (5.120)$$

To obtain $F_{m1} = 0$

$$F_{m1A} = F_{m1B} \qquad\qquad \beta = -\frac{\pi}{2} \qquad (5.121)$$

$$F_{m1} = \frac{1}{2}\sqrt{F_{m1A}^2 + F_{m1B}'^2 + 2F_{m1A}F_{m1B}\sin(\beta)}$$

$$= \frac{1}{2}\sqrt{2F_{m1A}^2 - 2F_{m1A}^2\sin(-90°)} = 0 \qquad (5.122)$$

The sum of MMFs $F_1(x,t)$ and $F_2(x,t)$ expressed by eqns (5.111) and (5.112) gives the resultant MMF of elliptic form — see eqn (5.113). An *ellipse* is a curve on a plane surrounding two focal points such that the sum of the distances to the two focal points is constant for every point on the curve. The ellipse equation is

$$\frac{y^2}{(F_{m1} + F_{m2})^2} + \frac{x^2}{(F_{m1} - F_{m2})^2} = 1 \qquad (5.123)$$

Fig. 5.36 shows how the elliptical MMF can be resolved into a forward-rotating circular magnetic field with amplitude F_{m1} and a backward-rotating circular magnetic field with amplitude F_{m2}.

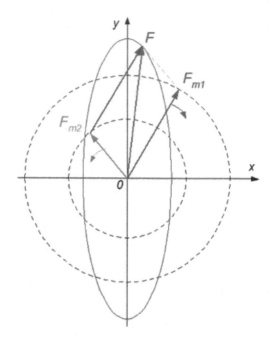

Fig. 5.36. The elliptical MMF is the superposition of two circular MMFs with different amplitudes $F_{m1} \neq F_{m2}$ rotating in opposite direction.

5.11 MMF of a single-phase winding

A single-phase winding can be distributed in slots (Fig. 5.37a) or consist of concentrated coils (Fig. 5.37b). In the case of a single-phase stator winding with distributed parameters (Fig. 5.37a) fed with sinusoidal current, the fundamental harmonic $\nu = 1$ of the MMF is described by the equation:

$$F_1(x, t) = F_{m1} \sin(\omega t) \cos\left(\frac{\pi}{\tau}x\right) \tag{5.124}$$

Making use of trigonometric identity (5.82) the MMF (5.124) can be replaced by superposition of two MMFs with the same amplitude and rotating in opposite direction, i.e.,

$$F_1(x, t) = \frac{1}{2}F_{m1} \sin\left(\omega t - \frac{\pi}{\tau}x\right) + \frac{1}{2}F_{m1} \sin\left(\omega t + \frac{\pi}{\tau}x\right) \tag{5.125}$$

The first term on the right-hand side of the above equation is the *forward-rotating MMF*, i.e., in the direction of the x-axis, and the second term is the *backward-rotating MMF*, i.e., in the opposite direction to the x-axis. The conclusion is that when feeding a single-phase winding with single-phase AC current with the frequency $f = \omega/(2\pi)$, a *pulsating magnetic field* is excited, which can be resolved into two fields that rotate in opposite direction.

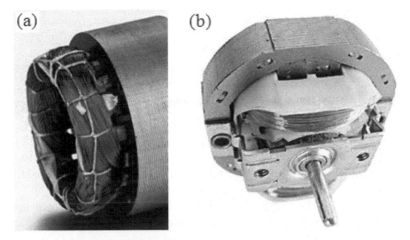

Fig. 5.37. Examples of single-phase stator windings: (a) winding distributed in slots; (b) winding with concentrated parameters of a two-pole motor with one coil per pole.

A concentrated-parameter single-phase winding in the case of a uniform air gap excites the MMF with its shape very similar to a rectangular waveform (Fig. 5.26) with amplitude

$$F_{m1} - \frac{4}{\pi}F_m \tag{5.126}$$

Higher space harmonics can be found by resolving the rectangular waveform (Fig. 5.26) into Fourier series.

Summary

The armature (stator) windings of AC electrical machines are made of coils, which create coil groups. Unlike DC machines, each phase winding of an AC machine has entry and exit terminals. A single-phase winding is a system of groups of coils connected in series or in parallel. In one phase of a three-phase winding consisting of $q_1 = $ integer coils, there is at least p coil groups (p is the number of pole pairs), in which the EMFs have the same amplitude and phase angle. Basic elements of a winding with distributed parameters are

- Turn;
- Coil;
- Coil group consisting of several coils.

Coils groups can be made of *concentric coils* or *diamond-shaped coils*. Basic parameters of AC windings are

- The number of slots per pole per phase $q_1 = s_1/(2pm_1)$, where s_1 is the number of stator slots, $2p$ is the number of poles and m_1 is the number of phases;
- The number of slots per pole $Q_1 = s_1/(2p)$;
- Coil span $y \leq \tau$, where the pole pitch measured in slots $\tau = Q_1$;
- Electrical angle between neighboring slots $\alpha_{el} = 360°p/s_1$;

The *electrical angle* α_{el} is p times greater than the *mechanical angle* α_m, i.e., $\alpha_{el} = p\alpha_m$. For a two-pole machine, i.e., $2p = 2$, the electrical angle is the same as the mechanical angle, i.e., $\alpha_{el} = \alpha_m$.

Dependent on the number of coil sides in slots, the windings are divided into *single layer* and *double layer* windings (Fig. 5.4).

With respect to the shape of end turns, the windings are divided into (a) double-tier, (b) three-tier and (c) made of former coils (Fig. 5.5).

A three-phase winding can be *Y-connected* or *Δ-connected* (Fig. 5.6).

Double-layer windings can be divided into *lap windings* and *wave windings* (Fig. 5.2).

With respect to the number of slots per pole per phase $q_1 = s_1/(2pm_1)$ the windings can be classified into *windings with q_1 an integer* and *windings with fractional q_1* (fractional slot winding).

With respect to coil span y the windings are divided into *full pitch windings* with $y = \tau$ and *chorded (short-pitch) windings* with $y < \tau$ (Fig. 5.7).

With respect to fabrication of coils the windings are divided into windings inserted in open slots, dropped in semi-closed slots, sewed up and pushed through a pairs of slots (Fig. 5.8).

The winding made of *elastic coils* is shown in Fig. 5.9 and the winding made of *stiff diamond-shaped coils* is shown in Fig. 5.10.

Since the EMFs induced by the magnetic rotating field in all conductors have the same amplitude and frequency, it is possible to represent the EMFs with the aid of phasors of the same module and uniformly shifted in phase (time). Such a diagram is called *star of slot EMFs* (Fig. 5.12).

Rules for designing a symmetrical, three-phase winding are (Figs 5.14, 5.15):

(a) EMFs of three phases should be equal and shifted symmetrically in phase by $2\pi/3$;
(b) A coil should be designed in such a way as to obtain EMFs in the left and right coil side shifted by an angle $\pi \equiv 180°$ (full-pitch coil) or slightly less (chorded coil);
(c) It is necessary to create a phase winding by connecting conductors (coil sides), in which EMFs are least shifted in phase.

The coil group within a one-phase winding can create one parallel path $a = 1$ or more than one parallel paths. For single layer windings, the maximum possible number of parallel paths of one phase is $a = p$. For double layer windings the maximum possible number of parallel paths of one phase is $a = 2p$.

The *rms* EMF per phase is given by eqn (5.24), i.e.,

$$E_1 = \pi\sqrt{2}fN_1k_{w1}\Phi \approx 4.44fN_1k_{w1}\Phi$$

where f is the stator current and magnetic flux frequency, N_1 is the number of turns in series per phase, k_{w1} is the winding factor for fundamental harmonic, and Φ is the magnetic flux given by eqn (5.20), i.e.,

$$\Phi = L \int_0^\tau \sin\left(\frac{\pi}{\tau}x\right) dx = \frac{2}{\pi}B_m\tau L$$

In the above equation, B_m is the peak value of the magnetic flux density in the air gap, $\tau = \pi D/(2p)$ is the pole pitch and L is the length of coil in the slot.

The winding factor k_{w1} of the stator winding for fundamental space harmonic $\nu = 1$ is the product of the distribution factor k_{d1} and pitch factor k_{p1}, i.e., $k_{w1} = k_{d1}k_{p1}$. The distribution factor k_{d1} (Fig. 5.21) is defined as the phasor sum-to-the-arithmetic sum ratio of EMFs induced in coils belonging to the same coil group (5.29), i.e.,

$$k_{d1} = \frac{\sin\left(q_1\frac{\alpha_{el}}{2}\right)}{q_1\sin\left(\frac{\alpha_{el}}{2}\right)} = \frac{\sin\left(\frac{\pi}{2m_1}\right)}{q_1\sin\left(\frac{\pi}{2m_1q_1}\right)}$$

where q_1 is the number of slots per pole per phase, $\alpha_{el} - p\alpha_m$ is the angle measured in electrical degrees, p is the number of pole pairs and m_1 is the number of stator phases. The pitch factor k_{p1} of the stator winding (Fig. 5.22) is defined as the ratio of the phasor sum-to-arithmetic sum of EMFs induced in two sides of a coil (5.32), i.e.,

$$k_{p1} = \sin\left(\frac{\beta}{2}\right) = \sin\left(\frac{w_c}{\tau}\frac{\pi}{2}\right)$$

where $\beta = \pi w_c/\tau$ is the coil span measured in radians or electrical degrees, w_c is the coil span measured in meters, and τ is the pole pitch.

In a three-phase AC machine with symmetrical phase winding fed with balanced three-phase voltage system, the third harmonic $\nu = 3$ and its multiples theoretically do not exist, i.e., $\nu = 3k = 0$ where $k = 1, 2, 3, \ldots$. The harmonics $\nu = 6k + 1 = 7, 13, 19, 25, \ldots$ rotate in the direction of the fundamental harmonic $\nu = 1$. The harmonics $\nu = 6k - 1 = 5, 11, 17, 23, \ldots$ rotate in the opposite direction. This applies both to the magnetic flux density and MMF. The frequency f_ν, pole pitch τ_ν, electrical angle $\alpha_{el\nu}$, winding factor k_w, magnetic flux ϕ_ν and the EMF $E_{1\nu}$ for higher space harmonics are different than those for the fundamental harmonic, i.e., $f_\nu = \nu f$, $\tau_\nu = \tau/\nu$, $\alpha_{el\nu} = \nu\alpha_{el}$,

$$k_{w1\nu} = k_{d1\nu} k_{p1\nu} \qquad k_{d1\nu} = \frac{\sin\left(q_1 \nu \frac{\alpha_{el}}{2}\right)}{q_1 \sin\left(\nu \frac{\alpha_{el}}{2}\right)} \qquad k_{p1\nu} = \sin\left(\nu \frac{w_c}{\tau} \frac{\pi}{2}\right)$$

$$\Phi_\nu = \frac{2}{\pi} B_{m\nu} \tau_\nu L = \frac{2}{\pi} B_{m\nu} \frac{\tau}{\nu} L$$

$$E_{1\nu} = \pi\sqrt{2} f_\nu k_{w1\nu} N_1 \Phi_\nu \approx 4.44 f_\nu k_{w1\nu} N_1 \Phi_\nu$$

where $B_{m\nu}$ is the amplitude of the νth harmonic of the magnetic flux density in the air gap. The winding distribution factor $k_{d1\nu}$ and pitch factor $k_{p1\nu}$ for higher space harmonics ν are expressed by eqns (5.37) and (5.38).

The *saturation factor* k_{sat} of magnetic circuit is defined as the total sum of reluctances-to-reluctance of air gaps (5.49), i.e.,

$$k_{sat} = \frac{2R_{\mu g} + R_{\mu Fe}}{2R_{\mu g}} = 1 + \frac{R_{\mu Fe}}{2R_{\mu g}}$$

where $R_{\mu g}$ is the reluctance of the air gap g and $R_{\mu Fe}$ is the reluctance of the ferromagnetic core.

Carter's coefficient k_C takes into account the effect of slot opening on the thickness of the air gap g. It is defined as the equivalent thickness of air gap with slot openings to the thickness of slotless air gap g (5.56).

The *equivalent air gap* g' with the effect of slot openings being included is

$$g' = k_C g \qquad \text{where} \qquad k_C \geq 1$$

The *MMF of one phase* is given by eqns (5.70) and (5.71), i.e.,

- For the fundamental harmonic $\nu = 1$

$$F_{m1} = \frac{2\sqrt{2}}{\pi} I_1 \frac{N_1 k_{w1}}{p} \approx 0.9 I_1 \frac{N_1 k_{w1}}{p}$$

- For higher harmonics $\nu > 1$

$$F_{m\nu} = \frac{2\sqrt{2}}{\pi} \frac{1}{\nu} I_1 \frac{N_1 k_{w1\nu}}{p} \approx 0.9 \frac{1}{\nu} I_1 \frac{N_1 k_{w1\nu}}{p}$$

A three-phase symmetrical winding fed with three-phase balanced sinusoidal current system produces *forward-rotating magnetic field* the rotational speed of which depends on the frequency f of the current and the number of pole pairs of the winding, i.e., $n_s = f/p$. Table 5.1 specifies synchronous speeds n_s as functions of the number of pole pairs p for power frequencies 50 Hz and 60 Hz.

The *resultant MMF of a three-phase winding* for the fundamental space harmonic $\nu = 1$ is given by eqn (5.85), i.e.,

$$F_1(x, t) = \frac{3\sqrt{2}}{\pi} I_1 \frac{N_1 k_{w1}}{p} \sin\left(\omega t - \frac{\pi}{\tau} x\right) \approx 1.35 I_1 \frac{N_1 k_{w1}}{p} \sin\left(\omega t - \frac{\pi}{\tau} x\right)$$

In general, for m_1-phase winding the numbers of higher space harmonics are $2m_1 k \pm 1$ where $k = 0, 1, 2, 3, \ldots$.

In a three-phase AC machine every harmonic rotates around the periphery of the machine with the speed

$$\frac{n_s}{\nu} \qquad \Rightarrow \qquad \frac{n_s}{1}, -\frac{n_s}{5}, \frac{n_s}{7}, -\frac{n_s}{11}, \frac{n_s}{13}, -\frac{n_s}{17}, \frac{n_s}{19}, \ldots$$

The "$-$" sign means that the given harmonic of the MMF and the magnetic flux density rotates in the direction opposite to the fundamental harmonic $\nu = 1$. In a three-phase AC machine, the 3rd harmonic and their multiples $\nu = 3, 9, 15, 21, \ldots$ do not excite the resultant magnetic field. The amplitudes of higher space harmonics of the magnetic field are inversely proportional to the harmonic number ν.

In terms of the *magnetic voltage drops*, the *saturation factor* of the magnetic circuit can be expressed as the total sum of magnetic voltage drops in the magnetic circuit-to-magnetic voltage drops across air gaps, as given by eqns (5.100) and (5.101).

A symmetrical *two-phase stator winding* (Fig. 5.34) fed with *balanced system of two-phase currents* excites a *circular magnetic rotating field*. The windings on the periphery of the stator are shifted by an electrical angle $\beta = 90°$ ($\pi/2$) and the currents are shifted in phase also by the same angle. The fundamental harmonics of MMFs are described by eqns (5.104) to (5.113).

In *two-phase machines*, in general, the windings are asymmetrical because the windings may have different numbers of turns $N_{1A} \neq N_{1B}$, different numbers of slots per pole $Q_{1A} \neq Q_{1B}$, different winding factors $k_{w1A} \neq k_{w1B}$ or different diameters of wires $d_{1A} \neq d_{1B}$. Two-phase machines can also be fed with currents with different *rms* values $I_{1A} \neq I_{1B}$ and different phase shift $\beta < \pi/2$. If the symmetry conditions for windings and currents are not met, the magnetic field in the air gap is not circular, but *elliptical* (Fig. 5.36).

A *single-phase winding* can be distributed in slots (Fig. 5.37a) or consist of concentrated coils (Fig. 5.37b). Single-phase windings can be analyzed similarly as two-phase windings.

Problems

1. Sketch the full diagram of a three-phase double-layer winding with $m_1 = 3$, $2p = 4$, $s_1 = 24$, $y = 5$ assuming the number of parallel paths $a = 1$ and $a = 2$.

Answer: $Q_1 = 6$, $q_1 = 2$, $\alpha_{el} = 30°$, distance from the left sided to the right return side of the coil $y = 5$ slots or $150°$, distance between beginning of phase A and B is 4 slots or $120°$.

2. The stator of a three-phase, 8-pole, 60-Hz, AC electrical machine has the inner diameter $D = 0.13$ m and the stack length $L = 0.15$ m. The number of stator slots is $s_1 = 72$ and the peak value of the magnetic flux density in the air gap $B_m = 0.711$ T. The number of turns per phase is $N_1 = 480$ and the coil span measured in slots is $y = 7$. Find the EMF E_1 per phase.

Answer: $k_{w1} = 0.902$, $E_1 = 400$ V.

3. For the stator of Problem 5.2 find the pole pitch and winding factors for higher space harmonics $\nu = 11$ and $\nu = 13$.

Answer: $\tau_{11} = 4.6$ mm, $\tau_{13} = 3.9$ mm, $k_{w11} = -0.136$, $k_{w13} = -0.038$.

4. The inner diameter of stator of a 2-pole electrical machine is $D = 0.104$ m, the stack length $L = 0.14$ m and the air gap $g = 0.9$ mm. If the saturation factor $k_{sat} = 1.08$ and Carter's coefficient $k_C = 1.13$, find the reluctance of the air gap and ferromagnetic core. What will be the air gap magnetic flux density, MMF for fundamental and for $\nu = 3, 5$ and 7 space harmonics if in the stator core a single coil consiting of $N_c = 120$ with rms current $I = 3.4$ is placed?

Answer: $R_{\mu g} = 1/H$, $R_{\mu Fe} = 5662\ 1/H$, $B = 0.33$ T, $F_{mc1} = 367.3$ A, $F_{mc3} = 122.4$ A, $F_{mc5} = 73.5$ A, $F_{mc7} = 52.5$ A.

5. If the stator of Problem 5.4 has $s_1 = 48$ slots that accommodate a three-phase winding consisting of full pitch coils, find the amplitude of the first harmonic of MMF of a single phase and amplitude of MMF of all three-phases.

Answer: MMF of single phase $F_{m1} = 2808.2$ A, MMF of all three phases $F_m = 4212.3$ A.

6. In an AC electrical machine with $2p = 2$ the inner diameter of stator core is $D = 0.48$ m, length of the stator core $L = 1.1$ m, number of stator slots $s_1 = 24$. A group of coils consisting of 8 series connected coils has been placed in slots. Each coil consists of $N_c = 6$ turns with coil span $y = 10$ measured in slots. The stator produces a circular magnetic field rotating with the synchronous speed $n_s = 3600$ rpm and magnetic flux $\Phi = 1.83$ Wb. Calculate the rms EMF in:
(a) Conductor (bar);
(b) Single coil;
(c) Coil group.

Answer: (a) $E_s = 243.9$ V, (b) $E_c = 2827.2$ V; (c) $E_g = 10\ 830.2$ V.

7. In an electrical machine the pole pitch is $\tau = 0.055$ m, the length of stack is $L = 0.15$ m and the air gap $g = 0.8$ mm. The magnetic flux is $\Phi = 0.0052$ Wb. Assuming the saturation factor of the magnetic circuit $k_{sat} = 1.1$, Carter's coefficient $k_C = 1.1$, find:

   (a) MVD across the air gap and in ferromagnetic parts of the magnetic circuit;

   (b) MMF per pole pair and magnetic flux density in the air gap.

   Answer: (a) $V_{\mu g} = 441.4$ A, $V_{\mu Fe} = 88.3$ A; (b) $F = 971.1$ A, $B_g = 0.63$ T.

8. In a two-phase, four-pole induction motor the number of series turns in phase A is $N_{1A} = 440$, the number of series turns in phase B is $N_{1B} = 220$, winding factors $k_{w1A} = k_{w1B} = 0.9$ and phase currents $I_{1A} = 1.2$ A, $I_{1B} = 0.8$ A. The phase shift of currents is $\beta = \pi/3$.

   (a) Find the resultant MMFs F_{m1}, F_{m2} of the forward and backward-rotating fields and angles γ and ϵ in Fig. 5.35;

   (b) What would be the MMFs F_{m1}, F_{m2} and angles γ and ϵ if the phase currents are shifted by $\beta = \pi/2$?

   Answer: (a) $F_{m1} = 142.8$ A, $F_{m2} = 80.3$ A, $\gamma = 7.4°$, $\epsilon = 13.2°$; (b) $F_{m1} = 146.6$ A, $F_{m2} = 73.3$ A, $\gamma = 0°$, $\epsilon = 0°$.

6

INDUCTION MACHINES

Induction motors operate on the principle of interaction of the stator magnetic rotating field on currents induced in the rotor winding. It can be a cage rotor winding, wound-rotor winding with slip rings or winding in the shape of high-current conducting sleeve, i.e., solid rotor coated with copper layer.

Three important inventors have made major contributions to the development of induction motors:

- 1985 - Galileo Ferraris, Italian professor, who invented self-starting single-phase induction motor;
- 1987 - Nicola Tesla, American inventor of Yugoslavian origin, who invented two-phase induction motor;
- 1989 - Michael Dolivo-Dobrovolsky, Polish engineer working in Germany, who invented three-phase cage induction motor. He has also developed a double-cage induction motor by 1893.

6.1 Construction

Generally, there are three-types of induction machines:

(a) Cage-rotor induction machines (Fig. 6.1);
(b) Wound-rotor (slip-ring) induction machines (Fig. 6.2);
(c) Solid-rotor induction machines (Fig. 6.3).

The stator consists of a laminated core (Fig 6.4) and three-phase winding (Fig 6.5) embedded in slots. This winding, when energized by a three-phase source of power, provides a *rotating magnetic field* (Section 5.8).

The rotor windings are also contained in slots in a laminated core which is mounted on the shaft. In small motors, the rotor-lamination stack is pressed directly on the shaft. In larger machines, the core is mechanically connected to the shaft through a set of spokes called a "spider."

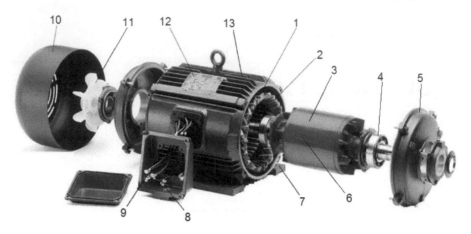

Fig. 6.1. Cage induction motor: 1 – stator core, 2 – stator winding, 3 – cage rotor, 4 – bearing, 5 – end plate (end bell), 6 – rotor bar, 7 – cast iron mounting feet, 8 – terminal box, 9 – terminal leads, 10 – fan cover, 11 – fan, 12 – nameplate, 13 – cast iron frame (housing).

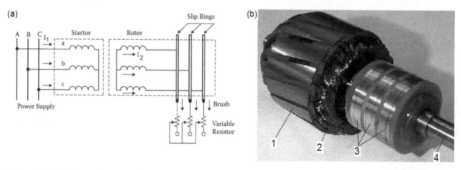

Fig. 6.2. Wound-rotor (slip-ring rotor): (a) stator and rotor connection diagram; (b) construction of rotor. 1 – rotor core, 2 – rotor winding, 3 – slip rings, 4 – shaft.

The cage–rotor winding (Fig 6.6) consists of solid bars of conducting material which are positioned in the rotor slots. These *rotor bars* are shorted together at the two ends of the rotor by *end rings*. In large machines, the rotor bars may be made of copper alloy, which is driven into the slots and then brazed to the end rings. Rotors up to about 0.5 m in diameter usually have die cast aluminum cage windings. The core laminations for such rotors are stacked in a mold, which is then filled with molten aluminum. In this industrial process, the rotor bars, the end rings and the cooling-fan blades are all cast at the same time (Fig. 6.6b).

Cage induction motors are cheaper and more reliable than comparable wound–rotor motors. Recently, copper casting technology has been matured. Induction motors with copper cast cage rotors (Fig. 6.7) have better per-

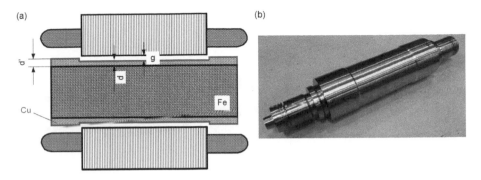

Fig. 6.3. Solid rotor induction motor: (a) longitudinal section; (b) solid steel rotor coated with cylindrical copper layer. Photo courtesy of *Sundyne Corporation*, Espoo, Finland.

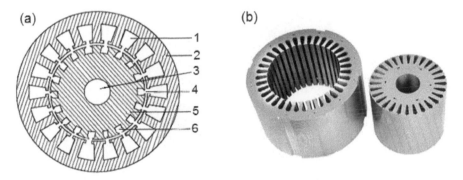

Fig. 6.4. Laminated cores of induction motors: (a) stator and rotor single laminations; (b) stator and rotor laminated stacks. 1 – stator slot, 2 – stator lamination, 3 – shaft, 4 – air gap, 5 – rotor lamination, 6 – shaft.

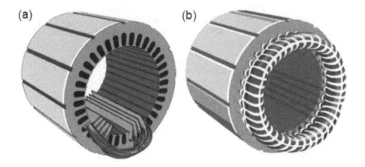

Fig. 6.5. Stator core with winding: (a) partially wound stator core; (b) stator winding completed.

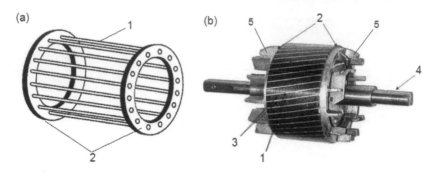

Fig. 6.6. Cage winding: (a) cage; (b) complete cage rotor. 1 – rotor bar, 2 – rotor end ring, 3 – rotor core, 4 – shaft, 5 – cooling fan blades.

Fig. 6.7. Copper cast rotor cage winding.

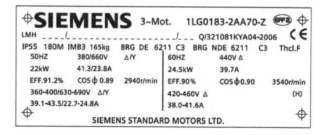

Fig. 6.8. Name plate of induction motor.

formance than those with aluminum cast cage rotors because the copper has higher electric conductivity ($\approx 57 \times 10^6$ S/m) than aluminum alloy ($\approx 30 \times 10^6$ S/m).

The winding of a wound rotor is a polyphase winding, consisting of coils. It is almost always a three-phase Y-connected winding. The three terminal leads are connected to *slip rings*, mounted on the shaft (Fig. 6.2). Carbon brushes riding on these slip rings are shorted together for normal operation. External resistances are inserted into the rotor circuit, via the brushes, which improves the motor's starting characteristics (Fig. 6.2a). As the motor accelerates, the external resistances are gradually reduced to zero. Another external resistance can also be used to control the speed (in a continuous duty cycle).

The name plate (Fig. 6.8) of an induction motor usually contains the following information:

- Serial number;
- Rated (nominal power);
- Class of insulation;
- Year of manufacturing;
- Number of phases;
- Frequency;
- Stator line-to-line voltage;
- Stator current;
- Stator winding conenction (Y or Δ);
- Power factor $\cos \varphi$;
- Internal protection IP;
- Duty cycle;
- Nominal (rated) speed;
- Ambient temperature of operation;
- Mass;
- Number of standard the machine meets.

6.2 Fundamental relationships

Induction motor operates on the principle of interaction of the stator magnetic rotating field on currents induced in the rotor winding (Fig. 6.9) The force on an elementary conductor dl with current I placed in the magnetic field with magnetic flux density B is created as according to eqn (1.1).

6.2.1 Slip

The *slip* (per unit) is the ratio of the *slip speed* ($n_s - n$) to the synchronous speed n_s of the rotating magnetic field, i.e.,

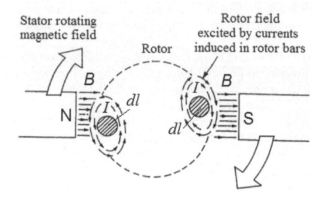

Fig. 6.9. Principle of operation of induction motor.

$$s = \frac{n_s - n}{n_s} = 1 - \frac{n}{n_s} \tag{6.1}$$

where

$$n_s = \frac{f}{p} \tag{6.2}$$

is the *synchronous speed* in rev/s, f is the input frequency, p is the number of pole pairs, and n is the rotor speed. The slip speed $(n_s - n)$ expresses the speed of the rotor relative to the rotating magnetic field of the stator.

6.2.2 Rotor speed

According to eqn (6.1), the *rotor speed* is a function of the input frequency f, the number of pole pairs p, and the slip s. Thus:

$$n = n_s(1 - s) = \frac{f}{p}(1 - s) \tag{6.3}$$

The current in the rotor creates its own magnetic field rotating with a speed that is given by

$$n_{rot} = \frac{sf}{p} = sn_s \tag{6.4}$$

Since $n_s(1 - s) + n_{rot} = n_s - sn_s + sn_s = n_s$, the stator and rotor magnetic fields rotate with the same speed.

6.2.3 Input power

The electrical *input active power* delivered to the motor is

$$P_{in} = m_1 V_1 I_1 \cos\varphi \tag{6.5}$$

where m_1 is the number of stator phases, V_1 is the input phase voltage, I_1 is the input phase current, $\cos\varphi$ is the power factor, and φ is the phase angle between current and voltage. For a three-phase system

$$P_{in} = 3V_1 I_1 \cos\varphi \tag{6.6}$$

Both the voltage V_1 and current I_1 are phase quantities. For the Y-connection

$$V_{1L} = \sqrt{3}V_1 \qquad I_{1L} = I_1 \tag{6.7}$$

For the Δ-connection

$$V_{1L} = V_1 \qquad I_{1L} = \sqrt{3}I_1 \tag{6.8}$$

6.2.4 Electromagnetic power

The *electromagnetic power* (or airgap power) is the active (true) power crossing the air gap from the stator to the rotor, written

$$P_{elm} = P_{in} - \Delta P_{1w} - \Delta P_{1Fe} \tag{6.9}$$

where the *stator winding* (or copper) losses are

$$\Delta P_{1w} = m_1 I_1^2 R_1 \tag{6.10}$$

and the *stator core losses* (comprising mainly hysteresis ΔP_{1h} and eddy–current losses ΔP_{1e}) are

$$\Delta P_{1Fe} = \Delta P_{1h} + \Delta P_{1e} \tag{6.11}$$

In the above equations R_1 is the AC stator winding resistance, hysteresis losses ΔP_{1h} are proportional to fB^2 and eddy–current losses ΔP_{1e} are proportional to $f^2 B^2$, where B is the magnetic flux density in the stator core's teeth or yoke. The hysteresis losses are given by eqn (1.7) and eddy current losses are given by eqn (1.8). In practical calculations it is better to use eqn (1.11) for stator core losses, which allows for separate calculations of losses in the teeth and yoke and includes increases in losses due to metallurgical and manufacturing processes. Eqn (1.11) is for calculation of losses in transformers. To calculate the core losses in induction machines, the losses in legs should be replaced by the losses in teeth.

Since the slip frequency sf under rated operation is very low ($s \approx 0.005 \ldots 0.02$), the rotor core losses proportional to the slip frequency due to the fundamental space and time harmonics are negligible.

6.2.5 Electromagnetic (developed) torque

The *electromagnetic torque* developed by the motor is

$$T_{elm} = \frac{P_{elm}}{\Omega_s} = \frac{P_{elm}}{2\pi n_s} \tag{6.12}$$

where the *synchronous angular speed*

$$\Omega_s = 2\pi n_s = 2\pi \frac{f}{p} \tag{6.13}$$

Another quantity, the *stator angular frequency*, is written as

$$\omega_s = 2\pi f = 2\pi n_s p \tag{6.14}$$

Thus, the following relationship exists between the synchronous angular speed Ω_s and the *stator angular frequency* ω_s:

$$\Omega_s = \frac{\omega_s}{p} \tag{6.15}$$

For $p = 1$ the angular speed Ω_s and angular frequency ω_s are equal, i.e., $\Omega_s = \omega_s$.

6.2.6 Mechanical power

The developed *mechanical power* is obtained by subtracting the rotor losses from the electromagnetic (air gap) power, to obtain

$$P_m = P_{elm} - \Delta P_{2w} - \Delta P_{2Fe} \approx P_{elm} - \Delta P_{2w} \tag{6.16}$$

Since the rotor hysteresis losses $\Delta P_{2h} \propto sf$, eddy current losses in the rotor core $\Delta P_{2e} \propto s^2 f^2$, and the rotor total core losses $P_{2Fe} \propto s(f + sf^2)$ are all negligible (except in the case of inverter-fed motors). In terms of the electromagnetic developed torque, T_{elm},

$$P_m = T_{elm}\Omega = 2\pi n T_{elm} \tag{6.17}$$

where

$$\Omega = 2\pi n = \Omega_s(1 - s) \tag{6.18}$$

is the *rotor angular speed*. Combining eqns (6.12) and (6.17) leads to

$$\frac{P_{elm}}{P_m} = \frac{n_s}{n} = \frac{1}{1 - s} \tag{6.19}$$

and thus

$$P_m = P_{elm}(1 - s) \tag{6.20}$$

6.2.7 Rotor winding losses

The *rotor winding losses* are calculated in the same way as those in the stator winding, namely as

$$\Delta P_{2w} = m_2 I_2^2 R_2 = m_1 (I_2')^2 R_2' \tag{6.21}$$

where m_2 is the number of rotor phases, I_2 is the rotor current, I_2' is the rotor current referred to the stator winding, R_2 is the AC rotor resistance and R_2' is the rotor AC resistance referred to as the stator winding.

6.2.8 EMF (voltage induced) in the stator winding

The *rms* EMF or *voltage induced* per phase in the stator winding is given by the equation

$$E_1 = 4\sigma_f f N_1 k_{w1} \Phi \tag{6.22}$$

where σ_f is the *form factor* of the EMF (i.e., the ratio of its *rms* value to mean value), f is the input frequency, N_1 is the number of stator turns per phase, k_{w1} is the stator winding factor (Section 5.4), and Φ is the magnetic flux. For sinusoids,

$$\sigma_f = \frac{\pi\sqrt{2}}{4} \approx 1.11 \tag{6.23}$$

and the stator winding EMF is then expressed as

$$E_1 = \pi\sqrt{2} f N_1 k_{w1} \Phi = 4.44 f N_1 k_{w1} \Phi \tag{6.24}$$

Eqn (6.23) can be derived from Faraday's law for electromagnetic induction, i.e.,

$$e_1 = -N_1 \frac{d\phi}{dt} \qquad\qquad e_1 = \sqrt{2} E_1 \sin(\omega t) \tag{6.25}$$

so that

$$d\phi = -\frac{1}{N_1} \sqrt{2} E_1 \sin(\omega t) \tag{6.26}$$

After integration

$$\phi = -\frac{1}{N_1 \omega} \sqrt{2} E_1 \cos(\omega t) = \frac{1}{N_1 \omega} \sqrt{2} E_1 \sin\left(\omega t + \frac{\pi}{2}\right) \tag{6.27}$$

The expression $1/(N_1\omega)\sqrt{2}E_1 = \Phi$ is the peak magnetic flux. Replacing N_1 with the effective number of turns $N_1 k_{w1}$, the stator *rms* EMF becomes

$$E_1 = \frac{\omega}{\sqrt{2}} N_1 k_{w1} \Phi = \frac{2\pi}{\sqrt{2}} N_1 k_{w1} \Phi = 4 \frac{\pi\sqrt{2}}{4} N_1 k_{w1} \Phi = \sigma_f N_1 k_{w1} \Phi \qquad (6.28)$$

The stator winding factor k_{w1} has been discussed in Section 5.4.

The *magnetic flux* can be obtained by integrating the cosinusoidal distribution of the magnetic flux density in the air gap over one pole pitch τ, i.e.,

$$\Phi = L_i \int_{-0.5\tau}^{0.5\tau} B_{mg} \cos\left(\frac{\pi}{\tau} x\right) dx = \frac{2}{\pi} \tau L_i B_{mg} \qquad (6.29)$$

where $\tau = \pi D/(2p)$ is the stator pole pitch, D is the stator core inner diameter, L_i is the effective (ideal) length of the stator core and B_{mg} is the peak value of the air gap magnetic flux density. Eqn (6.29) expresses also the first space harmonic of the magnetic flux in the case of nonsinusoidal distribution of the magnetic flux density. For nonsinusoidal waveforms instead of $2/\pi$, it is necessary to use the ratio of the *average to the peak value* of the air gap magnetic flux density α_i, i.e.,

$$\Phi = \alpha_i \tau L_i B_{mg} \qquad (6.30)$$

Once again, for sinusoids, $\alpha_i = 2/\pi \approx 0.637$.

6.2.9 EMF induced in the rotor winding

The slip-dependent EMF that is induced in the rotor winding per phase is

$$E_2(s) = 4\sigma_f s f N_2 k_{w2} \Phi \qquad (6.31)$$

where sf is the *slip frequency* (i.e., the frequency of the rotor current), N_2 is the number of rotor turns per phase, and k_{w2} is the rotor winding factor which can be calculated in the same way as shown in Section 5.4 for the stator winding. For a cage winding shown in Fig. 6.6 the number of turns $N_2 = 0.5$ and winding factor $k_{w2} = 1$. Eqn (6.31) can also be written in the form

$$E_2(s) = s E_{20} \qquad (6.32)$$

where

$$E_{20} = 4\sigma_f f N_2 k_{w2} \Phi \qquad (6.33)$$

is the EMF at standstill, i.e., $n = 0$ $(s = 1)$.

6.2.10 Rotor EMF referred to the stator system

Using the *turns ratio*

$$\frac{N_1 k_{w1}}{N_2 k_{w2}} \tag{6.34}$$

the induced rotor voltage (EMF) referred to as the stator winding is

$$sE'_{2o} = sE_{2o}\frac{N_1 k_{w1}}{N_2 k_{w2}} = sE_1 \tag{6.35}$$

For a transformer, the turns ratio is simply N_1/N_2. However, an AC electrical machine has its windings distributed in slots (Section 5.1) with coil pitch $w_c \leq \tau$ (5.5) and the turns ratio contains the *effective number of turns* in the stator and rotor, $N_1 k_{w1}$ and $N_2 k_{w2}$, respectively.

6.2.11 Rotor current referred to as the stator system

At speed $n = 0$ (locked rotor), i.e., $s = 1$, the magnetizing current can be neglected in comparison with the rotor current (see Section 2.1.5). Thus, the rotor MMF approximately counterbalances the stator MMF, or the apparent internal powers are equal, i.e.,

$$m_1 E_1 I_1 = m_2 E_2 I_2 \tag{6.36}$$

or

$$m_1 E'_2 I'_2 = m_2 E_2 I_2 \tag{6.37}$$

Using eqn (6.35) for $s = 1$ and eqns (6.36) and (6.37) the *rotor current* can be referred to the stator (primary) system in a similar way to finding the secondary current of a transformer, i.e.,

$$I'_2 = \frac{m_2 N_2 k_{w2}}{m_1 N_1 k_{w1}} I_2 \tag{6.38}$$

where m_1 is the number of stator phases, m_2 is the number of rotor phases, N_1 is the number of stator turns per phase, N_2 is the number of rotor turns per phase, k_{w1} is the stator winding factor, k_{w2} is the rotor winding factor, and I_2 is the current in the rotor conductors (bars).

For a cage rotor, $m_2 = s_2$ (the number of rotor slots), $N_2 = 0.5$ and $k_{w2} = 1$.

6.2.12 Rotor impedance

The slip-dependent *rotor impedance* is

$$\mathbf{Z}_2(s) = R_2 + jX_2(s) = R_2 + j2\pi s f L_2 = R_2 + jsX_2 \qquad (6.39)$$

where L_2 is the rotor winding inductance, and

$$X_2 = 2\pi f L_2 \qquad (6.40)$$

is the rotor reactance for $s = 1$.

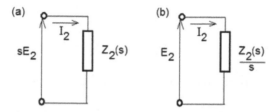

Fig. 6.10. Equivalent circuit of the rotor. The current I_2 in circuit (a) is the same as that in circuit (b).

The voltage across $\mathbf{Z}_2(s)$ is sE_2. Each phase of the rotor circuit can also be replaced with the impedance

$$\mathbf{Z}_2 = \frac{\mathbf{Z}_2(s)}{s} = \frac{R_2}{s} + jX_2 \qquad (6.41)$$

for the EMF (induced voltage) E_2. The rotor current (Fig. 6.10)

$$\mathbf{I}_2(s) = \frac{sE_2}{\mathbf{Z}_2(s)} = \frac{E_2}{\mathbf{Z}_2(s)/s} \qquad (6.42)$$

6.2.13 Rotor impedance referred to as the stator system

The rotor impedance, resistance and reactance referred to as the stator winding are:

$$\mathbf{Z}_2' = R_2' + jX_2' = \frac{m_1(N_1 k_{w1})^2}{m_2(N_2 k_{w2})^2} \mathbf{Z}_2 \qquad (6.43)$$

$$R_2' = \frac{m_1(N_1 k_{w1})^2}{m_2(N_2 k_{w2})^2} R_2 \qquad (6.44)$$

$$X_2' = \frac{m_1(N_1 k_{w1})^2}{m_2(N_2 k_{w2})^2} X_2 \qquad (6.45)$$

Eqn (6.44) is obtained by equating the rotor winding active losses

$$m_2 I_2^2 R_2 = m_1 (I_2')^2 R_2' \tag{6.46}$$

and then applying eqn (6.38). Similarly, eqn (6.45) is obtained by equating the rotor *reactive* losses, i.e.,

$$m_2 I_2^2 X_2 = m_1 (I_2')^2 X_2' \tag{6.47}$$

6.2.14 Output power

The shaft *output power* is

$$P_{out} = P_{in} - \Delta P_{1w} - \Delta P_{1Fe} - \Delta P_{2w} - \Delta P_{rot} - \Delta P_{str} = P_m - \Delta P_{rot} - \Delta P_{str} \tag{6.48}$$

where ΔP_{1w} are the stator winding losses, ΔP_{1Fe} are the stator core losses, ΔP_{2w} are the rotor winding losses, ΔP_{rot} are the rotational (mechanical) losses and ΔP_{str} are the stray load losses. The mechanical power

$$P_m = P_{in} - \Delta P_{1w} - \Delta P_{1Fe} - \Delta P_{2w} \tag{6.49}$$

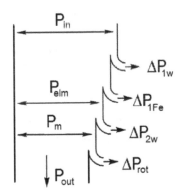

Fig. 6.11. Power (energy) balance diagram. The stray load losses ΔP_{str} can be added to rotational losses ΔP_{rot}.

In connection with the air gap power (electromagnetic power)

$$P_{elm} = P_{in} - \Delta P_{1w} - \Delta P_{1Fe} = P_m + \Delta P_{2w} \tag{6.50}$$

The rotor core losses ΔP_{2Fe} are very small and can be neglected at the rated speed, since the frequency of magnetic flux in the rotor, sf (the slip frequency), is very low.

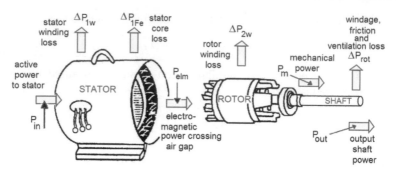

Fig. 6.12. Power flow diagram. Stray losses ΔP_{str} have been neglected.

Table 6.1. Average losses in two-pole and 4-pole low-power cage induction motors.

Losses	$2p = 2$ %	$2p = 4$ %	Factors affecting losses
Stator winding losses	26	34	Conductor cross-section, mean length of turn, current density
Stator core losses	19	21	Lamination material and thickness, frequency, magnetic flux density
Rotor winding losses	19	21	Material of cage, cross-section of rotor bars and end rings, current density
Rotational losses	25	10	Fan efficiency, bearings, lubrication
Stray load losses	11	14	Manufacturing, slot openings, core surfaces, air gap

The power (energy) balance diagram also called *Sankey's diagram* is given in Fig. 6.11. The other diagram of power flow is shown in Fig. 6.12. Average losses in two-pole and four-pole low-power cage induction motors are specified in Table 6.1.

6.2.15 Rotational (mechanical) losses

The speed-dependent *rotational losses* are

$$\Delta P_{rot} = \Delta P_{fr} + \Delta P_{wind} + \Delta P_{vent} \tag{6.51}$$

where ΔP_{fr} is the frictional loss (in the bearings and between the slip rings and brushes, if exist), ΔP_{wind} is the windage loss and ΔP_{vent} represents the ventilation losses (cooling fan losses).

6.2.16 Stray losses

Stray load losses ΔP_{str} (sometimes called *additional losses*) are due to higher harmonics. It is difficult to calculate the stray load losses. According to IEC (International Electrotechnical Commission) standards stray load losses are equal to 0.5% of the input power, i.e.,

$$\Delta P_{str} = 0.005 P_{in} \tag{6.52}$$

According to NEMA (National Electrical Manufacturers Association of the USA) standards, stray load losses are equal to 1.2% of P_{out} if $P_{out} < 2500$ hp (1865 kW) and to 0.9% of P_{out} if $P_{out} \geq 2500$ hp, i.e.,

$$\Delta P_{str} = 0.012 P_{out} \quad \text{if} \quad P_{out} < 1865 \text{ kW}$$

$$\Delta P_{str} = 0.009 P_{out} \quad \text{if} \quad P_{out} \geq 1865 \text{ kW} \tag{6.53}$$

6.2.17 Slip, electromagnetic power, and mechanical power

The slip can also be defined on the basis of power flow

$$s = \frac{P_{elm} - P_m}{P_{elm}} = \frac{2\pi n_s T_{elm} - 2\pi n T_{elm}}{2\pi n_s T_{elm}} = \frac{n_s - n}{n_s} \tag{6.54}$$

See also eqns (6.19) and (6.20). The electromagnetic torque T_{elm} is expressed by eqn (6.12). Since

$$P_{elm} - P_m \approx P_{2w},$$

the electromagnetic power is equal to the rotor winding losses divided by the slip, or

$$P_{elm} = \frac{\Delta P_{2w}}{s} \tag{6.55}$$

Similarly, the mechanical power may be approximated as

$$P_m = P_{elm}(1 - s) = \Delta P_{2w} \frac{1 - s}{s} \tag{6.56}$$

6.2.18 Efficiency

The *efficiency* is the ratio of the output P_{out} to input power P_{in}. Since $P_{out} < P_m$ and $P_{in} > P_{elm}$, this useful approximation to the efficiency of induction motors (see Fig. 6.13) can be deduced:

$$\eta = \frac{P_{out}}{P_{in}} \approx \frac{P_m}{P_{elm}} = 1 - s \tag{6.57}$$

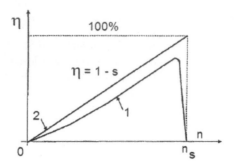

Fig. 6.13. Actual and approximated efficiency *versus* speed for induction motors: 1 – actual, 2 – approximated ($\eta \approx 1 - s$).

Eqn (6.57) is only valid for medium and large-power induction motors. The output power as a function of efficiency and *power factor* $\cos \varphi$ is

$$P_{out} = \eta P_{in} = m_1 V_1 I_1 \eta \cos \varphi \tag{6.58}$$

The product $\eta \cos \varphi$ characterizes the performance of induction machine.

6.2.19 Shaft torque

The *shaft torque* (load torque) is defined as the output power divided by the rotor angular speed $\Omega = 2\pi n$, i.e.,

$$T = \frac{P_{out}}{\Omega} = \frac{P_{out}}{2\pi n} \tag{6.59}$$

where n is the shaft rotational speed. The electromagnetic torque T_{elm} was expressed by eqn (6.12), in which the synchronous angular speed $\Omega_s = 2\pi n_s$. Since the rotor speed $n = n_s(1 - s)$, the rotor angular speed $\Omega = \Omega_s(1 - s)$.

6.3 Equivalent circuit

From Kirchhoff's current law, the *stator current* is

$$\mathbf{I}_1 = \mathbf{I}_0 + \mathbf{I}_2' \tag{6.60}$$

and the *exciting current* (in the vertical branch) is (Fig. 6.14)

$$\mathbf{I}_0 = I_{Fe} + jI_\Phi \tag{6.61}$$

where $I_{Fe} = E_1/R_{Fe}$ is the *core loss current* (the active component of I_0), and I_Φ is the *magnetizing current* (the reactive component of I_0). The stator core losses can be calculated as

$$\Delta P_{Fe} = m_1 I_{Fe}^2 R_{Fe} = m_1 \frac{E_1^2}{R_{Fe}} \tag{6.62}$$

where R_{Fe} is the *resistance representing the stator core losses* (or core loss resistance).

The *equivalent circuits* or *circuital models* per phase of an induction motor are shown in Figs 6.15 and 6.16. Figs 6.15a, 6.15b, 6.16a and 6.16b explain step by step how the T-type equivalent circuit has been derived.

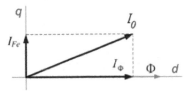

Fig. 6.14. Exciting current I_0, core loss current I_{Fe} and magnetizing current I_Φ.

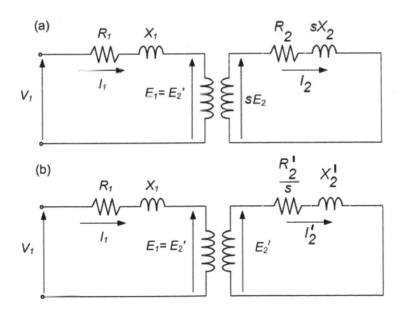

Fig. 6.15. Equivant circuit (circuital model) of an induction motor with magnetic coupling: (a) rotor winding EMF sE_2; (b) rotor winding EMF E_2'. R_1 – stator winding resistance, X_1 – stator winding leakage reactance, R_2' – rotor winding resistance referred to as the stator system, X_2' – rotor winding leakage reactance referred to as the stator system, V_1 – input phase voltage, I_1 – input (stator) phase current, I_2' – rotor current referred to as the stator winding.

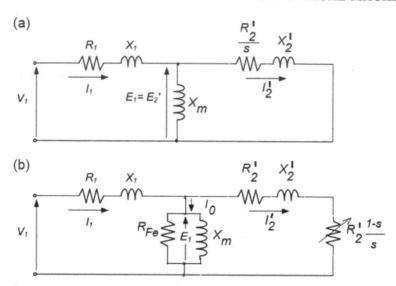

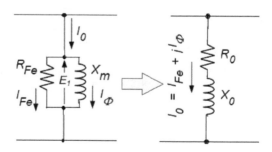

Fig. 6.16. T-type equivalent circuit without magnetic coupling of an induction motor: (a) without core loss resistance; (b) with core loss resistance R_{Fe} and separated rotor resistance R_2' for winding losses. R_1 – stator winding resistance, X_1 – stator winding leakage reactance, R_2' – rotor winding resistance referred to as the stator system, X_2' – rotor winding leakage reactance referred to as the stator system, R_{Fe} – core-loss resistance, X_m – mutual reactance, $\mathbf{V}_1$ – input phase voltage, $\mathbf{I}_1$ – input (stator) phase current, $\mathbf{I}_2'$ – rotor current referred to as the stator winding, $\mathbf{I}_0$ – exciting current, I_{Fe} – core loss current, I_Φ – magnetizing current.

Fig. 6.17. Transformation of parallel connection of R_{Fe}, X_m into series connection R_0, X_0.

Parallel connection of R_{Fe} and X_m in vertical branch can be replaced by a series connection of equivalent resistance R_0 and equivalent reactance X_0 (Fig. 6.17), i.e.,

$$\frac{R_{Fe}(jX_m)}{R_{Fe} + jX_m} = R_0 + jX_0$$

$$\frac{R_{Fe}X_m^2 + jR_{Fe}^2 X_m}{R_{Fe}^2 + jX_m^2} = R_0 + jX_0$$

From the above equation

$$R_0 = \frac{R_{Fe}X_m^2}{R_{Fe}^2 + X_m^2} \tag{6.63}$$

$$X_0 = \frac{R_{Fe}^2 X_m}{R_{Fe}^2 + X_m^2} = \frac{X_m}{1 + X_m^2/R_{Fe}^2} \approx X_m \tag{6.64}$$

$$\mathbf{Z}_0 = R_0 + jX_0 \qquad\qquad R_{Fe} \gg X_m \tag{6.65}$$

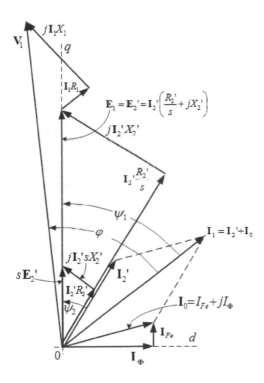

Fig. 6.18. Phasor diagram of an induction motor.

The *rms* line stator current can be evaluated by the following *rule of thumb*:
$I_1 \approx 2.2P_{out}$ at $V_1 = 380$ V and with $2p = 2$, where P_{out} is in kilowatts and I_1 is in amps.

Then, from Kirchhoff's voltage law

$$\mathbf{V}_1 = \mathbf{I}_1(R_1 + jX_1) + \mathbf{E}_1 \tag{6.66}$$

and

$$\mathbf{E}_1 = \mathbf{E}'_2 = \mathbf{I}'_2\left(\frac{R'_2}{s} + jX'_2\right) \tag{6.67}$$

so

$$\mathbf{E}'_2 = \mathbf{I}'_2(R'_2 + jX'_2) + \mathbf{I}'_2 R'_2 \frac{1-s}{s} \tag{6.68}$$

since

$$\frac{R'_2}{s} = R'_2 + R'_2 \frac{1-s}{s} \tag{6.69}$$

The phasor diagram of an induction motor is sketched in Fig. 6.18.

6.4 No-load and locked-rotor tests

The resistances and reactances of the equivalent circuit can be determined from the results of a no-load test, locked-rotor test and from measurement of the DC stator winding resistance. The connection diagram for the no-load and locked rotor tests on induction motors is shown in Fig. 6.19. As a laboratory power supply, usally a three-phase adjustable-ratio autotransformer, the so-called *variac* is used (Fig. 6.20). Nowadays, instead of ammeters, voltmeter and wattmeter, a three-phase electronic power analyzer (Fig. 6.21) is used.

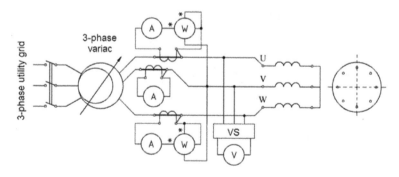

Fig. 6.19. Connection diagrams of induction motor, power supply and instrumentation for no-load and locked rotor tests. A – ammeter, V – voltmeter, W – wattmeter, VS – voltmeter switch.

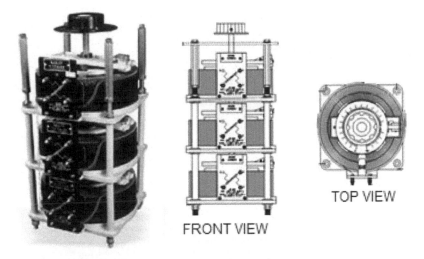

Fig. 6.20. Three-phase variac (adjustable-ratio autotransformer).

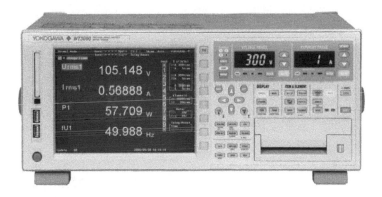

Fig. 6.21. Power analyzer.

6.4.1 No-load test

The *no-load test* on an induction motor is similar to the open-circuit test on a transformer. In this test the motor runs without any load. The input voltage V_1, input phase current I_{10}, input power P_{in0}, no-load speed n_0, stator winding resistance per phase R_1 and rotor winding resistance R_2 per phase (only for a slip ring motor) are measured. The rotor branch in the equivalent circuit of an induction motor, as shown in Fig. 6.22a, can be neglected because

$$n \to n_s \qquad s = 1 - \frac{n}{n_s} \to 0 \qquad \text{and} \qquad \frac{R_2'}{s} \to \infty \qquad (6.70)$$

The no-load parameters are found from the following equations:

- The no-load power factor

$$\cos \varphi_0 = \frac{P_{in0}}{m_1 I_{10} V_1} \qquad (6.71)$$

- The no-load stator winding losses

$$\Delta P_{1w0} = m_1 I_{10}^2 R_1 \qquad (6.72)$$

- The cage rotor resistance referred to as the stator system

$$R_2' \approx R_1 \qquad (6.73)$$

- The no-load reactance

$$X_1 + X_m = \frac{Q_{in0}}{m_1 I_{10}^2} = \frac{\sqrt{(m_1 I_{10} V_1)^2 - P_{in0}^2}}{m_1 I_{10}^2} \qquad (6.74)$$

- The no-load losses (core losses and rotational losses)

$$\Delta P_0 = P_{in0} - \Delta P_{1w0} \qquad (6.75)$$

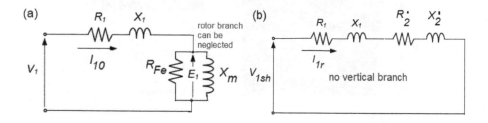

Fig. 6.22. Simplified equivalent circuits per phase of an induction motor for: (a) no-load test; (b) locked-rotor test.

The no-load current of induction motors

$$\Delta i_{0\%} = \frac{I_{10}}{I_{1n}} \times 100\% \qquad (6.76)$$

is in the range of 25 to 60% of the nominal (rated current) I_{1n}. The higher the power and lower the number of poles, the lower the no-load current.

6.4.2 Locked-rotor test

The *locked rotor test* on an induction motor corresponds to the short-circuit test on a transformer. In this test the rotor is blocked ($n = 0$, $s = 1$) and a reduced voltage V_{1sh} is applied to the motor to obtain the nominal current I_{1n} in the stator winding. The reduced input voltage V_{1sh}, input current I_{1n} and input power P_{insh} are measured (Fig. 6.19). The vertical branch in the equivalent circuit (Fig. 6.22b) can be neglected because for $s = 1$

$$R_2' \frac{1-s}{s} = 0 \qquad \text{and} \qquad R_2' + jX_2' << R_0 + jX_0 \qquad (6.77)$$

The locked-rotor parameters are calculated as follows:

- The locked-rotor power factor

$$\cos \varphi_{sh} = \frac{P_{insh}}{m_1 I_{1n} V_{1sh}} \qquad (6.78)$$

- The stator winding losses

$$\Delta P_{1w} = m_1 I_{1n}^2 R_1 \qquad (6.79)$$

- The locked-rotor impedance

$$\mid Z_{sh} \mid = \sqrt{R_{sh}^2 + Z_{sh}^2} = \frac{V_{1sh}}{I_{1n}} \qquad (6.80)$$

- The locked rotor resistance

$$R_{sh} = \frac{P_{insh}}{m_1 I_{1n}^2} \qquad (6.81)$$

- The rotor resistance referred to as the stator winding

$$R_2' = R_{sh} - R_1 \qquad (6.82)$$

- The locked rotor reactance

$$X_{sh} = \sqrt{Z_{sh}^2 - R_{sh}^2} \qquad (6.83)$$

- The stator and rotor leakage reactances

$$X_1 \approx 0.5 X_{sh}, \qquad X_2' \approx 0.5 X_{sh} \qquad (6.84)$$

- The electromagnetic torque developed at $s = 1$ and rated current

$$T_{elmsh} \approx \frac{m_1 I_{1n}^2 R_2'}{2\pi n_s} \qquad (6.85)$$

Example 6.1

No-load and locked-rotor tests have been performed on a three-phase, four-pole, 60-Hz, 10-kW, Y-connected, 208-V (line-to-line) cage induction motor, with the following results:

No-load test: input frequency $f = 60$ Hz, input voltage (line-to-line) $V_{10L} = 208$ V, no-load current $I_{10} = 6.49$ A, no-load power $P_{ino} = 332$ W, stator winding resistance per phase $R_1 = 0.25$ Ω.

Locked-rotor test (s=1): input frequency $f = 60$ Hz, input voltage (line-to-line) $V_{1shL} = 78.5$ V, input current $I_{1n} = 42$ A, input active power $P_{insh} = 2116.8$ W.

Solution

Calculations on the basis of the no-load test:

$$V_1 = \frac{V_{10L}}{\sqrt{3}} = \frac{208}{\sqrt{3}} = 120 \text{ V}$$

$$\cos\varphi_0 = \frac{P_{ino}}{m_1 I_{10} V_1} = \frac{332}{3 \times 6.49 \times 120} = 0.142, \quad \varphi_0 = 81.8^0$$

$$\Delta P_{1w0} = m_1 I_{10}^2 R_1 = 3 \times 6.49^2 \times 0.25 = 31.6 \text{ W}$$

$$X_1 + X_m = \frac{\sqrt{(m_1 I_{10} V_1)^2 - P_{ino}^2}}{m_1 I_{10}^2} = \frac{\sqrt{(3 \times 6.49 \times 120)^2 - 332^2}}{3 \times 6.49^2} = 18.3 \text{ } \Omega$$

$$\Delta P_0 = P_{ino} - \Delta P_{1w0} = 332 - 31.6 = 300.4 \text{ W}$$

Calculations on the basis of the locked-rotor test:

$$V_{1sh} = \frac{V_{1shL}}{\sqrt{3}} = \frac{78.5}{\sqrt{3}} = 45.3 \text{ V}$$

$$\cos\varphi_{sh} = \frac{P_{insh}}{m_1 I_{1n} V_{1sh}} = \frac{2116.8}{3 \times 42 \times 45.3} = 0.37, \quad \varphi_{sh} = 68.23^0$$

$$\Delta P_{1w} = m_1 I_{1n}^2 R_1 = 3 \times 42^2 \times 0.25 = 1323 \text{ W}$$

$$| Z_{sh} |= \frac{V_{1sh}}{I_{1n}} = \frac{45.3}{42} = 1.078 \ \Omega$$

$$R_{sh} = \frac{P_{insh}}{m_1 I_{1n}^2} = \frac{2116.8}{3 \times 42^2} = 0.4 \ \Omega$$

$$R_2' = R_{sh} - R_1 = 0.4 - 0.25 = 0.15 \ \Omega$$

$$X_{sh} = \sqrt{Z_{sh}^2 - R_{sh}^2} = \sqrt{1.078^2 - 0.4^2} = 1.0 \ \Omega$$

$$X_1 \approx 0.5 X_{sh} = 0.5 \times 1.0 = 0.5 \ \Omega, \qquad X_2' \approx 0.5 X_{sh} = 0.5 \times 1.0 = 0.5 \ \Omega$$

$$n_s = \frac{60}{2} = 30 \ \text{rev/s}$$

$$T_{elmsh} \approx \frac{m_1 I_{1n}^2 R_2'}{2\pi n_s} = \frac{3 \times 42^2 \times 0.15}{2\pi \times 30} = 4.21 \ \text{Nm}$$

6.5 Torque-speed characteristics

6.5.1 Equivalent circuit impedance

The impedance of the stator winding

$$\mathbf{Z}_1 = R_1 + jX_1 \tag{6.86}$$

The impedance of the rotor winding referred to as the stator

$$\mathbf{Z}_2'(s) = \frac{R_2'}{s} + jX_2' = R_2' + R_2'\frac{1-s}{s} + jX_2' \tag{6.87}$$

The impedance of vertical branch $\mathbf{Z}_0$ for series connection of R_0 and X_0 is given by eqn (6.65). The impedance of the rotor and vertical branch being in parallel

$$\frac{1}{\mathbf{Z}_{20}'} = \frac{1}{\mathbf{Z}_2'} + \frac{1}{\mathbf{Z}_0} \qquad \text{or} \qquad \mathbf{Z}_{20}' = \frac{\mathbf{Z}_2'\mathbf{Z}_0}{\mathbf{Z}_2' + \mathbf{Z}_0} \tag{6.88}$$

The resultant impedance $\mathbf{Z}$ (stator, rotor and vertical branch) per phase

$$\mathbf{Z} = \mathbf{Z}_1 + \mathbf{Z}_{20}' = \mathbf{Z}_1 + \frac{\mathbf{Z}_2'\mathbf{Z}_0}{\mathbf{Z}_2' + \mathbf{Z}_0} \tag{6.89}$$

6.5.2 Stator current derived from the equivalent circuit

On the basis of Ohms law, the stator phase current $\mathbf{I}_1$ is equal to the terminal phase voltage V_1 divided by the resultant impedance of the machine (6.89), i.e.,

$$\mathbf{I}_1 = \frac{V_1}{\mathbf{Z}} = \frac{V_1}{\mathbf{Z}_1 + \mathbf{Z}_{20}'} = \frac{V_1}{\mathbf{Z}_1 + \dfrac{\mathbf{Z}_2'\mathbf{Z}_0}{\mathbf{Z}_2' + \mathbf{Z}_0}} \tag{6.90}$$

It has been assumed that $\mathbf{V}_1 = |\mathbf{V}_1| = V_1$.

6.5.3 Rotor current derived from the equivalent circuit

The stator winding EMF per phase

$$\mathbf{E}_1 = \frac{\mathbf{V}_1}{\mathbf{Z}_1 + \mathbf{Z}_o\mathbf{Z}_2'/(\mathbf{Z}_o + \mathbf{Z}_2')} \frac{\mathbf{Z}_o\mathbf{Z}_2'}{\mathbf{Z}_o + \mathbf{Z}_2'} = \frac{\mathbf{V}_1\mathbf{Z}_o\mathbf{Z}_2'}{\mathbf{Z}_1\mathbf{Z}_o + \mathbf{Z}_1\mathbf{Z}_2' + \mathbf{Z}_o\mathbf{Z}_2'}$$

$$= \frac{\mathbf{V}_1}{\mathbf{Z}_1 + \mathbf{Z}_2' + \mathbf{Z}_2'\mathbf{Z}_1/\mathbf{Z}_o}\mathbf{Z}_2' = \mathbf{I}_2'\mathbf{Z}_2' \tag{6.91}$$

where the rotor current

$$\mathbf{I}_2' = \frac{\mathbf{V}_1}{\mathbf{Z}_1 + \mathbf{Z}_2' + \mathbf{Z}_2'\mathbf{Z}_1/\mathbf{Z}_o} \tag{6.92}$$

is referred to as the stator system.

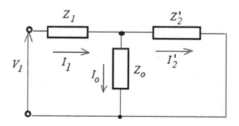

Fig. 6.23. T-type equivalent circuit per phase of an induction motor containing impedances $\mathbf{Z}_1$, $\mathbf{Z}_2'$ and $\mathbf{Z}_0$.

The T-type equivalent circuit containing impedances $\mathbf{Z}_1$, $\mathbf{Z}_2'$ and $\mathbf{Z}_0$ is given in Fig. 6.23. The denominator of the fraction in eqns (6.91) and (6.92) expressing $\mathbf{E_1}$ can be brought to the form

$$\mathbf{Z}_1 + \mathbf{Z}_2' + \mathbf{Z}_2'\frac{\mathbf{Z}_1}{\mathbf{Z}_o} = \mathbf{Z}_1 + \mathbf{Z}_2'(1 + \frac{\mathbf{Z}_1}{\mathbf{Z}_o}) = \mathbf{Z}_1 + \mathbf{Z}_2'(1 + \tau_1) \tag{6.93}$$

where the so-called *Heyland's coefficient* for the stator is

$$\tau_1 = \frac{\mathbf{Z}_1}{\mathbf{Z}_o} \approx \frac{X_1}{X_m} \tag{6.94}$$

Thus, the rotor current referred to as the stator can be expressed as

$$\mathbf{I}_2' = \frac{\mathbf{V}_1}{R_1 + jX_1 + (R_2'/s + jX_2')(1 + \tau_1)}$$

$$= \frac{\mathbf{V}_1}{R_1 + (R_2'/s)(1 + \tau_1) + j[X_1 + X_2'(1 + \tau_1)]} \tag{6.95}$$

The denominator of eqn (6.95) is equivalent to $\mathbf{Z}_1 + \mathbf{Z}_2'(1 + \tau_1)$. The rotor *rms* current referred to as the stator

$$I_2' = \frac{V_1}{\sqrt{[R_1 + (R_2'/s)(1 + \tau_1)]^2 + [X_1 + X_2'(1 + \tau_1)]^2}} \tag{6.96}$$

For $s \to \pm\infty$, the rotor current is obviously at its maximum, so

$$I_{2max}' = \lim_{s \to \pm\infty} I_2' = \frac{V_1}{\sqrt{R_1^2 + [X_1 + X_2'(1 + \tau_1)]^2}} \tag{6.97}$$

The stator and rotor current–speed characteristics are plotted in Fig. 6.24.

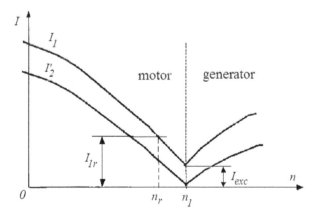

Fig. 6.24. Stator and rotor current versus speed.

6.5.4 Electromagnetic torque developed by an induction machine

The electromagnetic torque, first given by eqn (6.12), may now be expressed in terms of the equivalent circuit parameters as

$$T_{elm} = \frac{P_{elm}}{2\pi n_s} = \frac{m_1}{2\pi n_s}(I_2')^2\frac{R_2'}{s} \tag{6.98}$$

Putting the rotor current I_2' (6.96) referred to the stator

$$T_{elm} = \frac{m_1}{2\pi n_s}\frac{V_1^2(R_2'/s)}{[R_1 + (R_2'/s)(1+\tau_1)]^2 + [X_1 + X_2'(1+\tau_1)]^2} \tag{6.99}$$

6.5.5 Critical slip and maximum electromagnetic torque

To find the critical value of slip $s = s_{cr}$ which corresponds to the maximum (breakdown) electromagnetic torque T_{elmmax}, the first derivative of T_{elm} with respect to s is taken and then equated to zero (in other words, the first derivative test for maxima and minima is applied):

$$\frac{dT_{elm}}{ds} = 0 \tag{6.100}$$

The critical value of the slip is found to be

$$s_{cr} = \pm\frac{R_2'(1+\tau_1)}{\sqrt{R_1^2 + [X_1 + X_2'(1+\tau_1)]^2}} \tag{6.101}$$

The "+" sign signifies a machine in motor mode and the "−" sign denotes generator mode.

In conventional induction machines, R_1 is considerably less than $X_1 + X_2'(1+\tau_1)$. For this reason, $R_1^2 \ll [X_1 + X_2'(1+\tau_1)]^2$ and may therefore be disregarded. Thus

$$s_{cr} \approx \pm\frac{R_2'(1+\tau_1)}{X_1 + X_2'(1+\tau_1)} \approx \pm\frac{R_2'}{X_1 + X_2'} \tag{6.102}$$

The *maximum (breakdown) torque* is now found by re-substitution:

$$T_{elmmax} = T_{elm}(s = s_{cr}) = \pm\frac{m_1 V_1^2}{4\pi n_s(1+\tau_1)}\frac{1}{\sqrt{R_1^2 + [X_1 + X_2'(1+\tau_1)]^2} \pm R_1}$$

$$\approx \pm\frac{m_1 V_1^2}{4\pi n_s(1+\tau_1)}\frac{1}{X_1 + X_2'(1+\tau_1)} \tag{6.103}$$

In the denominator, "$+R_1$" is for $s_{cr} > 0$ (signifying a motor or brake) and "$-R_1$" is for $s_{cr} < 0$ (denoting a generator). The absolute value of maximum torque is slightly higher for the generator mode than for the motor mode, if the stator winding resistance R_1 is taken into account. In practice, only the leakage reactances X_1 and X_2 affect the maximum torque T_{elmmax}.

6.5.6 Starting torque

The *starting torque* is for the slip value $s = 1$ $(n = 0)$, and so:

$$T_{elmst} = T_{elm}(s = 1) = \frac{m_1 V_1^2}{2\pi n_s} \frac{R_2'}{[R_1 + R_2'(1 + \tau_1)]^2 + [X_1 + X_2'(1 + \tau_1)]^2}$$
(6.104)

At constant voltage V_1, the value of the starting torque mainly depends on the resistance R_2' and leakage reactance X_2' of the rotor.

6.5.7 Torque–speed and torque–slip curves

The torque–speed characteristic of an induction machine is plotted in Fig. 6.25, where it can be seen that the maximum torque T_{elmmax} for a generator is higher than that for a machine in motor mode. The torque–slip characteristic is plotted in Fig. 6.26, where five modes of operation, i.e.,

- Induction generator $(s < 0)$,
- Synchronous machine $(s = 0)$,
- Induction motor $(0 < s < 1)$,
- Transformer $(s = 1)$,
- Electromagnetic brake $(s > 1)$

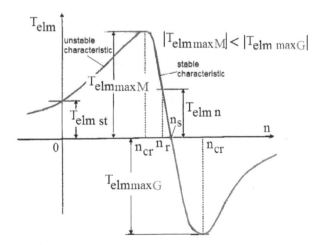

Fig. 6.25. Mechanical characteristic $T_{elm}(n)$ of an induction machine. $T_{elmmaxM}$ – maximum torque for motor, $T_{elmmaxG}$ – maximum torque for generator.

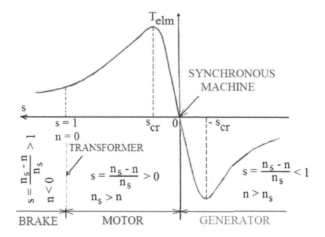

Fig. 6.26. Mechanical characteristic $T_{elm}(s)$ of an induction machine. Five modes of operation, i.e., induction generator, synchronous machine, induction motor, transformer and brake have been shown.

have been shown. Only wound-rotor induction machine can operate as a synchronous machine and a transformer. For synchronous operation, a DC current must be injected to the rotor. The electromagnetic torque developed by an induction machine depends, as do the maximum and starting torques, on the voltage square (see Figs 6.25 and 6.26). This is one of fundamental disadvatages of induction motors, because the torque is very sensitive to the voltage fluctuations (Fig. 6.27).

The ratio

$$OCF = \frac{T_{elmmax}}{T_{elmn}} \tag{6.105}$$

where T_{elmn} is the nominal (rated) developed torque, is called the *overload capacity factor* and the ratio

$$STR = \frac{T_{elmst}}{T_{elmn}} \tag{6.106}$$

is called the *starting torque ratio*. Similarly, the ratio

$$SCR = \frac{I_{1st}}{I_{1n}} \tag{6.107}$$

where I_{1n} is the nominal (rated) input current, is called the *starting current ratio*. The values of the OCF, STR and SCR arc given for various induction motor designs in Table 6.2.

Finally, the ratio of any torque T_{elm} to the breakdown torque T_{elmmax} is expressed by Kloss' formula:

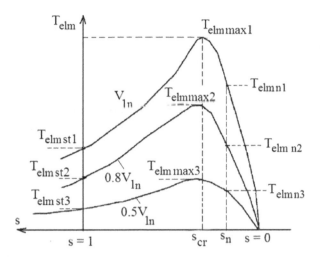

Fig. 6.27. The influence of the input voltage on the torque–slip characteristic of an induction machine.

Table 6.2. The OCF, STR, and SCR for single-cage, double-cage, and deep-bar induction motors

Motor	$OCF = T_{elmmax}/T_{elmn}$	$STR = T_{elmst}/T_{elmn}$	$SCR = I_{1st}/I_{1n}$
Single-cage	1.6 to 1.8	1.0 to 1.5	4.0 to 7.0
Double-cage	1.6 to 3.2	1.0 to 2.0	3.0 to 5.0
Deep-bar	1.6 to 3.2	0.2 to 1.0	3.0 to 5.0

$$\frac{T_{elm}}{T_{elmmax}} \approx \frac{2}{s_{cr}/s + s/s_{cr}} \tag{6.108}$$

Note that eqn (6.108) can be derived with the aid of eqns (6.99), (6.102), and (6.103).

Example 6.2

A three-phase, 12-pole ($2p = 12$), 420-V (line-to-line), Y-connected, 5.5-kW, 50-Hz cage induction motor has the following equivalent circuit parameters: $R_1 = 0.833$ Ω, $X_1 = 1.864$ Ω, $R'_2 = 0.833$ Ω, $X'_2 = 1.864$ Ω, $X_m = 36.25$ Ω. The machine operates with a slip of 0.03. The rotational loss is $\Delta P_{rot} = 250$ W and the stray loss is $\Delta P_{str} = 60$ W.

For an input frequency of 60 Hz, rated input voltage of 420 V, and rated slip of 0.03 find: (a) the rotor speed, (b) the stator and rotor currents at slip 0.03, (c) the stator and rotor winding losses, (d) the electromagnetic (air gap) power and electromagnetic (developed) torque, (e) the mechanical power and the output power assuming that the rotational loss is $\Delta P_{rot} = 250$ W and the stray loss is $\Delta P_{str} = 60$ W, (f) the input power, efficiency and power factor, (g) the starting torque T_{elmst} and starting torque ratio STR, (h) the breakdown torque T_{elmmax}, the critical slip s_{cr}, and the overload capacity factor OCF.

Assumptions: The core-losses are neglected ($R_{Fe} = 0$) and the skin effect in the rotor bars is also negligible.

Solution

(a) The rotor speed

- Synchronous speed
  At 50 Hz, $\qquad\qquad\qquad n_s = f/p = 50/6 = 8.33 \times 60 = 500$ rpm
  At 60 Hz, $\qquad\qquad\qquad n_s = f/p = 60/6 = 10.0 \times 60 = 600$ rpm

- Rotor speed
  At 50 Hz, $\qquad\qquad\qquad n = 500(1 - 0.03) = 485$ rpm
  At 60 Hz, $\qquad\qquad\qquad n = 600(1 - 0.03) = 582$ rpm

(b) The stator and rotor rated currents

- Reactances at 60 Hz

$$X_1 = X_2' = \frac{60}{50}1.864 = 2.237 \ \Omega$$

$$X_m = \frac{60}{50}36.25 = 43.5 \ \Omega$$

- The impedances of the vertical and rotor branches for the nominal slip

$$\frac{(R_2'/s + jX_2')jX_m}{R_2'/s + j(X_2' + X_m)} = \frac{(0.833/0.03 + j2.237)j43.5}{0.833/0.03 + j(2.237 + 43.5)} = (18.353 + j13.269)\Omega$$

- The stator (input) current

$$\mathbf{I}_1 = \frac{V_1}{R_1 + 18.353 + j(X_1 + 13.269)} = \frac{420/\sqrt{3}}{0.833 + 18.353 + j(2.237 + 13.269)}$$

$$= \frac{242.5}{19.186 + j15.506} = (7.645 - j6.179) \ \text{A}$$

$$|\ \mathbf{I}_1\ | = I_1 = \frac{242.5}{\sqrt{19.186^2 + 15.506^2}} = 9.83 \ \text{A}$$

- Heyland's coefficient

$$\tau_1 = \frac{X_1}{X_m} = \frac{2.237}{43.5} = 0.0514$$

- The rotor current referred to as the stator system

$$I_2' = \frac{242.5}{\sqrt{\left(0.833 + \frac{0.833}{0.03}1.0514\right)^2 + (2.237 + 2.237 \times 1.0514)^2}} = 7.98 \text{ A}$$

(c) The stator and rotor winding losses

- Stator winding losses

$$\Delta P_{1w} = 3I_1^2 R_1 = 3 \times 9.83^2 \times 0.833 = 241 \text{ W}$$

- Rotor winding losses

$$\Delta P_{2w} = 3(I_2')^2 R_2' = 3 \times 7.98^2 \times 0.833 = 159.1 \text{ W}$$

(c) The electromagnetic (air gap) power and electromagnetic torque

- Electromagnetic power crossing the airgap

$$P_{elm} = \frac{3(I_2')^2 R_2'}{s} = \frac{\Delta P_{2w}}{s} = \frac{159.1}{0.03} = 5.303 \text{ kW}$$

- Electromagnetic developed torque

$$T_{elm} = \frac{P_{elm}}{2\pi n_s} = \frac{5303.3}{2\pi 10} = 84.4 \text{ Nm}$$

(e) The mechanical power and output power

- The rotational loss is $\Delta P_{rot} = 250$ W and the stray loss is $\Delta P_{str} = 60$ W.
- Mechanical power

$$P_m = (1 - s)P_{elm} = (1 - 0.03)5303.3 = 5.144 \text{ kW}$$

- Output (shaft) power

$$P_{out} = P_m - \Delta P_{rot} - \Delta P_{str} = 5144.2 - 250 - 60 = 4.834 \text{ kW}$$

(f) The input power, efficiency and power factor

- The input current was found above to be

$$\mathbf{I}_1 = (7.645 - j6.179) \text{ A} \qquad \text{so} \qquad |\mathbf{I}_1| = I_1 = \sqrt{7.645^2 + 6.179^2} = 9.83 \text{ A}$$

- Apparent input power

$$S_{in} = 3V_1 I_1 = 3 \times \frac{420}{\sqrt{3}} \times 9.83 = 7150.95 \text{ VA}$$

- Active input power

$$P_{in} = 3V_1 \times 7.636 = 3 \times \frac{420}{\sqrt{3}} \times 7.636 = 5.555 \text{ kW}$$

or

$$P_{in} = P_{out} + \Delta P_{rot} + \Delta P_{str} + \Delta P_{2w} + \Delta P_{1w} =$$

$$4834.2 + 250.0 + 60.0 + 241.5 + 159.1 = 5.545 \text{ kW}$$

- Efficiency and power factor

$$\eta = \frac{P_{out}}{P_{in}} = \frac{4834.20}{5554.89} = 0.87 \qquad cos\varphi = \frac{P_{in}}{S_{in}} = \frac{5554.89}{7150.95} = 0.777$$

(g) The starting torque and starting torque ratio (STR)

- Starting electromagnetic torque $(s = 1)$

$$T_{elmst} = \frac{3V_1^2}{2\pi n_s} \frac{R_2'}{[R_1 + R_2'(1 + \tau_1)]^2 + [X_1 + X_2'(1 + \tau_1)]^2}$$

$$= \frac{3(420/\sqrt{3})^2}{2\pi \times 10} \frac{0.833}{(0.833 + 0.833 \times 1.0514)^2 + (2.237 + 2.237 \times 1.0514)^2}$$

$$= 95.97 \text{ Nm}$$

- Starting torque ratio

$$STR = \frac{T_{elmst}}{T_{elmn}} = \frac{95.97}{84.40} = 1.137$$

(h) The breakdown torque, critical slip, and overload capacity factor (OCF)

- Critical slip

$$s_{cr} = + \frac{R_2'(1 + \tau_1)}{\sqrt{R_1^2 + [X_1 + X_2'(1 + \tau_1)]^2}}$$

$$\frac{0.833 \times 1.0514}{\sqrt{0.833^2 + (2.237 + 2.237 \times 1.0514)^2}} = 0.1877$$

- Kloss' formula

$$\frac{T_{elmn}}{T_{elmmax}} \approx \frac{2}{s_{cr}/s + s/s_{cr}} = \frac{2}{0.1877/0.03 + 0.03/0.1877} = 0.312$$

- Maximum electromagnetic torque

$$T_{elmmax} = \frac{T_{elmn}}{0.312} = \frac{84.4}{0.312} = 270.9 \text{ Nm}$$

- Overload capacity factor (OCF)

$$OCF = \frac{270.9}{84.4} = 3.21$$

6.5.8 Influence of rotor resistance on torque–speed characteristics

The maximum torque according to eqn (6.103) is independent of both the stator and rotor circuit resistances (Fig. 6.27). However, the critical slip (6.102) is directly proportional to the rotor resistance, R_2'. Thus, if R_2' increases then s_{cr} increases too, while $T_{elmmax} = const$ (Fig. 6.28).

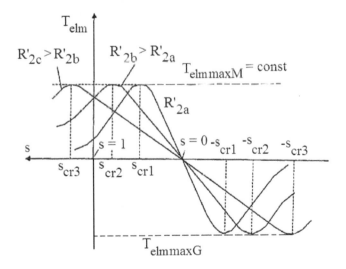

Fig. 6.28. The influence of the rotor circuit resistance on the torque–slip characteristic of an induction machine. $T_{elmmaxM}$ – maximum torque for motor, $T_{elmmaxG}$ – maximum torque for generator.

6.5.9 Load characteristics

Load characteristics are steady-state performance curves plotted against the load torque T or input current I_1 at constant input voltage V_1, usually equal to nominal (rated) voltage, i.e., $V_1 = V_{1n} = const$. Fig. 6.29 shows the rotor speed n, efficiency η, power factor $cos\varphi$, input power P_{in} and output power P_{out} as functions of the load torque T at $V_1 = V_{1n} = const$.

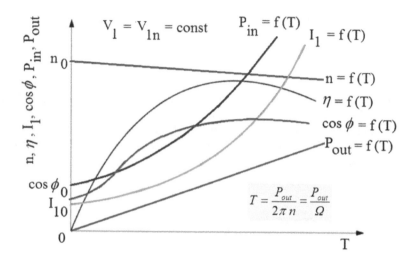

Fig. 6.29. Load characteristics of induction motor at $V_1 = V_{1n} = const$.

6.6 Starting

An induction motor is not a self-starting motor. Wound–rotor (slip–ring) induction motors are started with the aid of additional rotor resistance. Cage–rotor induction motors are started using one of the following methods:

- Direct on-line starting – only for small motors;
- Star–delta switching;
- Stator impedance starting;
- Autotransformer starting;
- Solid state soft starters.

6.6.1 Slip–ring motors

An induction motor with slip rings is started by connecting the stator to the power line, with some additional three-phase resistance R_{st} being fully cut-in in the rotor circuit, in the form of a starting rheostat (Fig. 6.30). The electromagnetic torque T_{elm} according to eqn (6.99) must be the same for $s = 1$ and $R_{st} > 0$ as it is for $0 < s < 1$ and $R_{st} = 0$. So,

$$\frac{(R_2'/s)}{[R_1 + (R_2'/s)(1 + \tau_1)]^2 + [X_1 + X_2'(1 + \tau_1)]^2}$$

$$= \frac{R_2' + R_{st}'}{[R_1 + (R_2' + R_{2st}')(1 + \tau_1)]^2 + [X_1 + X_2'(1 + \tau_1)]^2}$$

where R_{st}' is the rotor starting resistance referred to the stator winding. Thus

$$\frac{R_2'}{s} = R_2' + R_{st}' \qquad \text{and} \qquad R_1 + \frac{R_2'}{s}(1 + \tau_1) = R_1 + (R_2' + R_{st}')(1 + \tau_1)$$

or

$$\frac{R_2}{s} = R_2 + R_{st}$$

The starting rheostat resistance

$$R_{st} = \frac{R_2}{s} - R_2 = R_2 \frac{1 - s}{s} \tag{6.109}$$

gives the necessary starting torque and small starting current (as shown in Fig. 6.31). For example, if the rotor resistance is $R_2 = 0.1 \ \Omega$ and the rotor rated speed is $n = 975$ rpm, then the corresponding slip is $s = (1000 - 975)/1000 = 0.025$ and the starting rheostat resistance should be $R_{st} = 0.1(1 - 0.025)/0.025 = 3.9 \ \Omega$. After switching on the stator windings, the rheostat is gradually cut-out until the rotor winding is short-circuited.

Starting rheostats are usually made of metal and have oil or liquid cooling. To reduce the rotor circuit resistance and in order to decrease the friction losses of the brushes on the slip rings, wound-rotor induction motors are frequently provided with devices for short-circuiting the rings while running and for further lifting the brushes (Fig. 6.30a). Torque–speed and current–speed characteristics of a slip-ring induction motor are plotted in Fig. 6.31.

6.6.2 Cage–rotor motors

Direct on-line starting

A small cage induction motor may be switched "direct-on-line," meaning that it is switched directly onto normal voltage, momentarily taking several times the full-load current at a low power factor. Direct switching may be the subject of Supply Authority regulations. However, there is usually no restriction on small induction motors up to 5.5 kW.

(a)

(b)

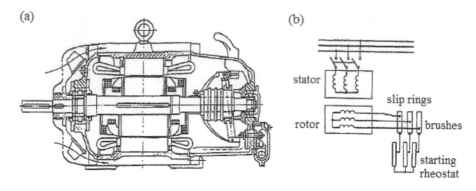

Fig. 6.30. A slip-ring induction motor: (a) longitudinal section with slip rings and mechanism for lifting the brushes and shorting the slip rings; (b) connection of three-phase starting rheostat to brushes.

(a)

(b)

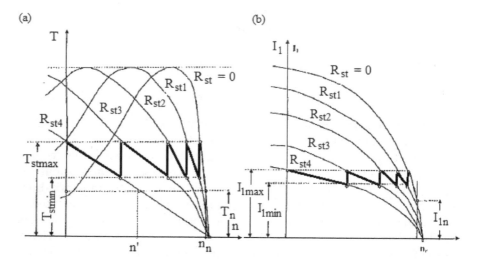

Fig. 6.31. Characteristics of a slip-ring induction motor with starting rheostat: (a) starting torque–speed curve; (b) starting input current–speed curve.

Star–delta switching

For star–delta $(Y-\Delta)$ switching (see Fig. 6.32), a motor must be built with a Δ-connected stator winding. Let V_{1L} be the line voltage, and let V_{1Y} and $V_{1\Delta}$ be the voltages per phase for Y and Δ connection of the windings. Next let I_{1stLY}, $I_{1stL\Delta}$, I_{1stY}, and $I_{1st\Delta}$ represent the starting current in the line and in the phases of the stator winding when it is Y- and Δ-connected, respectively.

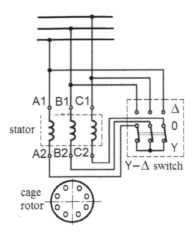

Fig. 6.32. Star–delta switching.

Finally, $\mid \mathbf{Z}_{sh} \mid = Z_{sh}$ is the blocked-rotor (short-circuit) impedance for $s = 1$ of the motor, per phase. Then, in the case of the Y–connection,

$$I_{1stLY} = I_{1stY} = \frac{V_{1Y}}{Z_{sh}} = \frac{V_{1L}}{\sqrt{3}Z_{sh}} \tag{6.110}$$

and for the motor with its stator winding Δ-connected

$$I_{1stL\Delta} = \sqrt{3}I_{1st\Delta} = \frac{\sqrt{3}V_{1\Delta}}{Z_{sh}} = \frac{\sqrt{3}V_{1L}}{Z_{sh}} \tag{6.111}$$

By comparing the above two equations, it is seen that

$$\frac{I_{1stLY}}{I_{1stL\Delta}} = \frac{1}{3} \tag{6.112}$$

Thus, *the starting current when the stator winding is Y-connected is one-third of that when it is Δ-connected.* Meanwhile, *the starting torque T_{st} also decreases by a factor of three.* Hence

$$T_{stY} \propto V_{1Y}^2 = \frac{1}{3}V_{1L}^2 \qquad \text{and} \qquad T_{st\Delta} \propto V_{1\Delta}^2 = V_{1L}^2$$

Combining these expressions now gives

$$\frac{T_{stY}}{T_{st\Delta}} = \frac{1}{3} \tag{6.113}$$

The characteristics of an induction motor for $Y - \Delta$ switching are shown in Fig. 6.33.

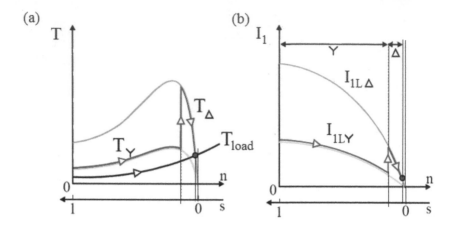

Fig. 6.33. Characteristics of $Y - \Delta$ switching: (a) torque–speed curve, (b) stator current–speed curves.

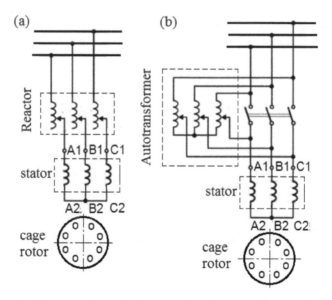

Fig. 6.34. Starting with the aid of variable inductive element: (a) reactor; (b) autotransformer.

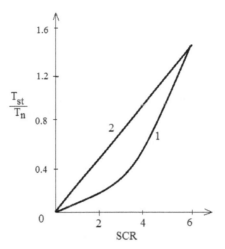

Fig. 6.35. Relation $T_{st}/T_n = f(SCR)$ for starting by means of reactor (1) and autotransformer (2).

Stator impedance starting

The inclusion of a three-phase resistor or of an inductive reactor (see Fig. 6.34a) in each of the stator input lines reduces the stator terminal voltage. Suppose that the starting current in the external circuit of a Y-connected motor is limited to I_{1stL}, so one can write

$$I_{1stL} = SCR \times I_{1n} \tag{6.114}$$

where SCR is the permissible starting current ratio and I_{1n} is the input nominal (rated) current. Assuming $I_1 \approx I_2'$, the starting torque is then

$$T_{st} \approx \frac{m_1 I_{1stL}^2 R_2'}{2\pi n_s} = SCR^2 \frac{m_1 I_{1n}^2 R_2'}{2\pi n_s} \propto SCR^2 \tag{6.115}$$

In other words, when starting an induction motor with a reactor, the starting torque depends on the square of the value of SCR. The rated torque is

$$T_n \approx \frac{m_1 I_{1n}^2 R_2'}{2\pi n_s} \frac{1}{s_n} \tag{6.116}$$

where s_n is the nominal (rated) slip. Consequently

$$\frac{T_{st}}{T_n} = SCR^2 s_n \tag{6.117}$$

The relation $T_{st}/T_r = f(SCR)$ is plotted in Fig. 6.35.

Autotransformer starting

Suppose that the autotransformer in Fig. 6.34b is used to reduce the stator applied voltage. If ϑ is the voltage ratio of the autotransformer and if Z_{sh} is the impedance of one motor phase for $s = 1$, then, neglecting for simplicity the autotransformer impedance, the starting voltage V_{1LY} across the motor input terminals and the starting motor current I_{1stY} are:

$$V_{1LY} = \frac{1}{\vartheta} V_{1L} \qquad \text{and} \qquad I_{1stY} = \frac{V_{1LY}}{Z_{sh}} = \frac{V_{1L}}{\vartheta Z_{sh}} \qquad (6.118)$$

where V_{1L} is the line voltage of the external circuit. The starting line current in the autotransformer winding (external circuit) is

$$I_{1L} = \frac{1}{\vartheta} I_{1stY} = \frac{1}{\vartheta^2} \frac{V_{1L}}{Z_{sh}} = \frac{1}{\vartheta^2} I_{1sh} \qquad (6.119)$$

where $I_{1sh} = V_{1L}/Z_{sh}$ is the short-circuit current of the motor at rated voltage. As compared with the stator impedance starting that was described in the previous subsection, this method of starting has a considerable advantage with respect to the starting torque. Indeed,

$$T_{st} = \frac{m_1 I_{1stY}^2 R_2'}{2\pi n_s} = \frac{m_1 (I_{1L}\vartheta)^2 R_2'}{2\pi n_s} \qquad (6.120)$$

and the nominal (rated) torque is given by eqn (6.116). Consequently, in this case,

$$\frac{T_{st}}{T_n} = \frac{I_{1L}^2 \vartheta^2}{I_{1n}^2} s_n = \frac{I_{1L}}{I_{1n}} \frac{\vartheta^2 I_{1L}}{I_{1n}} s_n = SCR \frac{\vartheta^2 I_{1L}}{I_{1n}} s_n \qquad (6.121)$$

where SCR is the starting current ratio in the external circuit and s_n is the nominal slip. Now, according to eqn (6.119),

$$\vartheta^2 = \frac{I_{1sh}}{I_{1L}}$$

and so

$$\frac{T_{st}}{T_n} = SCR \frac{I_{1sh}}{I_{1n}} s_n \qquad (6.122)$$

In this case, therefore, the relation $T_{st}/T_n = f(SCR)$ is a straight line, since for given values of I_{1sh}/I_{1n} and s_n the torque $T_{st} \propto SCR$ (see Fig. 6.35).

Solid state soft starters

A typical induction motor, designed according to IEC standards, produces approximately 140% of its normal full-load torque almost instantly when full voltage is applied. The same motor draws about six times its full load current at start-up. As has been emphasized, substantial damage to the motor and auxiliary equipment can occur as a result of this. Reducing the voltage at start-up reduces the starting torque, current surge, electrodynamic forces on the motor windings, and also the mechanical and electrical shock transmitted to the equipment.

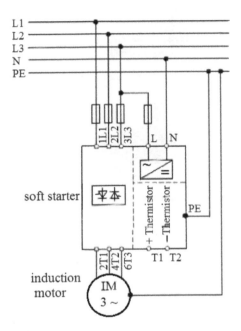

Fig. 6.36. Connection diagram of a three-phase induction motor with the aid of solid soft starter. L1, L2, L3 – line wires, N – neutral wire, PE – protective earth.

The voltage at the first instant of starting can be smoothly reduced with the aid of semiconductor devices, using the so-called *solid state soft starter*. The power circuit of a solid state soft starter is shown in Fig. 6.36. Practical connection diagram is shown in Fig. 6.37. At the heart of a soft starter one usually finds a microprocessor, which provides the control logic for the starting process. The load characteristics of an induction motor for direct on-line starting, Y-Δ switch and solid soft starter are shown in Fig. 6.38.

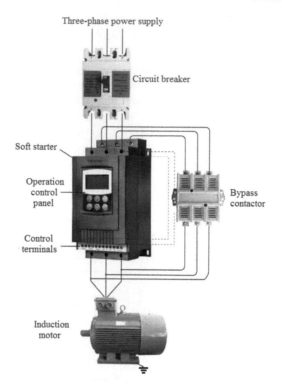

Fig. 6.37. Three-phase induction motor, solid soft starter, circuit breaker and by-pass contactor.

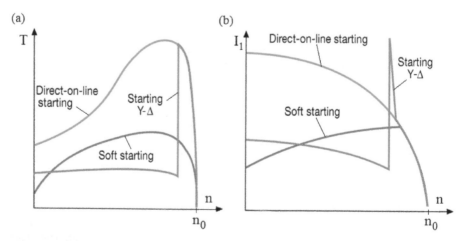

Fig. 6.38. The load characteristics of an induction motor for direct on-line starting, Y-Δ switch, and solid soft starter: (a) torque $T = f(n)$; (b) stator line current $I_1 = f(n)$.

6.7 Induction motors that use skin effect in the rotor winding

To obtain increased rotor resistance at $s = 1$ (rotor frequency $sf = f$) and low resistance at $s = s_n$ (very low rotor frequency $s_n f$), the so-called *skin effect* is utilized. Two rotor designs use the skin effect:

- Deep-bar rotor;
- Double-cage rotor.

The skin effect arises when the eddy currents flowing in the conductor produce electromagnetic fields which oppose the primary field. The skin effect causes non-uniform distribution of current in the conductor. The closer to the air gap, the higher the current density in the rotor bar.

The *depth of penetration* of electromagnetic wave into a conductive material is defined as the depth at which the amplitude of the current density falls to $1/e \approx 0.368$ of its surface value. The equivalent depth of penetration of electromagnetic field is expressed as [38]

$$\Delta = \frac{1}{\sqrt{\pi f \mu_0 \mu_r \sigma}} \text{ m} \tag{6.123}$$

where $\mu_0 = 0.4\pi \times 10^{-6}$ H/m is the magnetic permeability of free space, μ_r is the relative magnetic permeability and σ is the electric conductivity, S/m. For example, if $s = 1$, $f = 50$ Hz, $sf = 50$ Hz and if $s = 0.05$, $f = 50$ Hz, $sf \rightarrow 2.5$ Hz. The equivalent depth of penetration Δ is much lower at $s = 1$ (starting) than at nominal operation $s = 0.05$. Consequently, the resistance R_2 and inductance L_2 at starting is much higher than that at nominal operation and the starting current is reduced.

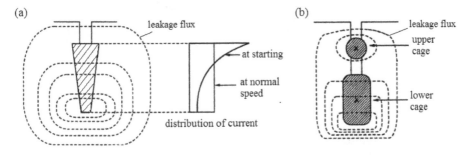

Fig. 6.39. Rotor slots of induction motors that use skin effect in the rotor winding: (a) deep-bar rotor; (b) double-cage rotor.

6.7.1 Deep bar motors

Principle of operation

The *deep bar rotor* (Fig. 6.39) of an induction motor has slots and bars with increased radial depth toward the shaft and has better starting characteristics compared with the conventional single-cage motor. Bars with cross-sections that are other than rectangular are also used, examples being trapezoidal or bottle-shaped. In further discussion only bars with a rectangular cross-section will be considered, since they are the main ones and of the simplest shape with respect to design and manufacture.

In deep bar motors, use is made of the *skin-effect* that is induced in the rotor winding bars by the slot leakage fluxes.

Let us first consider the phenomena at starting. At the initial instant, when $s = 1$, the frequency in the rotor is equal to the input (stator) frequency. The slot leakage flux paths within the rotor under these conditions are depicted in Fig. 6.39a. Bar portions of different height are linked by different numbers of leakage flux lines: the lower parts by the greatest number of flux lines, and the upper parts by the least number. For this reason, the maximum leakage EMFs are induced in the lower parts of the bars, and the minimum EMFs in the upper parts.

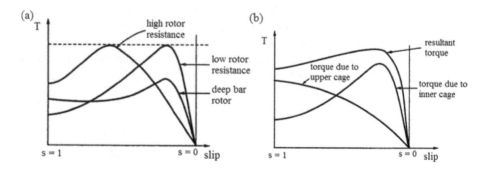

Fig. 6.40. Torque–slip curves for induction motors with: (a) deep-bar rotor; (b) double-cage rotor.

It can therefore be seen that the rotor leakage flux EMF, E_{2l}, is directed oppositely to the rotor main EMF, E_2, but, in accordance with the previous discussion, it is greater in the lower parts of the conductor than in its upper parts. Consequently, less current should flow through the lower parts than through the upper ones; in other words, the current is forced to the outside of the conductor (producing a skin effect). Hence the current density is distributed along the conductor height as shown in Fig. 6.39a.

The skin effect takes place in all types of motors. With the usual conductor height being 10 to 12 mm and the fundamental harmonic of input frequency being 50 to 60 Hz, however, it is almost unnoticeable. Nevertheless, in deep bars that are 20 to 50 mm high, the skin effect is very strong and appreciably changes the rotor parameters. Torque–slip curves for deep-bar rotor are shown in Fig. 6.40a.

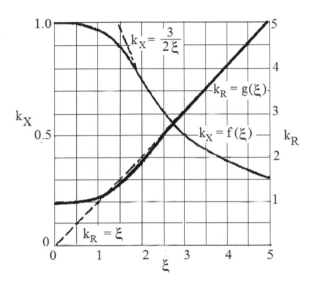

Fig. 6.41. Coefficients k_R and k_X versus ξ.

Rotor resistance and inductive reactance

For all practical purposes, the skin effect takes place only in that part of the conductor which lies in the slot, and is absent in the end connections of the winding. Therefore, the resistance R'_2 and inductive reactance X'_2 of the rotor winding can be expressed as follows:

$$R'_2 = k_R R'_{2sl} + R'_{2e} \qquad \text{and} \qquad X'_2 = k_X X'_{2sl} + X'_{2e} \qquad (6.124)$$

where R'_{2sl} is the resistance of the slot part of the rotor winding assuming a uniform current distribution along the conductor cross-section, k_R is the factor allowing for the increase of the resistance R'_{2sl} due to the skin effect, R'_{2e} is the constant-value resistance of the rotor end rings, X'_{2sl} and X'_{2e} are the leakage inductive reactances of the slot and end rings of the rotor winding with uniform current distribution over the conductor cross-section and with a frequency f ($s = 1$), and k_X is the factor allowing for the decrease of the inductive reactance X'_{2sl} due to the skin effect. Coefficients k_R and k_X depend on the shape and size of the rotor bar, the matarial of the conductor and the frequency of current in the rotor. They are defined as:

$$k_R = \frac{\text{AC resistance of bar}}{\text{DC resistace of bar}} \qquad (6.125)$$

$$k_X = \frac{\text{AC leakage reactance of bar}}{\text{DC leakage reactance of bar}} \qquad (6.126)$$

Analysis of this problem shows that

$$k_R = \xi \frac{\sinh(2\xi) + \sin(2\xi)}{\cosh(2\xi) - \cos(2\xi)} \qquad (6.127)$$

and

$$k_X = \frac{3}{2\xi} \frac{\sinh(2\xi) - \sin(2\xi)}{\cosh(2\xi) - \cos(2\xi)} \qquad (6.128)$$

where

$$\xi = h_{2b} \sqrt{\pi \mu_0 s f \sigma_2 \frac{b_{2b}}{b_2}} \qquad (6.129)$$

and where h_{2b} is the height of the rotor bar, $\mu_0 = 0.4\pi \times 10^{-6}$ H/m is the magnetic permeability of free space, sf is the rotor slip frequency, σ_2 is the conductivity of the rotor winding, b_{2b} is the width of the rotor bar, and b_2 is the width of the rotor slot. Knowing ξ, it is possible to determine the factors k_R and k_X (see Fig. 6.41). For values of $\xi > 2$, the skin-effect coefficients can be simply expressed as:

$$k_R = \xi \qquad \text{and} \qquad k_X = \frac{3}{2\xi} \qquad (6.130)$$

6.7.2 Double-cage motors

The *double-cage induction motor* was invented by M. Dolivo-Dobrovolsky in 1893, and it is one of the variations of the cage induction motor that uses the skin-effect phenomenon in the rotor winding to improve the starting properties. The stator of such a motor does not differ from that of a conventional AC motor, but its rotor consists of two cages (see Fig. 6.42). The upper cage nearer to the air gap, called the *starting cage,* is made of a somewhat resistive material such as brass, aluminum, bronze, etc., while the lower cage, called the *operating cage,* is made of copper. In small induction motors both of the two cages, together with the end rings, are made of aluminum alloy.

At starting, the frequency of the rotor current is high and equal to the frequency of the source. The current amplitude is distributed between the upper

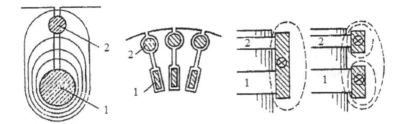

Fig. 6.42. Design of rotor windings of a double-cage motor: 1 – operating cage, 2 – starting cage.

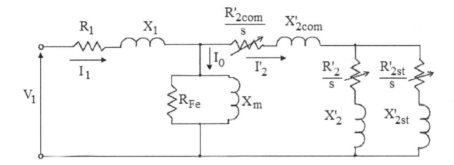

Fig. 6.43. Equivalent circuit per phase of a double-cage induction motor: R_2', X_2' – resistance and leakage reactance of the operating cage, R_{2st}', X_{2st}' – resistance and leakage reactance of the starting cage, R_{2com}', X_{2com}' – common resistance and reactance. All other symbols are according to Fig. 6.16.

and lower windings in inverse proportion to their impedances. Since the lower cage has a very high inductive reactance, its impedance is several times that of the upper cage, whose reactance is practically zero. The current in the lower cage is thus appreciably smaller than that in the upper cage. Furthermore, owing to the high leakage inductive reactance of the lower cage, the current in it lags by a large angle behind the EMF induced by the mutual inductance flux, and the winding consequently produces a small torque. Conversely, the current in the upper cage at starting is not only of considerable magnitude, but is also nearly in phase with the mutual inductance EMF because of the negligible inductive inductance and high resistance of the cage, owing to which this winding produces a very high torque. The torque during the starting period is hence developed mainly by the upper cage, which, in consequence, is referred to as the *starting cage*.

As the motor speed increases, the frequency in the rotor winding begins to decrease. This results in a reduction of the inductive reactance and an increase in the current in the lower cage, accompanied by a corresponding decrease in the phase angle between the current and the voltage. Consequently, this cage gradually begins to develop a greater and greater torque. When the motor reaches full speed and has a very small slip s, the inductive reactance of the lower cage becomes negligible in comparison with its resistance. The total current of the whole rotor winding will be divided between two cage windings in inverse proportion to their resistances, and, since the upper cage winding has a resistance which is 5 to 6 times that of the lower cage winding, the current in the upper cage becomes considerably smaller than that in the lower cage. Hence the torque is mainly developed by the lower cage, which is often called the *operating cage*.

The equivalent circuit per phase of a double-cage induction motor is shown in Fig. 6.43. The common resistance R'_{2com} exists only if the double-cage winding has common end rings. The common reactance X'_{2com} represents the linkage flux between the upper and lower cages.

Fig. 6.40b shows approximate torque curves for the upper and lower cages, as well as the resultant torque curve of both cage windings.

6.8 Speed control

Eqn (6.3) provides the basis for several methods of speed control for induction motors, which can be grouped under the headings of changing:

- The input frequency f;
- The number of stator pole pairs p;
- The slip s, i.e., by adjusting the rotor circuit resistance of wound-rotor induction motors or the input voltage of cage induction motors.

6.8.1 Frequency changing for speed control

An excellent way to control the speed of an induction motor is to vary the input frequency (Fig. 6.44) using solid–state converters. Assuming $R_1 \ll X_1 + X'_2(1 + \tau_1)$ and $1 + \tau_1 \approx 1$, the breakdown torque according to eqn (6.103) is

$$T_{elmmax} \approx \pm \frac{m_1 V_1^2}{4\pi n_s} \frac{1}{2\pi f (L_1 + L'_2)} = \pm \frac{m_1 p}{8\pi^2} \frac{1}{L_1 + L'_2} \left(\frac{V_1}{f}\right)^2 \qquad (6.131)$$

where $L_1 = X_1/(2\pi f)$ and $L'_2 = X'_2/(2\pi f)$ are, respectively, the stator winding inductance and the rotor winding inductance (referred to as the stator winding).

With a constant V_1/f ratio (constant magnetic flux Φ), an induction motor develops a constant maximum torque T_{elmmax} (Fig. 6.45), except at low frequencies. If $V_1 = const$ and frequency f is variable, an increase in frequency f reduces the maximum torque T_{elmmax} (Fig. 6.44). For a frequency k times the rated frequency 50 Hz and voltage k times the rated phase voltage

$$\frac{V_1}{f} = \frac{kV_{1n}}{k50} = const \qquad (6.132)$$

the maximum electromagnetic torque according to eqn (6.103) is

$$T_{elmmax} = \frac{m_1}{4\pi k n_s(1+\tau_1)} \times \frac{(kV_{1n})^2}{\sqrt{R_1^2 + k^2[X_1 + X_2'(1+\tau_1)]^2} + R_1}$$

$$= \frac{m_1}{4\pi n_s(1+\tau_1)} \times \frac{V_{1n}^2}{\sqrt{(R_1/k)^2 + [X_1 + X_2'(1+\tau_1)]^2} + R_1/k} \qquad (6.133)$$

Before the power electronics era, eddy-current couplings were frequently used to build a variable-speed drive with an induction motor fed from a constant-frequency line. Although the dynamic response was poor, sufficient speed stability could be obtained in closed-loop eddy-current coupling drives.

The critical slip according to eqn (6.102) is

$$s_{cr} \approx \pm \frac{R_2'}{2\pi f(L_1 + L_2')} \qquad (6.134)$$

Note that *both the breakdown torque and critical slip are inversely proportional to the input frequency*. Hence, if the input frequency f decreases while V_1 =const, the speed will decrease too (via an increase in slip), and the overload capacity factor OCF will increase according to eqn (6.105) (see Fig. 6.44a). The magnetic flux Φ will also increase, since the stator EMF $F_1 \approx V_1$ is expressed by eqns (6.22) and (6.24). The product $f\Phi$ must be constant at V_1 =const. Obviously, at a reduced input frequency the stator core losses increase due to an increase in the magnetic flux density. Changing the input frequency, the speed is usually controlled by keeping $V/f = const$ (see Fig. 6.45 and eqn (6.132)). This ensures approximately constant magnetic flux density $\Phi = const$ in the air gap.

A schematic power circuit diagram of a solid–state converter is shown in Fig. 6.46. It consists of a *rectifier* (AC to DC converter), an *intermediate circuit* (filter), and an *inverter* (DC to AC converter, each of which employs *power semiconductor devices* also called *solid state switches*. Placing in the DC link a capacitor C provides voltage supply of the inverter (Fig. 6.47). The output voltage is a sequence of rectangular pulses that arise from the cyclic connection of the output phases with positive and negative terminal of the intermediate circuit. For proper operation of the inverter the freewheeling diodes are necessary that are plugged antiparallel to each solid state switch (Fig. 6.47). The principle of switching the solid-state switch so that it ceases to conduct is termed *commutation*.

The sequence of the firing pulses that are sent as inputs to the inverter circuit can be controlled by a pulse generator in any desired manner, and the semiconductor switches, e.g., thyristors, are generally fired in diagonal pairs,

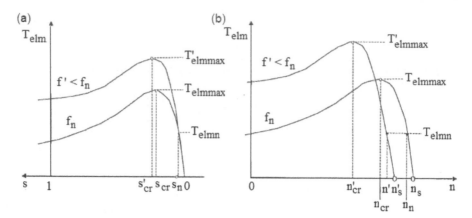

Fig. 6.44. Influence of the input frequency on the characteristics: (a) torque–sleep; (b) torque–speed ($V_1 = const$).

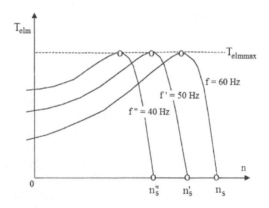

Fig. 6.45. Torque–speed characteristics of an induction motor for various input frequencies at $V_1/f = const$.

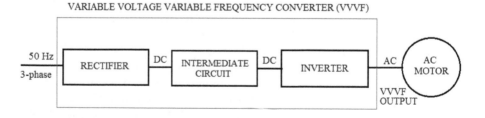

Fig. 6.46. Frequency conversion circuitry.

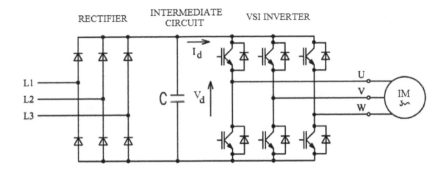

Fig. 6.47. Solid state converter with a voltage source inverter (VSI). The control of output parameters (voltage and frequency) is only in the control circuit of the inverter.

resulting in a stepped waveform such as the one shown in Fig. 6.48a. The fundamental of this waveform is the frequency that defines the speed at which the induction motor being driven by the circuit will operate. An alternative type of output from an inverter circuit is obtained by *high frequency chopping*, which results in a stepped PWM (*pulse width modulation*) waveform (as in Fig. 6.48b). Principle of PWM modulation is explained in Fig. 3.15. Again, the fundamental frequency imposes the synchronous speed.

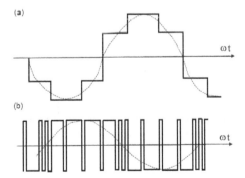

Fig. 6.48. Inverter output voltage waveforms: (a) six-pulse square-wave, (b) PWM waveform.

6.8.2 Pole changing for speed control

The number of pole pairs in the stator can be changed as follows:

- By placing one winding on the stator and changing the number of poles $2p$ by successively reconnecting the parts of this winding;

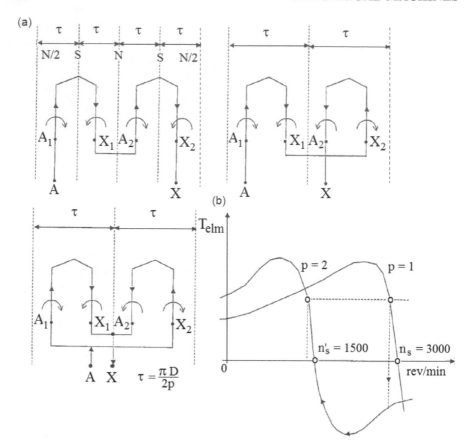

Fig. 6.49. Pole changing for speed control: (a) diagrams, (b) torque–speed characteristics.

- By placing two independent windings on the stator;
- By providing two independent stator windings, each with reconnection of the poles.

Double-speed motors are usually made with one winding on the stator, the number of poles being changeable in the ratio 1:2. Three- and four-speed motors are provided with two windings on the stator, one or both of which can have the number of its poles changed. So, for example, if it is desired to obtain a motor for four synchronous speeds, such as 1500, 1000, 750, and 500 rpm, two windings should be placed on the stator, one of which gives $p = 2$ and $p = 4$ pole pairs, the other $p = 3$ and $p = 6$.

If the motor has a wound rotor, the number of pole pairs must be changed simultaneously on the stator and rotor. This complicates the design of the rotor. Therefore, motors with a changing number of poles usually have a

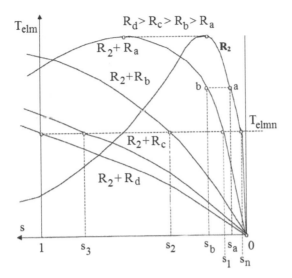

Fig. 6.50. Speed control by changing the slip (external rheostat resistance) of a wound-rotor motor.

short-circuited rotor with a cage winding. Such a rotor can operate without any reconnection and with any number of stator poles.

There exist several methods of switching over the pole pairs of a winding. The one most frequently used is the method of changing the direction of the current in the separate halves of each phase winding or, more simply, in the half-windings. Schematic diagrams of the half-winding commutation which changes the number of poles in the ratio of 2:1 are given in Fig. 6.49a. When the number of poles is changed, the air gap flux density changes, and so do the torque–speed curves (Fig. 6.49b).

6.8.3 Speed control by voltage variation

The speed of an induction motor may be continuously controlled by controlling the input voltage applied to the stator winding: the electromagnetic torque is then proportional to the square of the input voltage. This method of speed control changes the output (shaft) torque, and hence the efficiency falls off drastically with the reduction in speed, making the method unsuitable for most applications.

6.8.4 Changing the resistance in the rotor circuit

In motors with slip rings the speed can be controlled with the aid of a rheostat in the rotor circuit. The control diagram does not differ from the ordinary diagram of an induction motor with a wound rotor (see Fig. 6.30). In practice, the control rheostats are similar to starting rheostats, but are bigger and are designed for continuous operation. The torque-speed characteristics are shown in Fig. 6.50.

6.9 Inverter-fed induction motor capabilities

As it has been discussed, in inverter-fed induction motors the speed can be controlled by varying:

(a) Input frequency f,
(b) Input voltage V_1,
(c) Both input frequency and voltage (VVVF),

keeping the air gap magnetic flux constant, i.e.,

$$\pi\sqrt{2}N_1 k_{w1}\Phi \approx \frac{V_1}{f} = const \tag{6.135}$$

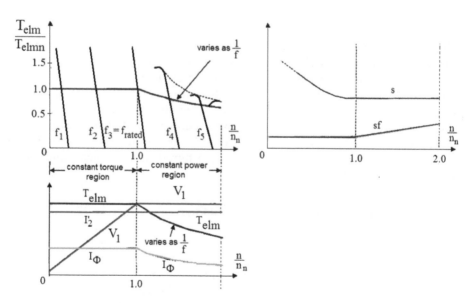

Fig. 6.51. Induction motor characteristics and capabilities.

Speed control by means of VVVF also allows the motor to operate not only at speeds below the rated speed, but also at above the rated speed (Fig. 6.51). Most induction motors can be operated under such conditions without mechanical and thermal problems.

In the region of low speed, below its rated value, the lines in Fig. 6.51a show the electromagnetic (developed) torque T_{elm} versus speed for low slip frequencies sf. The flux Φ is kept constant by controlling V_1/f. The voltage is decreased from its nominal (rated) value V_{1n} approximately in proportion to f. The electromagnetic torque developed

$$T_{elm} = \frac{P_{elm}}{\Omega_s} = \frac{m_1 E_1 I_2' \cos \Psi_2}{\Omega_s} = \frac{1}{\sqrt{2}} m_1 p N_1 k_{w1} \Phi I_2' \cos \Psi_2 \qquad (6.136)$$

is constant if $\Phi = const$, $I_2' = const$ and $\cos \Psi_2 = const$. The angle Ψ_2 between sE_1 and I_2' is approximately constant since

$$\tan \Psi_2 = \frac{sX_2'}{R_2'} \qquad (6.137)$$

is very small for a cage rotor ($sX_2' = 2\pi s f L_2 << R_2'$) and $\cos \Psi_2 \approx 1$. The magnetizing current $I_\Phi = const$ and the slip s increases for $sf = const$ as the frequency f decreases (Fig. 6.51).

The speed can be increased beyond its rated value by increasing the input frequency f above its nominal value and keeping $V_1 = const$ (very often equal to V_{1n}). According to eqn (6.135) the flux Φ is inversely proportional to the frequency ($\Phi \propto (1/f)$) at $I_2' = const$ and $\Psi_2 = const$, so the electromagnetic torque (6.136) will also be inversely proportional to the frequency. The rotor current I_2' is approximately constant if $V_1 = const$ and $s = const$. Since the frequency increases, the slip frequency sf also increases at $s = const$ and the magnetizing current I_Φ decreases since Φ decreases. The mechanical power $P_m - T_{elm}\Omega$ can be held constant.

Example 6.3

A 3-phase, 7.5-kW, 380-V, 1450-rpm, 50-Hz, Δ-connected cage induction motor has the following equivalent circuit parameters: $R_1 = 2.144\ \Omega$, $R_2' = 1.323\ \Omega$, $X_1 = 2.891\ \Omega$, $X_2' = 5.487\ \Omega$, $X_m = 116.3\ \Omega$. The motor is controlled by a voltage source inverter (VSI) at $V_1/f = const$. The VSI output frequency varies from 5 to 75 Hz. Find: (a) the starting current and starting developed torque at rated frequency, (b) the starting current and starting developed torque at minimum and maximum frequency, and (c) the breakdown torque as a function of frequency.

Solution

This is a four-pole motor ($2p = 4$). The synchronous speed at rated frequency is

$$n_s = \frac{f}{p} = \frac{50}{2} = 25\ \text{rev/s}$$

Heyland's coefficient according to eqn (6.94)

$$\tau_1 \approx \frac{X_1}{X_m} = \frac{2.891}{116.3} = 0.0248$$

(a) The starting current and starting torque at rated frequency

$$I_{1st} \approx \frac{V_{1n}}{\sqrt{(R_1 + R_2')^2 + (X_1 + X_2')^2}} = \frac{380}{\sqrt{(2.144 + 1.323)^2 + (2.891 + 5.487)^2}}$$

$$= 41.91 \text{ A}$$

$$T_{elmst} = \frac{m_1 V_{1n}^2}{2\pi n_s} \times \frac{R_2'}{[R_1 + R_2'(1 + \tau_1)]^2 + [X_1 + X_2'(1 + \tau_1)]^2}$$

$$= \frac{3 \times 380^2}{2\pi \times 25} \times \frac{1.323}{[2.144 + 1.323 \times (1 + 0.0248)]^2 + [2.891 + 5.487 \times (1 + 0.0248)]^2}$$

$$= 43.06 \text{ Nm}$$

(b) The starting current and starting developed torque for minimum and maximum frequency

For a frequency k times the rated frequency and $V_1/f = kV_{1n}/(k \times 50) = const$ the equations for starting current and starting developed torque are

$$I_{1st} \approx \frac{kV_{1n}}{\sqrt{(R_1 + R_2')^2 + k^2(X_1 + X_2')^2}} = \frac{V_{1n}}{\sqrt{(R_1 + R_2')^2/k^2 + (X_1 + X_2')^2}}$$

$$T_{elmst} = \frac{m_1 V_{1n}^2}{2\pi n_s} \times \frac{R_2'/k}{[R_1 + R_2'(1 + \tau_1)]^2/k^2 + [X_1 + X_2'(1 + \tau_1)]^2}$$

For $f = 5$ Hz or $k = 0.1$

$$I_{1st5} = \frac{380}{\sqrt{(2.144 + 1.323)^2/0.1^2 + (2.891 + 5.487)^2}} = 10.65 \text{ A}$$

$$T_{elmst5} = \frac{3 \times 380^2}{2\pi \times 25} \times \frac{1.323/0.1}{(2.144 + 1.323 \times 1.0248)^2/0.1^2 + (2.981 + 5.487 \times 1.0248)^2}$$

$$= 28.12 \text{ Nm}$$

For $f = 75$ Hz or $k = 1.5$

$$I_{1st75} = \frac{380}{\sqrt{(2.144 + 1.323)^2/1.5^2 + (2.891 + 5.487)^2}} = 43.72 \text{ A}$$

$$T_{elmst75} = \frac{3 \times 380^2}{2\pi \times 25} \times \frac{1.323/1.5}{(2.144 + 1.323 \times 1.0248)^2/1.5^2 + (2.981 + 5.487 \times 1.0248)^2}$$

$$= 31.21 \text{ Nm}$$

The motor is started at the minimum available frequency 5 Hz. Both the starting current and starting torque are lower than that at $V_{1n} = 380$ V and $f = 50$ Hz. The 5 Hz to 50 Hz starting current and starting torque ratios

$$\frac{I_{1st5}}{I_{1st}} = \frac{10.65}{41.91} = 0.254 \qquad \frac{T_{elmst5}}{T_{elmst}} = \frac{28.12}{43.06} = 0.653$$

(c) The breakdown torque as a function of frequency

According to eqn (6.133)

$$T_{elmmax} = \frac{3}{4\pi \times 25 \times (1 + 0.0248)}$$

$$\times \frac{380^2}{\sqrt{(2.144/k)^2 + [2.891 + 5.487 \times (1 + 0.0248)]^2} + 2.144/k}$$

$$= \frac{1345.549}{\sqrt{4.5967/k^2 + 72.489} + 2.144/k}$$

The breakdown torque for various frequencies at $V_1/f = const$ can be calculated on the basis of the above equation. The results tabulated below show for a small induction motor (7.5 kW) that the breakdown torque T_{elmmax} decreases as the frequency decreases. Assuming $R_1 = 0$ (large motors), the breakdown torque is idependent of frequency and equal to 158 Nm.

Table 6.3. Breakdown torque versus frequency (Example 6.3)

$k =$	5.0	3.0	2.0	1.5	1.2	1.0	0.8	0.6	0.4	0.1	
$f =$	250	150	100	75	60	50	40	30	20	5	Hz
$T_{elmmax} =$	154.1	145.33	139.39	133.72	128.32	123.17	105.07	87.26	55.12	30.23	Nm

6.10 Braking

Simple friction brakes may serve for hoists, lifts and cranes but electromagnetic methods are adopted for more sophisticated drives, especially where precision in braking time or load positioning is called for. However, these electrical methods may impose severe duty cycles and may involve both thermal and mechanical stresses.

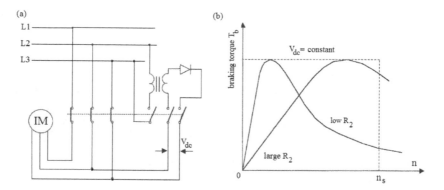

Fig. 6.52. Direct current injection braking: (a) circuit diagram, (b) braking characteristics.

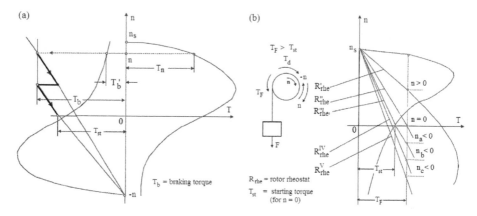

Fig. 6.53. Speed–torque characteristics for plugging obtained: (a) by reversal of two stator terminals, and (b) by lifting a mass, when $T_{st} < T_F$ and the rotor resistance is large.

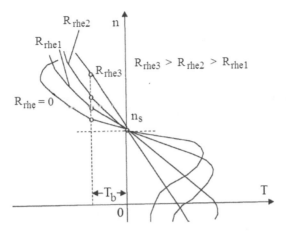

Fig. 6.54. Speed–torque charactersitics of a wound-rotor induction motor for re-generative braking. R_{rhe} – resistance of regeneration braking rheostat.

6.10.1 Direct current injection (dynamic) braking

In this method, the stator winding, immediately after its disconnection from the 3-phase supply, is excited with direct current from a rectifier and (in the case of slip-ring machines) a resistance is introduced into the rotor circuit. The DC excitation establishes a stationary gap flux, the magnitude and position of which depends on the resultant combined MMF of the DC stator and induced rotor currents. A rotor EMF, E_2, proportional to the flux and to the speed, is generated in the rotor winding, the braking effect being the result of the $I_2^2 R_2$ loss. A typical connection diagram is given in Fig. 6.52a. The torque-speed characteristics are shown in Fig. 6.52b.

6.10.2 Plugging

Braking is obtained by reversal of the stator connections while the motor is running, so reversing the direction of the rotating-wave air gap field. The slip s is then greater than unity, and the machine develops a braking (i.e., reversed) torque. The stator and rotor currents are both large. Cage motors up to 20 kW are plugged direct, using the star connection if a $Y-\Delta$ switch is provided. Larger cage motors require stator resistors. Slip-ring motors employ rotor resistance for current limitation. Some speed-torque characteristics obtained for plugging are shown in Fig. 6.53.

6.10.3 Regenerative braking

The motor, overdriven by the load into a hyper-synchronous speed (so that it exhibits negative slip $s < 0$), becomes an induction generator. In a slip-ring machine, braking can be maintained by a moderate rotor external resistance R_{rhe} while useful energy is still returned to the supply. The relevant speed-torque characteristics are shown in Fig. 6.54.

6.11 Connection of a three-phase motor to a single-phase power supply

Typical induction motors with symmetrical three-phase windings can work when connected to a single-phase power supply. To obtain a starting torque, appropriate winding connections with additional elements R, L or C are necessary (see Fig. 6.55). The output power of a three-phase induction motor fed from single-phase mains is *always lower* than that of a motor fed with three-phase mains, and it is equal to between 55% and 70% of the rated power of a three-phase motor.

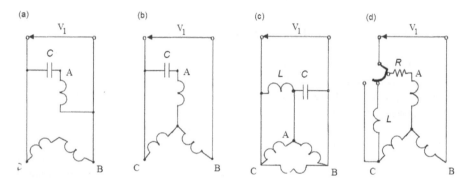

Fig. 6.55. Stator windings connection of a three-phase induction motor fed from single-phase mains.

6.12 Induction motors with copper cage rotor

Until recently, *die cast aluminum cage* rotors have been manufactured because copper pressure die-casting was unproven. Lack of a durable and cost-effective mold material has been the technical barrier preventing manufacture of the *copper cast cage* rotor. The high melting point of copper (1083°C) causes rapid deterioration in dies made of traditional tool steels.

Studies conducted by the International Copper Research Association (IN-CRA) in the 1970s have identified tungsten and molybdenum as good candidate materials for copper casting. They have not found use in the industry largely because of fabrication costs. In the late 1990s, several new materials have been developed for high temperature applications. Two promising candidates for parts of the die caster are the nickel-based superalloys and the beryllium-nickel alloys. None of these materials has the low expansion of tungsten or molybdenum, but they do retain exceptional strength at high temperatures. Inconel 617, 625 and 718 alloys (1260 to 1380°C melting temperature range) are very promising die materials for die casting of copper motor rotors.

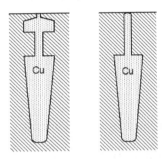

Fig. 6.56. Shapes of copper cage bars.

Table 6.4. Test data and performance of 1.1, 5.5, 11 and 37 kW IMs. Standard efficiency series aluminum rotor models compared to high efficiency copper rotor designs at 400 V, 50 Hz [24].

Rated power, kW	1.1		5.5		11		37	
Rotor conductor	Al	Cu	Al	Cu	Al	Cu	Al	Cu
Rated current, A	2.68	2.45	11	10.9	21.8	21.9	67.1	67.5
Power factor	0.77	0.79	0.83	0.83	0.83	0.81	0.87	0.85
Speed, rpm	1418	1459.5	1424	1455.7	1437	1460	1468	1485
Rated torque, Nm	7.4	7.21	36.9	36.15	73	71.9	240	237.7
Slip	0.055	0.027	0.051	0.0295	0.042	0.0267	0.021	0.01
Input power, W	1435	1334	6485	6276	12590	12330	40700	39900
Stator winding losses, W	192.6	115.1	427.4	372.4	629	521	1044	975
Core losses, W	63.6	51	140.8	101	227	189	749	520
Stray load losses, W	9.5	6.7	100.3	31.4	163	171	699	200
Rotor losses, W	64.1	31.4	299.2	170.4	483	311	837	451
Windage and friction, W	15.9	25	17.5	36	63	56.5	304	203
Efficiency, %	75.9	82.8	84.8	88.12	87.6	89.9	91.1	93.2
Temperature rise, K	61.1	27.8	80.0	61.3	75	62.1	77.0	70.4

The use of copper in place of aluminum for the cage rotor windings of induction motors reduces the rotor winding losses and improves the motor efficiency. The overall manufacturing cost and motor weight can also be reduced.

The higher conductivity of copper increases the skin effect and therefore makes deep bar and double cage effects more pronounced. To increase the bar resistance at high frequency (unity slip at starting) the cross-section of the rotor bar should be as shown in Fig. 6.56.

Comparison of induction motors with aluminum cage rotors and copper cage rotors is shown in Table 6.4 [24].

Nowadays, motors with copper cage rotors are being produced in sizes ranging from 100 W to almost 100 kW.

6.13 Abnormal operating conditions

6.13.1 Increase in voltage, $P_{out} = const$

1. Rotor and stator currents dependent on the load fall almost inversely with voltage — see eqn (6.5);
2. Magnetizing current, flux density, saturation of the magnetic circuit and core losses rise. The temperature rise in the stator core due to the increased core losses and the quickly-rising magnetizing current limit the voltage increase;
3. The starting current rises at about the same rate;
4. The starting and breakdown torques increase as the square of the voltage;
5. At the same output, the power factor decreases due to the increased magnetizing current and the smaller active current;
6. The temperature rise in the rotor windings is reduced. The overall heating effect on the motor depends on whether the effect of the temperature rise in the core or in the windings is predominant. For normal voltage fluctuations it will hardly vary;
7. Efficiency will change little, rising or falling with the change in winding or core losses, respectively;
8. Owing to the lower rotor losses, the speed will rise slightly.

6.13.2 Decrease in voltage $P_{out} = const$

1. Both the stator and rotor currents increase – see eqn (6.5);
2. Magnetizing current, flux density, core losses, and hence the temperature rise in the core, all fall;
3. The starting current drops approximately in proportion to the voltage;
4. The starting and breakdown torques fall as the square of the voltage;
5. The power factor improves;

6. The rotor winding losses and in general the stator losses also increase, and as a rule heating increases;
7. Efficiency is hardly affected;
8. Speed drops off a little.

6.13.3 Change in frequency

A change in frequency results in proportional change in speed. Both the critical slip (6.102) and breakdown torque (6.103) are inversely proportional to the input frequency. The power changes approximately in proportion to the frequency, although at lower frequencies it falls at a greater rate due to the deterioration in cooling.

6.14 Single-phase induction motors

A *single-phase induction motor* is designed in a similar way as a three-phase induction motor. The *main winding* occupies 2/3 of the stator slots (Fig. 5.37a. The *auxiliary starting winding* or *auxiliary operating winding* occupies 1/3 of the stator slots. The rotor has a cage winding, most often die cast aluminum alloy cage winding (Fig. 6.6b). The current in the main stator winding produces pulsating magnetic flux density, i.e.,

$$B(x,t) = B_m \sin(\omega t) \cos\left(\frac{\pi}{\tau}x\right) \tag{6.138}$$

According to trigonometric identity (5.82), the pulsating magnetic field (6.138) can be resolved into two magnetic fields with the same amplitude and rotating in opposite directions, i.e.,

$$F(x,t) = \frac{1}{2}B_m \sin\left(\omega t - \frac{\pi}{\tau}x\right) + \frac{1}{2}B_m \sin\left(\omega t + \frac{\pi}{\tau}x\right) \tag{6.139}$$

The above eqns (6.138) and (6.139) are similar to eqns (5.124) and (5.124) that express the time-space distribution of the MMF. Thus, a single-phase induction motor can be considered as two polyphase induction motors with a common shaft and magnetic fluxes that rotate in opposite directions (Fig. 6.57). The electromagnetic starting torque of such a motor is $T_{elmst} = 0$.

Since a single-phase induction motor does not produce a starting torque, an auxiliary winding is necessary for starting and sometimes also to achieve better performance. The most popular single-phase induction motor with auxiliary starting winding are the following:

(a) Split-phase induction motor;
(b) Capacitor-start induction motor;
(c) Permanent–split capacitor induction motor;
(d) Capacitor–start capacitor-run induction motor;
(e) Shaded–pole induction motor.

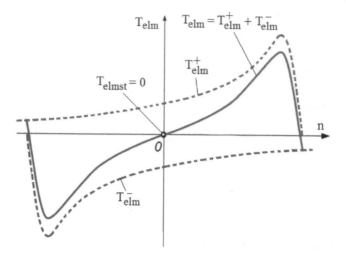

Fig. 6.57. Torque–speed characteristics of a single-phase induction motor. T_{elm}^{+} – electromagnetic torque produced by forward-rotating magnetic field, T_{elm}^{-} – electromagnetic torque produced by backward-rotating magnetic field, $T_{elm} = T_{elm}^{+} + T_{elm}^{-}$ – resultant electromagnetic torque, $Telmst = 0$ – electromagnetic starting torque.

6.14.1 Split-phase induction motor

The stator of a *split-phase induction motor* has main winding and auxiliary winding displaced in space by 90° electrical (Fig. 6.58). The auxiliary winding is wound of thin wire with higher resistance than the main winding. The currents I_{main} in the main winding and I_{aux} in the auxiliary winding lag behind the supply voltage V_1 and the current I_{aux} leads the current I_{main} (Fig. 6.58b). Thus, at starting, the motor becomes a two-phase unbalanced induction motor. The elliptical rotating magnetic field (Fig. 5.36) produces a starting torque and the motor becomes a self-starting motor.

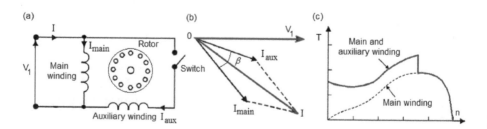

Fig. 6.58. Split-phase single-phase induction motor: (a) connection diagram; (b) phasor diagram; (c) torque-speed characteritics.

After the motor starts, the auxiliary winding is disconnected usually by means of a centrifugal switch that operates at about 75% of the synchronous speed n_s. Under normal operation the motor runs only with the main winding being energized.

6.14.2 Capacitor-start induction motor

The *capacitor-start induction motor* has the starting winding identical to the main winding and a capacitor C connected in series with the starting winding (Fig. 6.59). The capacitance C is chosen in such a way that the current I_{aux} leads the main current I_{main} by approximately 90° electrical to obtain maximum starting torque (Fig. 6.58b). After starting, the auxiliary winding is opened by a centrifugal switch when the motor develops about 75% of the synchronous speed n_s. The motor then accelerates until it reaches the nominal speed and operates only with the main winding connected to the single-phase AC power supply.

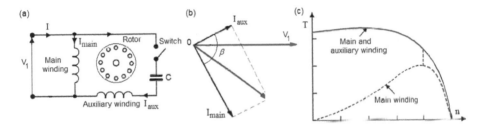

Fig. 6.59. Capacitor-start single-phase induction motor: (a) connection diagram; (b) phasor diagram; (c) torque-speed characteritics.

6.14.3 Permanent–split capacitor induction motor

The *permanent–split capacitor induction motor* has the capacitor C in series with the auxiliary winding all the time (Fig. 6.60a). No centrifugal switch is needed (reduction of cost). The torque (Fig. 6.60b), efficiency and power factor are improved because the induction motor operates as a two-phase motor. Typically, the capacitance $C = 20\ldots50\ \mu\text{F}$.

6.14.4 Capacitor–start capacitor-run induction motor

There are two capacitors in the *capacitor–start capacitor-run induction motor*, both in series with the auxiliary winding (Fig. 6.61a). The *starting capacitor* C_{st} is disconnected usually with the aid of a centrifugal switch when the motor attains a speed of 75 to 80% of the synchronous speed n_s. The *running*

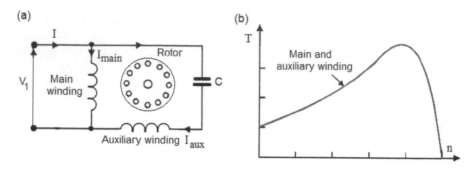

Fig. 6.60. Permanent–split capacitor single-phase induction motor: (a) connection diagram; (b) torque-speed characteritic.

capacitor $C_r < C_{st}$ is connected permanently. Theoretically, with two capacitors, optimum starting and running performance can be achieved (Fig. 6.61b) because the phase angle difference between currents in the main and auxiliary windings is almost 90°. Typical values of these capacitors for a 375-W motor are $C_s = 300$ μF and $C_r = 40$ μ F.

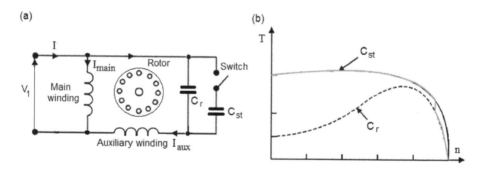

Fig. 6.61. Capacitor–start capacitor-run single-phase induction motor: (a) connection diagram; (b) torque-speed characteritic.

6.14.5 Shaded–pole induction motor

The stator of the *shaded–pole single-phase induction motor* has salient poles divided into two unequal halves (Fig. 6.62). The smaller portion carries the shorted copper coil and is called as shaded portion of the pole.

When a single-phase supply is given to the stator of shaded–pole induction motor an alternating flux is produced. This change of flux induces EMF in the shaded coil. Since this shaded portion is short circuited, the current is produced in it in such a direction to oppose the main flux. The flux in

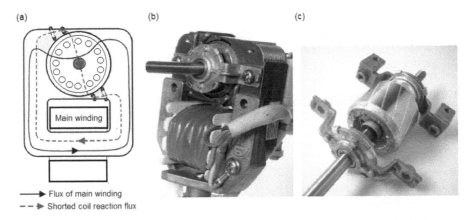

(a)   (b)   (c)

Main winding

→ Flux of main winding
- - → Shorted coil reaction flux

Fig. 6.62. Shaded–pole single-phase induction motor: (a) magnetic fluxes in a 2-pole rotor; (b) assembled motor; (c) cage rotor.

shaded portion of the pole lags behind the flux in unshaded portion of the pole. The phase difference between these two fluxes produces resultant rotating magnetic field and electromagnetic torque. The rotor is made as a die cast aluminum alloy cage rotor (Fig. 6.62c). The efficiency of a shaded-pole induction motor is very small and usually does not exceed 10%. The main advantage of shaded–pole motors is very low manufacturing cost. These motors are in mass production for portable cooling fans, portable fan heaters and domestic appliances.

Summary

There are three-types of induction machines:

(a) Cage-rotor induction machine (Fig. 6.1);
(b) Wound-rotor (slip-ring) induction machine (Fig. 6.2);
(c) Solid-rotor induction machine (Fig. 6.3).

The *stator of induction machines* consists of a laminated core (Fig. 6.4) and three-phase winding (Fig. 6.5) embedded in slots. This winding, when energized by a three-phase source of AC power, provides a rotating magnetic field (5.8).

The rotor windings in most induction machines are also contained in slots in a laminated core which is mounted on the shaft (Figs 6.1 and 6.2). The only exception is a solid rotor induction machine, in which the rotor winding has a form of high-electric conductivity sleeve placed on solid steel cylinder (Fig. 6.3). The winding embedded in slots may have a form of a cage winding that consists of axial bars and end rings or a three-phase winding wound of copper wire the terminals of which are connected to three slip rings. In wound-rotor induction motors a three-phase external resistance (rheostat) is used to reduce the inrush current at starting and/or control the speed. The cage winding is a shorted winding and usually made of aluminum alloy, brass, bronze or copper.

The electromagnetic torque is created as a result of interaction of the stator rotating magnetic field on rotor currents induced by the stator (Fig. 6.9).

The name plate (Fig. 6.8) of an induction motor provides the following information:

- Serial number;
- Rated (nominal power);
- Class of insulation;
- Year of manufacturing;
- Number of phases;
- Frequency;
- Stator line-to-line voltage;
- Stator current;
- Stator winding conenction (Y or *Delta*);
- Power factor $\cos \varphi$;
- Internal protection IP;
- Duty cycle;
- Rated (nominal speed);
- Ambient temperature of operation;
- Mass;
- Standard.

The *slip* s (6.1) is the ratio of the slip speed $(n_s - n)$ to the synchronous speed n_s of the rotating magnetic field, i.e.,

$$s = \frac{n_s - n}{n_s} = 1 - \frac{n}{n_s}$$

where $n_s = f/p$ is the synchronous speed (speed of the stator magnetic rotating field), n is the rotor speed, f is the frequency of the stator current and magnetic flux and p is the number of pole pairs.

The input power (6.5) is expressed as

$$P_{in} = m_1 V_1 I_1 \cos \phi$$

where m_1 is the number of phases, V_1 is the *rms* phase voltage, I_1 is the *rms* phase current and $\cos \varphi$ is the power factor.

The electromagnetic power P_{elm} crossing the air gap (6.9) is

$$P_{elm} = P_{in} - \Delta P_{1w} - \Delta P_{1Fe} = \frac{\Delta P_{2w}}{s}$$

where the stator winding losses are $P_{1w} = m_1 I_1^2 R_1$, the rotor winding losses are $\Delta P_{2w} = m_2 I_2^2 R_2$, and the stator core losses ΔP_{1Fe} consist of hysteresis losses, eddy current losses, excess losses and additional losses in the ferromagnetic core due to metallurgic and manufacturing processes (Section 1.3). The electromagnetic torque T_{elm} (6.12) is defined as the electromagnetic power P_{elm} crossing the air gap divided by the angular synchronous frequency $\Omega_s = 2\pi n_s$, i.e.,

$$T_{elm} = \frac{P_{elm}}{\Omega_s} = \frac{P_{elm}}{2\pi n_s}$$

The developed mechanical power P_m is electromagnetic (air gap) power (6.50) less than rotor losses, i.e.,

$$P_m = P_{elm} - \Delta P_{2w} - \Delta P_{2Fe} \approx -\Delta P_{2w} = (1-s)P_{elm} = \frac{1-s}{s}\Delta P_{2w}$$

where $\Delta P_{2w} = m_2 I_2^2 R_2$ are losses in the rotor winding, m_2 is the number of rotor phases, I_2 is the rotor current and R_2 is the rotor winding resistance. The core losses in the rotor core ΔP_{2Fe} can be neglected because the frequency of the rotor flux sf is very small.

The slip can also be defined on the basis of the electromagnetic P_{elm} and mechanical P_m power (6.54), i.e.,

$$s = \frac{P_{elm} - P_m}{P_{elm}} = \frac{2\pi n_s T_{elm} - 2\pi n T_{elm}}{2\pi n_s T_{elm}} = \frac{n_s - n}{n_s}$$

The *rms* EMF or voltage induced per phase (6.22) in the stator winding is given by the equation

$$E_1 = 4\sigma_f f N_1 k_{w1} \Phi$$

where the magnetic flux (6.29)

$$\Phi = \frac{2}{\pi}\tau L_i B_{mg}$$

In the above equations σ_f is the form factor of the EMF (i.e., the ratio of its *rms* value to average value), f is the input frequency, N_1 is the number of stator turns per phase, k_{w1} is the stator winding factor (Section 5.4), α_i is the ratio of the average to the peak value of the air gap magnetic flux density, $\tau = D/(2p)$ is the pole pitch, L_i is the effective (ideal) length of the stator core and B_{mg} is the peak value of the air gap magnetic flux density. For sinusoids

$$\sigma_f = \pi\sqrt{2}/4 \approx 1.11 \qquad\qquad \alpha_i = \frac{2}{\pi} \approx 0.637$$

The slip-dependent *rms* EMF induced in the rotor phase winding is given by eqn (6.31), i.e.,

$$E_2(s) = 4\sigma_f sf N_2 k_{w2} \Phi = sE_{20} \qquad\qquad E_{20} = 4\sigma_f f N_2 k_{w2} \Phi$$

where N_2 is the number of rotor turns per phase and k_{w2} is the rotor winding factor for fundamental harmonic. For a cage winding $N_2 = 0.5$ and $k_{w2} = 1$.

The rotor parameters are referred to the stator winding as follows:

- EMF (6.35)

$$sE'_{2o} = sE_{20} \frac{N_1 k_{w1}}{N_2 k_{w2}} = sE_1$$

  where $N_1 k_{w1}/N_2 k_{w2}$ is the turns ratio.
- Current (6.38)

$$I'_2 = \frac{m_2 N_2 k_{w2}}{m_1 N_1 k_{w1}} I_2$$

- Resistance (6.44)

$$R'_2 = \frac{m_1 (N_1 k_{w1})^2}{m_2 (N_2 k_{w2})^2} R_2$$

- Reactance (6.45)

$$X'_2 = \frac{m_1 (N_1 k_{w1})^2}{m_2 (N_2 k_{w2})^2} X_2$$

where E_{20}, I_2, R_2 and X_2 are the rotor EMF for $s = 1$, rotor current, rotor resistance and rotor reactance, respectively. For a cage winding the number of rotor phases m_2 is equal to the number of rotor slots s_2, i.e., $m_2 = s_2$.

The power (energy) balance diagram called Sankey's diagram is given in Fig. 6.11, while power flow diagram for induction motor is given in Fig. 6.12.

The following power losses are in induction motor:

- Stator winding losses $\Delta P_{1w} = m_1 I_1^2 R_1$;
- Stator core losses ΔP_{1w} (please see Section 1.3);
- Rotor winding losses $\Delta P_{2Fe} = m_2 I_2^2 R_2 = m_1 (I'_2)^2 R'_2$
- Rotor core losses, for fundamental harmonic $\Delta P_{2Fe} \approx 0$;
- Rotational (mechanical) losses $\Delta P_{rot} = \Delta P_{fr} + \Delta P_{wind} + \Delta P_{vent}$, i.e., windage, friction in bearing and ventilation fan losses;
- Stray load losses ΔP_{str} due to higher harmonics equal to about 0.5% of the input power according to IEC.

The efficiency of induction machine is the ratio of the output P_{out} to input power P_{in} (6.57), i.e.,

$$\eta = \frac{P_{out}}{P_{in}} = \frac{P_{out}}{P_{out} + \sum \Delta P}$$

where $\sum \Delta P$ is the sum of all losses. For medium and large-power induction motors (6.57)

$$\eta = \frac{P_{out}}{P_{in}} \approx \frac{P_m}{P_{elm}} = \frac{(1-s)P_{elm}}{P_{elm}} = 1 - s$$

The output power (6.58)

$$P_{out} = \eta P_{in} = m_1 V_1 I_1 \eta \cos \varphi$$

where $\cos \varphi$ is the power factor, φ is the angle between the current I_1 and voltage V_1.

The following equations are the basis for the construction of the equivalent circuit (Figs 6.15, 6.16) and phasor diagram (Fig. 6.18)

- Stator current (6.60)

$$\mathbf{I}_1 = \mathbf{I}_0 + \mathbf{I}_2'$$

- Exciting current in vertical branch of the equivalent circuit (6.61)

$$\mathbf{I}_0 = I_{Fe} + jI_\Phi$$

where I_{Fe} is the core loss current (the active component of I_0) and I_Φ is the magnetizing current (the reactive component of I_0).
- Input phase voltage (6.66)

$$\mathbf{V}_1 = \mathbf{I}_1(R_1 + jX_1) + \mathbf{E}_1$$

- Stator EMF (6.67)

$$\mathbf{E}_1 = \mathbf{E}_2' = \mathbf{I}_2' \left(\frac{R_2'}{s} + jX_2' \right)$$

- Rotor EMF E_2' referred to the stator winding (6.68)

$$\mathbf{E}_2' = \mathbf{I}_2'(R_2' + jX_2') + \mathbf{I}_2'\frac{1-s}{s}R_2'$$

- Slip-dependent rotor resistance referred to the stator winding (6.69)

$$\frac{R_2'}{s} = R_2' + R_2'\frac{1-s}{s}$$

The last equation corresponds to the following power balance equation of the rotor

$$P_{elm} = \Delta P_{2w} + P_m$$

because $P_{elm} = \Delta P_{2w}/s$, $P_m = \Delta P_{2w}(1-s)/s$ and $\Delta P_{2w} = m_2 I_2^2 R_2$.

The *resistances and reactances* of the equivalent circuit can be determined from the results of a *no-load test* and *locked-rotor tests*. The connection diagrams for the no-load and locked rotor tests on induction motors are shown in Fig. 6.19.

The *no-load test* on an induction motor is similar to the open-circuit test on a transformer. In this test the motor runs without any load. The input voltage V_1, input phase current I_{10}, input power P_{ino}, no-load speed n_0, stator winding resistance per phase R_1 and rotor winding resistance R_2 per phase (only for a slip ring motor) are measured. The rotor branch in the equivalent circuit of an induction motor, as shown in Fig. 6.22a, can be neglected. The no load parameters can be found from eqns (6.71) to (6.75).

The no-load current (6.76) of induction motors

$$\Delta i_{0\%} = \frac{I_{10}}{I_{1n}} \times 100\%$$

is in the range of 25 to 60% of the nominal (rated current) I_{1n}. The higher the power and lower the number of poles, the lower the no-load current.

The *locked rotor test* on an induction motor corresponds to the short-circuit test on a transformer. In this test the rotor is blocked ($n = 0$, $s = 1$) and a reduced voltage V_{1sh} is applied to the motor to obtain the nominal (rated) current I_{1n} in the stator winding. The reduced input voltage V_{1sh}, input current I_{1n} and input power P_{insh} are measured (Fig. 6.19). The vertical branch in the equivalent circuit (Fig. 6.22b) can be neglected because for $s = 1$ $R_2' \frac{1-s}{s} = 0$ and $R_2' + jX_2' << R_0 + jX_0$. The locked-rotor parameters can be found from eqns (6.78) to (6.85).

The *electromagnetic torque calculated on the basis of the equivalent circuit* is given by eqn (6.99), i.e.,

$$T_{elm} = \frac{m_1}{2\pi n_s} \frac{V_1^2(R_2'/s)}{[R_1 + (R_2'/s)(1+\tau_1)]^2 + [X_1 + X_2'(1+\tau_1)]^2}$$

where $\tau_1 \approx X_1/X_2$ is the so-called *Heyland's coefficient* for the stator. The electromagnetic torque is proportional to the voltage square V_1^2.

The *critical slip* that corresponds to the maximum torque is given by eqns (6.101) and (6.102), i.e.,

$$s_{cr} = \pm \frac{R_2'(1+\tau_1)}{\sqrt{R_1^2 + [X_1 + X_2'(1+\tau_1)]^2}} \approx \pm \frac{R_2'}{X_1 + X_2'}$$

The "+" sign signifies a machine in motor mode and the "−" sign denotes generator mode.

The *maximum (breakdown) torque* corresponding to the critical slip s_{cr} is given by eqn (6.103), i.e.,

$$T_{elmmax} = T_{elm}(s = s_{cr}) = \pm \frac{m_1 V_1^2}{4\pi n_s (1 + \tau_1)} \frac{1}{\left\{ \sqrt{R_1^2 + [X_1 + X_2'(1 + \tau_1)]^2} \pm R_1 \right\}}$$

$$\approx \pm \frac{m_1 V_1^2}{4\pi n_s (1 + \tau_1)} \frac{1}{X_1 + X_2'(1 + \tau_1)}$$

In the denominator, '$+R_1$' is for $s_{cr} > 0$ (signifying a motor or brake) and '$-R_1$' is for $s_{cr} < 0$ (denoting a generator). The absolute value of maximum torque is slightly higher for the generator mode than for the motor mode, if the stator winding resistance is taken into account.

The *starting torque* is for the slip value $s = 1$ ($n = 0$) and is given by eqn (6.104), i.e.,

$$T_{elmst} = T_{elm}(s = 1) = \frac{m_1 V_1^2}{2\pi n_s} \frac{R_2'}{[R_1 + R_2'(1 + \tau_1)]^2 + [X_1 + X_2'(1 + \tau_1)]^2}$$

At constant voltage V_1, the value of the starting torque mainly depends on the resistance R_2' and leakage reactance X_2' of the rotor.

The ratio of any electromagnetic torque T_{elm} to breakdown torque T_{elmmax} is expressed by *Kloss' formula* (6.108), i.e.,

$$\frac{T_{elm}}{T_{elmmax}} \approx \frac{2}{s_{cr}/s + s/s_{cr}}$$

The *torque-speed characteristic* and *torque-slip characteristic* of an induction machine are plotted in Figs 6.25 and 6.26. There are five modes of operation of induction machines, i.e.,

- Induction generator ($s < 0$),
- Synchronous machine ($s = 0$),
- Induction motor ($0 < s < 1$),
- Transformer ($s = 1$),
- Brake ($s > 1$)

The behavior of an induction motor can be characterized by the following three parameters:

- Overload capacity factor (6.105)

$$OCF = \frac{T_{elmmax}}{T_{dn}}$$

- Starting torque ratio (6.106)

$$STR = \frac{T_{elmst}}{T_{elmn}}$$

- Starting current ratio (6.107)

$$SCR = \frac{I_{1st}}{I_{1n}}$$

where T_{elmmax} is the maximum electromagnetic torque, T_{elmn} is the nominal electromagnetic torque, T_{elmst} is the starting electromagnetic torque, I_{1st} is the starting current and I_{1n} is the nominal stator current. The values of OCF, STR and SCR for various designs of induction motors are given in Table 6.2.

The influence of the *input voltage on torque-slip characteristic* is shown in Fig. 6.27.

The influence of the *rotor resistance R'_2 on the torque-slip characteristics* is plotted in Fig. 6.28.

The load characteristics of induction motor, i.e., speed n, effciency η, power factor $\cos\varphi$, input power P_{in} and output power P_{out} as functions of the load torque T at $V_1 = V_{1n} = const$ are plotted in Fig. 6.29.

An *induction motor is not a self-starting motor*. Wound-rotor (slip-ring) induction machines are started with the aid of additional rotor resistance. Cage-rotor induction motors are started using one of the following methods:

- Direct on-line starting (only for small motors subject to supply authority regulations);
- Star-delta switching;
- Stator impedance starting (variable reactor);
- Autotransformer starting;
- Solid state soft starters.

Induction motors that use skin effect in the rotor winding are

- Deep-bar rotor;
- Double-cage rotor.

Skin effect causes non-uniform distribution of current in the conductor. Current density in the rotor bar is higher closer to the air gap. Skin effect is used to reduce the starting current and increase the starting torque.

Eqn (6.3), i.e., $n = (f/p)(1 - s)$ provides the basis for several methods of *speed control* for induction motors, which can be grouped under the headings of changing:

- The input frequency f using solid state converters (Figs 6.46 and 6.47);
- The number of stator pole pairs p using one winding with changeable poles (Fig. 6.49) or two windings;
- The slip s, i.e., by adjusting the rotor circuit resistance of wound-rotor induction motors or the input voltage of cage induction motors.

Inverter-fed induction motor characteristics and capabilities are shown in Fig. 6.51. In inverter-fed induction motors the speed can be controlled by varying: (a) input frequency f; (b) input voltage V_1; (c) both input frequency and voltage (VVVF), keeping the air gap magnetic flux constant (6.135), i.e.,

$$\pi\sqrt{2}N_1 k_{w1}\Phi \approx \frac{V_1}{f} = const$$

There are, in general, three methods of *electromagnetic braking* of induction motors:

- Direct current injection (dynamic) braking (stator excited with DC current);
- Plugging (reversal of the stator connections while the motor is running);
- Regenerative braking (overridden motor at negative slip becomes an induction generator).

A *three-phase induction motor can be connected to a single-phase power supply* with the aid of additional elements R, L or C (Fig. 6.55).

Traditionally, die cast aluminum cage rotors have been manufactured because copper pressure die-casting was unproven. In the 1970s tungsten and molybdenum were identified as good candidate materials for *copper casting*. The *use of copper in place of aluminum* for the cage rotor windings of induction motors reduces the rotor winding losses and improves the motor efficiency. The overall manufacturing cost and motor weight can also be reduced. The higher conductivity of copper increases the skin effect and therefore makes deep bar and double cage effects more pronounced. Comparison of induction motors with aluminum cage rotors and copper cage rotors is shown in Table 6.4.

When the *input voltage of an induction motor increases* at $P_{out} = const$, the rotor and stator currents dependent on the load fall almost inversely with voltage, while the magnetizing current, core losses, stator temperature, starting current, starting torque and breakdown torque increase.

When the *input voltage of an induction motor decreases* at $P_{out} = const$, both the stator and rotor currents increase, while the magnetizing current, core losses, temperature, starting current, starting torque and breaking torque decrease.

A *change in input frequency* of an induction motor results in proportional change in speed. Both the critical slip (6.101), (6.102) and breakdown torque (6.103) are inversely proportional to the input frequency f. The power changes approximately in proportion to the frequency, although at lower frequencies it falls at a greater rate due to the deterioration in cooling.

A *single-phase induction motor* is designed in a similar way as a three-phase induction motor. The *main winding* occupies 2/3 of the stator slots (Fig. 5.37a). The *auxiliary starting or auxiliary operating winding* occupies 1/3 of the stator slots. The rotor has a cage winding, most often die cast aluminum alloy cage winding (Fig. 6.6b). The current in the main stator winding produces pulsating magnetic field.

Since a single-phase induction motor does not produce a starting torque, an auxiliary winding is necessary for starting and sometimes also for better performance. The most popular single-phase induction motor with auxiliary starting winding are the following:

(a) Split-phase induction motor (Fig. 6.58);
(b) Capacitor-start induction motor (Fig. 6.59);
(c) Permanent–split capacitor induction motor (Fig. 6.60);
(d) Capacitor=-start capacitor-run induction motor (Fig. 6.61);
(e) Shaded-pole induction motor (Fig. 6.62).

Problems

1. The following data is read on the name plate of a three-phase cage induction motor: output power $P_{out} = 3$ kW, input frequency $f = 50$ Hz, speed $n = 2800$ rpm, voltage $V_1 = 380$ V (line-to-line), efficiency $\eta = 80\%$, power factor $\cos\phi = 0.92$. If the stator windings are Y-connected, find: (a) the number of poles $2p$, and the slip s, (b) the input current I_1, and input power P_{in}, (c) the mechanical power P_m and electromagnetic power P_{elm}, given rotational losses $\Delta P_{rot} = 120$ W, (d) the shaft torque T and the electromagnetic torque T_{elm}.

   Answer: (a) $2p = 2$, $s = 0.0666$, (b) $I_1 = 6.18$ A, $P_{in} = 3750$ W, (c) $P_m = 3120$ W, $P_{elm} = 3343$ W, (d) $T = 10.23$ Nm, $T_{elm} = 10.64$ Nm.

2. An induction motor, operating from a 60-Hz line, develops the output (shaft) power $P_{out} = 7.5$ kW at 1745 rev/min. At what speed will it run if the load torque is reduced to one-half ? What will be the output power at half torque ?

   Answer: $n = 1772.5$ rpm, $P_{out} \approx 3.8$ kW

3. A three-phase, six-pole (2p=6), 3000 V, 250 kW, 50 Hz induction motor has $s_1 = 90$ stator slots, and $s_2 = 80$ rotor slots. The number of stator turns per phase is $N_1 = 210$, and each coil has a span of 12 slots. The squirrel-cage rotor resistance is $R_2 = 13.6 \times 10^{-5}$ Ω and the leakage reactance is $X_2 = 51.0 \times 10^{-5}$ Ω. Find the rotor impedance referred to as the stator winding for $s = 1$.

   Answer: $\mathbf{Z}_2' = (0.75 + j2.8)\Omega$

4. Determine the stator rated current I_1, voltage drop across the stator winding impedance for rated operating conditions, blocked-rotor current I_2' and voltage drop across the rotor impedance assuming $s = 1$ for the motor of Problem 3 ($P_{out} = 250$ kW, $V_{1L} = 3000$ V, $m_1 = 3$, $f = 50$ Hz, Y

connection, $R_1 = 0.72$ Ω, $R'_2 = 0.75$ Ω, $X_1 = 2.78$ Ω, $X'_2 = 2.8$ Ω, $\cos \phi = 0.9$ lagging, $\eta = 0.89$).

Answer: $I_1 = 60$ A, $I_1 Z_1 = 172.3$ V, $I'_2 \approx 561$ A at $s = 1$, $I'_2 Z'_2 = 1626.3$ V at $s = 1$

5. A four-pole induction motor draws 25 A from a 460 V (line-to-line), 50 Hz, three-phase line at a power factor of 0.85, lagging. The stator winding loss is $\Delta P_{1w} = 1000$ W, and the rotor winding loss is $\Delta P_{2w} = 500$ W. The rotational losses are $\Delta P_{rot} = 250$ W, core loss $\Delta P_{Fe} = 800$ W, and stray load loss $\Delta P_{str} = 200$ W. Calculate:
   (a) The electromagnetic (air gap) power, P_{elm};
   (b) The mechanical power, P_m;
   (c) The output power, P_{out};
   (d) The efficiency, η;
   (e) The slip, s, and operating speed, n;
   (f) The electromagnetic torque, T_{elm};
   (g) The shaft (output) torque, T.

   Answer: (a) $P_{elm} = 15,131$ W, (b) $P_m = 14,631$ W, (c) $P_{out} = 14,181$ W, (d) $\eta = 0.838$, (e) $s = 0.033$, $n \approx 1450$ rpm, (f) $T_{elm} = 96.3$ Nm, (g) $T = 93.4$ Nm.

6. No-load and locked-rotor tests have been performed on a three-phase, 500 kW, 50 Hz, Y-connected, 6 kV (line-to-line), 57 A, 980 rpm cage induction motor, with the following results:

   No-load test: input frequency $f = 50$ Hz, input voltage (line-to-line) $V_{10L} = 6$ kV, no-load current $I_{10} \approx I_{1exc} = 17$ A, no-load losses $\Delta P_0 = 14$ kW, rotational losses $\Delta P_{rot} = 3.5$ kW.

   Locked-rotor test (s=1): input frequency $f = 50$ Hz, input voltage (line-to-line) $V_{1shL} = 380$ V, line current $I_{1sh} = 15$ A, input active power $P_{insh} = 1$ kW, stator winding resistance per phase $R_1 = 0.8$ Ω.

   Find the equivalent circuit resistances and reactances.

   Answer: $X_1 \approx X'_2 = 7.3$ Ω, $R_{Fe} = 3670$ Ω, $X_m = 205$ Ω.

7. Find the overload capacity factor and starting torque ratio for the following cage induction motor: $m_1 = 3$, $P_{out} = 250$ kW, $V_{1L} = 3000$ V, $f = 50$ Hz, Y connection, $2p = 6$, $n = 972$ rev/min, $R_1 = 0.72$ Ω, $R'_2 = 0.75$ Ω, $X_1 = 2.78$ Ω, $X'_2 = 2.8$ Ω.

Answer: $OCF \approx 2.3$, $STR = 0.662$.

8. A three-phase, four-pole, 210-kW, 50-Hz, 500-V (line-to-line), Y-connected cage induction motor is operating at its rated nominal of $n_n = 1485$ rpm, rated voltage, and rated frequency. The critical slip corresponding to the breakdown torque is $s_{cr} = 0.044$. Find: (a) the starting torque T_{st} and (b) the starting current ratio SCR.

   Assumption: The no-load (exciting) current, Heyland's coefficient, and the stator winding resistance have all been neglected.

   Answer: $T_{st} = 274.13$ Nm, $SCR = 4.5$.

9. A three-phase, 50-Hz, 75-kW, 380-V (line-to-line) Y-connected, 1460-rpm, wound-rotor induction motor is operating at rated speed, rated voltage, and rated frequency with a three-phase rotor rheostat $R_{rhe} = 0.5$ Ω. The critical slip for $R_{rhe} = 0$ is $s_{cr} = 0.13$ and the rotor resistance is $R_2 = 0.05$ Ω. Find: (a) the starting torque T_{st} and (b) the starting current ratio SCR.

   Assumption: The no-load (exciting) current, Heyland's coefficient, and stator winding resistance are neglected.

   Answer: $T_{st} = 1170$ Nm, $SCR = 2.85$.

10. A three-phase, six-pole, 50-Hz, 2.8-kW, 220-V, Δ-connected, 950-rpm, cage induction motor has the overload capacity factor $OCF = 1.9$. This motor is connected to a three-phase, 50 Hz, 220 V line using a $Y - \Delta$ switch. The external shaft torque is $T = 0.15T_n$ where T_n is the nominal torque. Find: (a) the starting current I_{1stY} and starting torque T_{stY} when the stator windings are Y-connected, (b) the steady-state speed n_Y and current I_{1Y} when the stator windings are Y-connected and (c) the electromagnetic torque $T^*_{elm\Delta}$ and line current $I^*_{1L\Delta}$ when the speed $n^*_\Delta = n_Y$ and the stator windings are Δ-connected.

   Assumption: The no-load (exciting) current, Heyland's coefficient, and stator winding resistance are neglected.

   Answer: (a) $I_{1stY} = 9.65$ A, $T_{stY} = 6.08$ Nm, (b) $n_Y = 978.1$ rpm, $I_{1Y} = 1.17$ A, (c) $I^*_{1L\Delta} = 3.5$ A, $T^*_{elm\Delta} = 12.67$ Nm.

11. A three-phase, 10-kW, 380-V (line-to-line), 1450-rpm cage induction motor is fed from a three-phase AC voltage regulator. Both the motor and regulator are Δ-connected and the motor current is sinusoidal. At nominal (rated) load the efficiency is $\eta = 0.82$ and the power factor is $\cos \phi = 0.86$.

Find: (a) the *rms* current rating of the thyristor, (b) the peak voltage rating of the thyristor, (c) the firing angle α to obtain sinusoidal motor current.

Answer: (a) 8.8 A, (b) 537.4 V, (c) $\alpha = 30.68°$.

12. A three-phase, 380-V (line-to-line), 50-Hz, 1450-rpm Y-connected cage induction motor is fed from a VVVF three-phase inverter. The maximum input frequency cannot exceed 50 Hz and the maximum to minimum required speed ratio is 10:1. The inverter DC input voltage is supplied from a fully controlled three-phase rectifier with three 380 V input. Find the maximum and minimum DC link voltage and corresponding firing angles of the rectifier.

Answer: 465.4 V DC, $\alpha = 24.9^0$; 46.5 V DC, $\alpha = 84.8°$.

13. A three-phase, 3.0-kW, 380-V, 710-rpm, 50-Hz, Δ-connected cage induction motor has the following equivalent circuit parameters: $R_1 = 4.9$ Ω, $R_2' = 0.27$ Ω, $X_1 = 14.0$ Ω, $X_2' = 0.66$ Ω, $X_m = 30.0$ Ω. The motor is controlled by a VSI at $V_1/f = const$. The inverter output frequency varies from 10 to 100 Hz. Find: (a) the starting current and starting electromagnetic torque at rated frequency, (b) the starting current and starting electromagnetic torque at minimum and maximum frequency and (c) the breakdown torque as a function of frequency.

Answer: (a) $I_{1st} = 24.45$ A, $T_{elmst} - 5.9$ Nm; (b) $I_{1st10} = 12.8$ A, $T_{elmst10} = 8.05$ Nm, $I_{1st100} = 25.53$ A, $T_{elmst100} = 3.22$ Nm; (c) $k = 2$, $f = 100$ Hz, $T_{elmmax} = 156.55$ Nm; $k = f/f_n = 1$, $f = 50$ Hz, $T_{elmmax} = 133.56$ Nm; $k = f/f_n = 0.2$, $f = 10$ Hz, $T_{elmmax} = 51.83$ Nm.

7

SYNCHRONOUS MACHINES

7.1 Construction

A *synchronous machine* operates at a constant speed in absolute synchronism with the line frequency. This means that the rotor speed is the same as that of the rotating magnetic field excited by the stator (or armature) AC winding. The magnetic flux of the armature rotates synchronously with the field excitation flux, i.e.,

$$n_s = \frac{f}{p} = n \tag{7.1}$$

where n_s is the speed of the armature field called the "synchronous speed" and n is the speed of the field excitation flux, in practice, the mechanical speed of the rotor.

There is no essential difference between the stators of polyphase synchronous and induction machines of comparable rating. The stator (armature) is made of stacked-up electrotechnical steel laminations, and the stator slots accommodate a three-phase distributed winding that sets up a rotating magnetic field.

The prinicple of operation of a synchronous machine at no load, in motoring mode and in generating mode is explained in Fig. 7.1

There are two types of rotors of wound-field synchronous machines: *non-salient pole rotor* (Fig. 7.2a) also called *cylindrical rotor* and *salient-pole rotor* (Fig. 7.2b). Non-salient pole rotor (Fig. 7.2a), usually two-pole ($2p = 2$) or four-pole ($2p = 4$) is made of solid steel. Salient pole core and pole shoes (or pole faces) are laminated (Fig. 7.2b) and mounted on a special "spider" construction, not shown in Fig. 7.2b. Salient pole shoes are equippped with a cage winding, the so-called *damper* or *amortisseur* placed near the surface of the pole shoes. In generators, this winding is necessary to dump oscillation and backward component of the armature magnetic rotating field. In motors, this winding can serve as a cage winding for asynchronous starting.

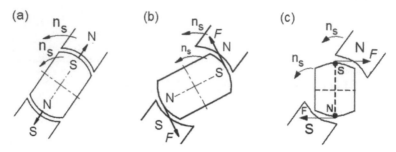

Fig. 7.1. Principle of operation of a synchronous machine: (a) no-load; (b) machine loaded with breaking torque (operation as a motor); (c) machine driven with external torque (operation as a generator).

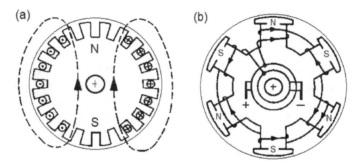

Fig. 7.2. Rotor of wound-field synchronous machines: (a) non-salient pole rotor ($2p = 2$); (b) salient pole machine ($2p = 6$).

7.2 Classification of synchronous machines

Synchronous machines can be classified as follows:

(a) Turboalternators (turbogenerators). Driven by steam turbines at speed from 1500 to 3600 rpm and with power ratings up to 1300 MVA ($2p = 2$) or 1700 MVA ($2p = 4$, nuclear power stations).

(b) Hydroalternators. Driven by hydro-power turbines at speeds from 60 to 1000 rpm and in power ratings up to 824 MVA (Itaipu Power Station, Brazil-Paraguay).

(c) Combustion engine-driven generators. Various forms of prime-mover of the internal combustion types with Diesel engines being predominant. Speed up to 3600 rpm, power rating up to 200 MVA.

(d) Gas turbine driven-generators. Speed in the range from a few thousand up to several thousand rpm, power ratings up to 500 MVA.

(e) Microturbines. Generators integrated with gas turbines. Output power up to 200 MW, speed up to 120,000 rpm.

(f) Wind turbine generators. Driven by wind turbines at very low speeds from 12 to 30 rpm, maximum power 5 MW (trends toward increasing power up to 10 MW).

(g) Synchronous motors. Wound field, PM, reluctance and hysteresis motors. The largest wound-field synchronous motor for ship propulsion is rated at 44 MW, 144 rpm (aboard *Queen Elizabeth 2*). Larger wound field synchronous motors rated up to 80 MW, 3600 rpm are built for liquefied natural gas (LNG) plant compressors. The largest PM brushless motor for ship propulsion is rated at 36.5 MW, 127 rpm (DRS Technologies, Parsippany, NJ, USA).

(h) Synchronous condensers. Self-driven synchronous machines consuming real power and delivering capacitive reactive power. Speed 3000 to 3600 rpm, power up to 500 MVAR.

7.2.1 Turboalternators

Synchronous generators driven by steam turbines are non-salient pole machines, also called cylindrical rotor machines (Fig. 7.2a). The power density (output power–to–mass) is 0.4 to 7.0 kVA/kg. The turboalternator is shown in Fig. 7.3 and the rotor of a turboalternator is shown in Fig. 7.4.

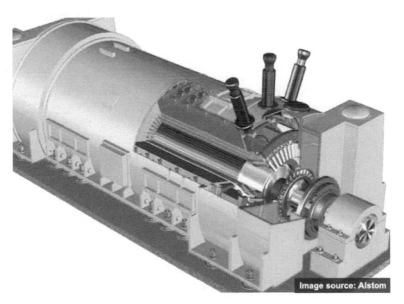

Fig. 7.3. Computer image of a large turboalternator with non-salient pole rotor.

In power-frequency two-pole machines the linear speed at the rotor surface is from 200 to 250 m/s. Centrifugal forces at these high speeds cause high

Fig. 7.4. Non-salient pole rotor of a large turboalternator. The end turns are protected against centrifugal and electrodynamic forces with the aid of sleeves made of nonferromagnetic material.

mechanical radial stresses on rotor parts. Therefore, rotor cores are manufactured as solid steel cylinders. The outside diameter of the rotor of a 2-pole alternator is limited to $D = 1.33$ m, i.e.,

$$v = \pi D n_s = \pi \times 1.33 \times (50\ldots60) = 209\ldots251 \text{ m/s} \qquad (7.2)$$

The length of the rotor

$$L = (2.5\ldots6.0)D \;[\text{m}] \qquad (7.3)$$

To protect the end turns of the rotor winding against deformation under action of centrifugal forces, end turns are covered with the aid of metal sleeves. To reduce the leakage magnetic flux and eddy-current losses in these sleeves, the sleeves are made of nonferromagnetic austenic steel (18% Mn, 3% Cr, 0,5% C) so maximum allowable stresses can be up to 1150 MN/m^2 (1150 MPa).

7.2.2 Hydroalternators

Hydroalternators (hydrogenerators) are low speed salient-pole synchronous machines (Fig. 7.2b) with a large number of poles (Fig. 7.5). Hydrogenerators have large diameter. The rotor diameter D is up to 15 m and in most cases the rotor diameter–to–rotor length ratio is

$$\frac{D}{L} = 0.15\ldots0.20 \qquad (7.4)$$

The number of poles is usually $2p = 10\ldots100$, which corresponds to synchronous speed

$$n_s = \frac{f}{p} = \frac{50}{5\ldots50} = 1\ldots10 \text{ rev/s} = 60\ldots600 \text{ rpm} \qquad (7.5)$$

Fig. 7.5. Salient-pole rotor of a hydrogenerator. Below the rotor on the floor, there are salient poles with cage damper winding.

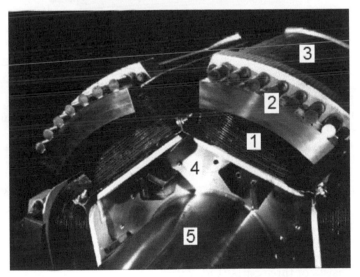

Fig. 7.6. Construction of a salient-pole rotor. 1 – field excitation winding, 2 – cage damper winding, 3 – pole shoe, 4 – spider construction, 5 – shaft.

at frequency $f = 50$ Hz. The rotor has salient poles and cage damper winding placed in pole faces (Figs 7.5 and 7.6).

7.2.3 Combustion engine driven synchronous generators

Combustion engine-driven synchronous generators have a salient-pole rotor, conventional stator and brushless exciter. Synchronous generator combustion engine systems are mostly designed as mobile units (Fig. 7.7). Typically, the rotor is a wound-field rotor. Modern generators are designed with permanent magnet (PM) rotor.

Fig. 7.7. Construction of synchronous generators driven by Diesel engine. 1 - Diesel engine, 2 - cooler (radiator), 3 - dry air filter, 4 - guard of turbocompressor (if exists), 5 - guard of fan, 6 - front shield of radiator, 7 - generator, 8 - control panel, 9 - lifting hook, 10 - support frame, 11 - integrated fuel tank, 12 - grounding terminal, 13 - batteries for starting, 14 - name plate, 15 - antivibration gasket.

Fig. 7.8 shows a cross-section and magnetic field distribution of a 55-kW, 2200-rpm, 12-pole PM brushless generator driven by a combustion engine.

7.2.4 Gas turbine driven generators

Construction of *gas turbine engines* is, theoretically, very simple. A gas turbine engine consists of three parts (Fig. 7.9):

- Compressor to compress the incoming air to high pressure;
- Combustion chamber to burn the fuel and produce high pressure, high velocity gas;
- Turbine to extract the energy from the high pressure, high velocity gas flowing from the combustion chamber.

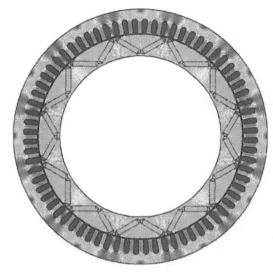

Number of phases	3
Number of poles	12
Number of stator slots	72
Electric power, kW	55
Frequency, Hz	220
Line-to-line voltage, V	400
Current, A	84
Efficiency, %	94
Power factor	1.0
Current density, A/mm^2	7.5
Outer diameter, mm	400
Stator stack length, mm	60
Mass (winding, core, PMs), kg	38.5
PMs	NdFeB

Fig. 7.8. Cross-section of a PM brushless generator rated at 55 kW, 2200 rpm, 220 Hz driven by a combustion engine.

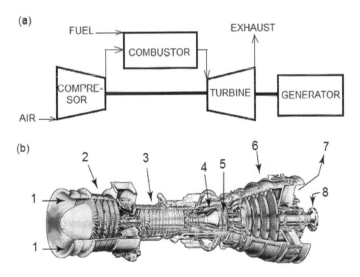

Fig. 7.9. Construction of gas turbine: (a) schematic; (b) 3D image. 1 - air inlet, 2 - low pressure compressor, 3 - high pressure compressor, 4 - combustor, 5 - high pressure turbine, 6 - low pressure turbine, 7 - exaust, 8 - power shaft.

Synchronous generators driven by gas turbines are usually high-speed generators rated up to 500 MW.

7.2.5 Microturbines

A *microturbine* is a high-speed small electrical machine the rotor of which and the rotor of turbine have common shaft (Fig. 7.10). The active (true) power is typically from 30 to 200 kW and rotational speed from $20,000$ to $120,000$ rpm.

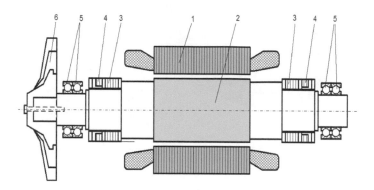

Fig. 7.10. High-speed PM generator for microturbine. 1 – stator stack with three-phase winding, 2 – PM rotor with retaining sleeve, 3 – rotor laminated stack of radial magnetic bearing, 4 – rotor of magnetic bearing sensor, 5 – additional rolling bearings, 6 – rotor of microturbine.

Technologies used in high-speed turbines:

- Small gas turbines operate in termodynamic Brayton cycle;
- Mini cogeneration (combined heat and power, CHP) steam power plant;
- Power plants operating in termodynamic organic Rankine cycle (ORC).

Water is not suitable fluid for small-scale power plants because it must be overheated up to 600°C. By using organic working fluids it is possible to design ORC plants that require a minimum of maintenance and can operate unattended for extensive periods of time. Organic substances that can be used below a temperature of 400°C do not need to be overheated.

7.2.6 Wind generators

Although most *wind turbines* use induction generators, PM brushless generators have become popular in the last few years. The power of wind generators is from hundreds of watts up to 5 MW, recently up to 10 MW. A wind generator is located in a special tower (Fig. 7.11) and driven by wind turbine with adjustable pitch propeller using a step-up gear or direct gearless drive. The speed of the wind turbine is typically from 12 to 24 rpm.

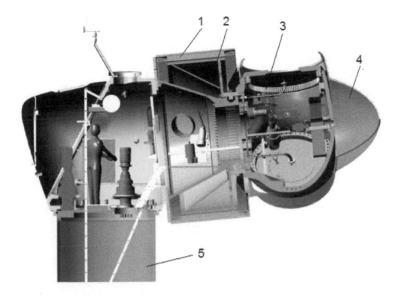

Fig. 7.11. Large power gearless wind turbine with PM brushless generator. 1 – stator stack with three-phase winding, 2 – PM rotor, 3 – mechanism for adjustment of pitch of propeller, 4 – nacelle, 5 – tower. Courtesy of Zephyros, ABB, Helsinki Factory, Finland.

7.3 Electromotive force induced in armature winding

Fig. 7.12 shows the stator (armature) phase windings in A, B, C stationary coordinate system and d and q rotor axes, while Fig. 7.13 shows the rotor field excitation flux Φ_f and EMF E_f induced in a single phase of armature winding.

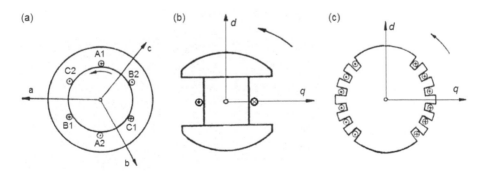

Fig. 7.12. Simplified cross-section of synchronous machines: (a) stator phase windings in A, B, C stationary coordinate system, (b) d, q-axes of salient pole rotor, (c) d, q-axes of non-salient pole rotor.

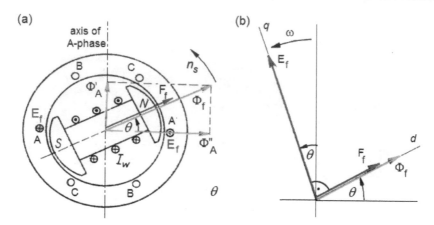

Fig. 7.13. Rotor field excitation flux Φ_f and EMF E_f induced in a single phase winding of the stator (armature): (a) rotor flux components and three-phase stator armature windings; (b) phasor diagram per phase.

The instantaneous magnetic flux of phase A is

$$\phi_f(t) = \Phi_f \sin(\alpha) = \Phi_f \sin(\omega t) \tag{7.6}$$

This magnetic flux $\phi_f(t)$ is excited by the field current I_f. The instantaneous value of EMF induced in a single turn by the flux $\phi_f(t)$

$$e_f(t) = \frac{d\phi_f(t)}{dt} = \omega \Phi_f \cos(\omega t) = 2\pi f \Phi_f \cos(\omega t) \tag{7.7}$$

The EMF induced in a single phase with N_1 number of turns

$$e_f(t) = 2\pi f N_1 \Phi_f \cos(\omega t) \tag{7.8}$$

If the winding is distributed in slots

$$e_f(t) = 2\pi f N_1 k_{w1} \Phi_f \cos(\omega t) \tag{7.9}$$

where k_{w1} is the winding factor given by eqn (5.23). The winding factor is the product of the distribution factor k_{d1} expressed by eqn (5.29) and pitch factor expressed by eqn (5.31). Since $\cos(\omega t) = \sin(90° - \omega t)$ the instantaneous phase EMF is

$$e_f(t) = 2\pi f N_1 k_{w1} \Phi_f \sin(90° - \omega t) = E_{fm} \sin(90° - \omega t) \tag{7.10}$$

The *rms* EMF per phase

$$E_f = \frac{E_{fm}}{\sqrt{2}} = \frac{2\pi}{\sqrt{2}} f N_1 k_{w1} \Phi_f = \pi \sqrt{2} f N_1 k_{w1} \Phi_f \tag{7.11}$$

This is similar equation to eqn (5.24). The magnetic flux excited by the rotor field winding

$$\Phi_f = \alpha_i \tau L_i B_{mg1} \tag{7.12}$$

where B_{mg1} is the peak value (amplitude) of the first space harmonic of the magnetic flux density in the air gap, L_i is the effective (ideal) length of the stator (armature) core and the *pole shoe width* b_p *–to–pole pitch* τ ratio is

$$\alpha_i = \frac{b_p}{\tau} \tag{7.13}$$

For non-salient rotor synchronous machines $\alpha_i = 2/\pi$ and for salient-pole synchronous machines $\alpha_i = 0.65 \ldots 0.80$. The total magnetic flux ϕ_{ft} of the rotor field winding

$$\Phi_{ft} = \Phi_f + \Phi_{f\sigma} \tag{7.14}$$

where $\Phi_{f\sigma}$ is the *leakage flux*. The magnetic flux leakage factor

$$\sigma_l = \frac{\Phi_{ft}}{\Phi_f} = 1 + \frac{\Phi_{f\sigma}}{\Phi_f} \tag{7.15}$$

7.4 Armature reaction

The armature MMFs and currents are sketched in Fig. 7.14. The armature MMFs and currents have the d and q axis components. According to Lentz's rule, the armature reaction MMF in the d-axis is in oppoite direction to the magnetizing MMF of the wound-field rotor. This component

$$F_{ad} = F_a \sin \psi \tag{7.16}$$

demagnetizes the rotor. The q-axis component

$$F_{aq} = F_a \cos \psi \tag{7.17}$$

causes cross armature reaction. The magnitude of the fundamental harmonic of the MMF excited by the armature

$$F_a = \frac{m_1 \sqrt{2}}{\pi} \frac{N_1 k_{w1}}{p} I_a \tag{7.18}$$

where I_a is the *rms* armature current and $F_a = F_{a1}$ is the fundamental space harmonic ($\nu = 1$) of the armature MMF. The magnitude of the fundamental harmonic of the MMF according to eqn (7.18) is identical as the magnitude of MMF in eqn (5.85) derived in Section 5.8. Similar equations to (7.16) and (7.17) can be written for currents, i.e.,

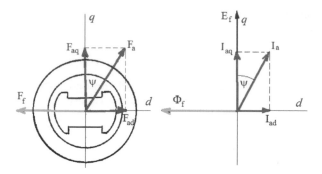

Fig. 7.14. Armature MMFs and and armature currents of a synchronous machine.

$$I_{ad} = I_a \sin\psi \tag{7.19}$$

$$I_{aq} = I_a \cos\psi \tag{7.20}$$

In phasor notation

$$\mathbf{F}_a = \mathbf{F}_{ad} + \mathbf{F}_{aq} \tag{7.21}$$

$$\mathbf{I}_a = \mathbf{I}_{ad} + \mathbf{I}_{aq} \tag{7.22}$$

The base impedance (nominal impedance) of a synchronous machine is the ratio of the nominal armature voltage to nominal armature current, i.e.,

$$Z_n = \frac{V_{1n}}{I_{an}} \tag{7.23}$$

7.5 Generator and motor operation

Fig. 7.15 shows the position of the phasor of the armature current in the $d-q$ coordinate system. Dependent on which quadrant is the armature current, the synchronous machine can operate as *overexcited* and *underexcited generator* or *overexcited* and *underexcited motor*.

7.6 Operation at no load

At no load the field winding is excited with the DC current and the armature terminals are open. The output terminal voltage V_1 can be controlled by changing the DC field excitation current I_f.

Fig. 7.16a shows the connection diagram for measurement of open-circuit (no-load) characteristics. The synchronous machine SG is driven by the prime

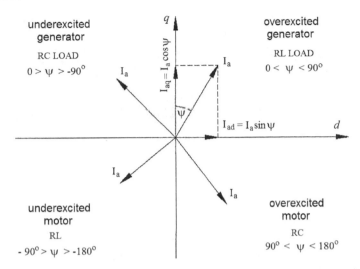

Fig. 7.15. Position of armature current I_a in the $d - q$ coordinate system and four modes of operation of a synchronous machine.

mover PM. The open-circuit (no-load) characteristic (Fig. 7.17a) is the output phase voltage V_1 curve plotted against the field excitation current I_f at constant rotor speed $n = n_s = const$ and open armature winding terminals, i.e., the load impedance $Z_L \to \infty$. At no load $V_1 = E_f$.

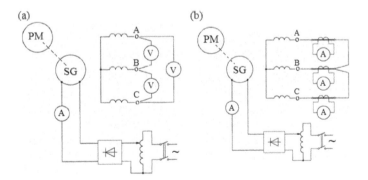

Fig. 7.16. Connection diagrams for laboratory tests on synchronous generators: (a) at no-load $Z_L \to \infty$, (b) at short-circuit $Z_L = 0$. SG – synchronous generator, PM – prime mover.

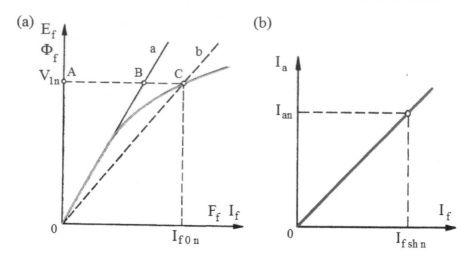

Fig. 7.17. Steady-state characteristics of synchronous generator: (a) open-circuit (no-load) characteristic $E_f = f(I_f)$ at $n = n_s = const$ and $Z_L \to \infty$; (b) short-circuit characteristic $I_a = f(I_f)$ at $n = n_s = const$, $Z_L = 0$, $V_1 = 0$.

7.7 Operation at short circuit

The short-circuit armature current is determined by the phase nominal voltage V_{1n} and the d-axis synchronous reactance X_{sd}, i.e.,

$$I_{ash} = \frac{V_{1n}}{\sqrt{R_1^2 + X_{sd}^2}} \approx \frac{V_{1n}}{X_{sd}} \tag{7.24}$$

For medium-power and large-power synchronous machine the armature winding resistance R_1 can be neglected.

Fig. 7.16b shows the connection diagram for measurement of *short-circuit characteristics*. The synchronous machine SG with shorted armature terminals is driven by the prime mover PM. The short-circuit characteristic (Fig. 7.17b) is the output armature current I_a curve plotted against the field excitation current I_f at constant rotor speed $n = n_s = const$ and shorted armature winding terminals, i.e., the load impedance $Z_L = 0$. At short-circuit the armature terminal voltage $V_1 = 0$.

The *short circuit ratio* is defined as

$$k_{sh} = \frac{I_{f0}}{I_{fsh}} = \frac{I_{ash}}{I_{an}} \tag{7.25}$$

where I_{f0} is field excitation current corresponding to rated voltage V_{1n} at no load, I_{fsh} is the field excitation current corresponding to rated armature current I_{an} at shorted armature terminals and I_{ash} is the armature current corresponding to I_{f0} read on short circuit characteristic and given by eqn (7.24).

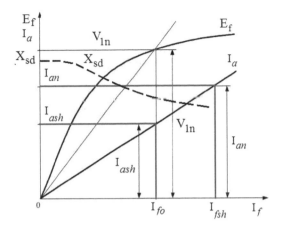

Fig. 7.18. No-load characteristic $E_f = f(I_f)$ ($Z_L \to \infty$), short-circuit characteristic $I_a = f(I_f)$ ($Z_L = 0$, $V_1 = 0$) and d-axis synchronous reactance curve $X_{sc} = f(I_f)$ at $n = n_s = const$ of a synchronous generator.

The no-load characteristic $E_f = f(I_f)$ at $n = n_s = const$ and $Z_L \to \infty$ and short-circuit characteristic $I_a = f(I_f)$ at $n = n_s = const$, $Z_L = 0$, $V_1 = 0$ are plotted in Fig. 7.17. These characteristics allow for the transformation of short-circuit ratio (7.25) to the following form:

$$k_{sh} = \frac{I_{f0}}{I_{fsh}} = \frac{V_{1n}}{X_{sd}I_{an}} = \frac{Z_n}{X_{sd}} = \frac{1}{X_{sd}/Z_n} = \frac{1}{x_{od}} \propto g \qquad (7.26)$$

where the nominal (base) impedance is according to eqn (7.23), g is the air gap (mechanical clearance between the stator and rotor cores), $R_1 = 0$ has been neglected and

$$x_{sd} = \frac{X_{sd}}{Z_n} \qquad (7.27)$$

is per-unit d-axis synchronous reactance.

The d-axis synchronous reactance curve $X_{sd} = f(I_f)$ at $n = n_s = const$, no-load characteristic $E_f = f(I_f)$ ($Z_L \to \infty$) and short-circuit characteristic $I_a = f(I_f)$ ($Z_L = 0$, $V_1 = 0$) are plotted in Fig. 7.18.

Typical values of short-circuit ratio are

- For non-salient pole synchronous turboalternators

$$k_{sh} = 0.4 \ldots 0.7 \qquad (7.28)$$

- For salient pole synchronous hydrogenerators

$$k_{sh} = 1.0 \ldots 1.4 \qquad (7.29)$$

Synchronous machines with low value of short-circuit ratio show greater change in voltage with the fluctuation of load, are less stable under parallel operation and show lower charging current, when the generator is loaded only with the capacitance of the transmission line (no load). Such a machine is cost-effective.

$$g \uparrow X_{sd} \downarrow k_{sh} \uparrow \qquad\qquad g \downarrow X_{sd} \uparrow k_{sh} \downarrow \qquad\qquad (7.30)$$

Utilization of active materials in a synchronous machine with low value of short circuit ratio (smaller air gap g) is better than in a machine with high value of short circuit ratio . The larger the air gap g, the heavier the machine and worse performance characteristics.

7.8 Phasor diagram of synchronous machine with non-salient pole rotor and unsaturated magnetic circuit

The equivalent circuit per phase of a synchronous machine with non-salient pole rotor is shown in Fig. 7.19. The stator (armature) winding is represented by the following impedance:

$$\mathbf{Z}_1 = R_1 + j(X_1 + X_a) \qquad\qquad (7.31)$$

where R_1 is the resistance of a single phase of armature winding, X_1 is the leakage reactance of a single phase of armature winding and X_a is the *armature reaction reactance* also called *mutual reactance*.

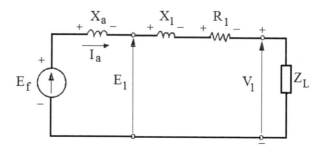

Fig. 7.19. Equivalent circuit per phase of a synchronous machine with non-salient pole rotor.

The *synchronous reactance X_s* of a non-salient pole synchronous machine is the sum of the leakage reactance X_1 and armature reaction reactance X_a, i.e.,

$$X_s = X_1 + X_a \qquad\qquad (7.32)$$

For the equivalent circuit shown in Fig. 7.19 the following equations result from Kirchhoff's voltage law

- The EMF excited by the resultant flux Φ (with the armature reaction being included)

$$\mathbf{E}_1 = \mathbf{V}_1 + \mathbf{I}_a(R_1 + jX_1) \tag{7.33}$$

- The EMF excited by the useful flux Φ_f of the rotor

$$\mathbf{E}_f = \mathbf{E}_1 + jX_a\mathbf{I}_a = \mathbf{V}_1 + \mathbf{I}_a(R_1 + jX_s) \tag{7.34}$$

- The EMF excited by the armature reaction flux Φ_a

$$\mathbf{E}_a = \mathbf{E}_f - \mathbf{E}_1 \tag{7.35}$$

The *rms* EMF E_1 (7.33), E_f (7.34) and E_a (7.35) can also be expressed as functions of magnetic fluxes Φ and Φ_f, i.e.,

$$E_1 = |\mathbf{E}_1| = \pi\sqrt{2}fN_1k_{w1}\Phi \tag{7.36}$$

$$E_f = |\mathbf{E}_f| = \pi\sqrt{2}fN_1k_{w1}\Phi_f \tag{7.37}$$

$$E_a = |\mathbf{E}_a| = \pi\sqrt{2}fN_1k_{w1}\Phi_a \tag{7.38}$$

See also eqns (5.24)and (7.11). The armature reaction magnetic flux due to the fundamental harmonic of the armature reaction magnetic flux density B_{ma1} (peak value) is

$$\Phi_a = \frac{2}{\pi}\tau L_i B_{ma1} \tag{7.39}$$

The resultant flux is the field excitation flux Φ_f reduced by the armature reaction flux Φ_a, i.e.,

$$\mathbf{\Phi} = \mathbf{\Phi}_f - \mathbf{\Phi}a \tag{7.40}$$

or

$$\Phi = |\mathbf{\Phi}| = \frac{2}{\pi}\tau L_i B_{mg1} \tag{7.41}$$

Since $R_1 << X_1$ and $X_s = X_1 + X_a$, the armature current can be expressed with the aid of a simple equation

$$I_a = \frac{\mathbf{E}_f - \mathbf{V}_1}{R_1 + jX_s} \approx -j\frac{\mathbf{E}_f - \mathbf{V}_1}{X_s} \tag{7.42}$$

The magnetizing MMF of the field winding which excites the same magnetic flux as the armature reaction flux is

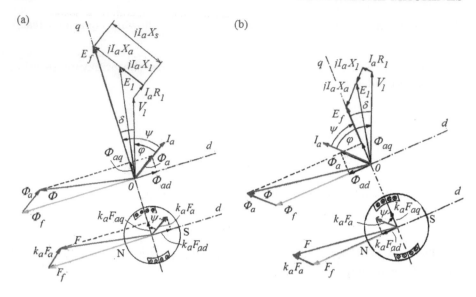

Fig. 7.20. Phasor diagrams of non-salient pole synchronous generators for: (a) RL load; (b) RC load.

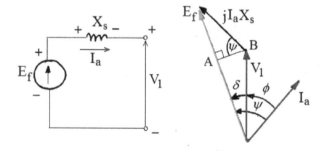

Fig. 7.21. Simplified equivalent circuit and phasor diagram for non-salient rotor synchronous generator under inductive load RL ($R_1 = 0$).

$$F'_f = k_a F_{feq} = \frac{m_1 \sqrt{2}}{\pi} \frac{N_1 k_{w1}}{p} k_a I_a \qquad (7.43)$$

See also eqn (5.85). The coefficient k_a for reducing the armature reaction MMF to the field winding MMF can be found on the basis of design specification of a non-salient pole synchronous machine. According to classical literature, e.g., [25]

$$k_a = \frac{\pi^2 \gamma}{8 \sin(0.5\pi\gamma)} \qquad (7.44)$$

where γ is the ratio of the rotor circumference occupied by slots to total circumference of the rotor. Physically, the reciprocal of k_a is the shape factor of the field excitation [7].

The equivalent circuit sketched in Fig. 7.19 and eqns (7.31) to (7.42) allow for construction of phasor diagrams for inductive (RL) and capacitive (RC) loads (Fig. 7.20).

Under inductive load RL the armature reaction magnetic flux Φ_a is de-magnetizing, i.e., it reduces the field excitation flux Φ_f (Fig. 7.20a).

Under capacitive load RC the armature reaction magnetic flux Φ_a is mag-netizing, i.e., it magnifies the field excitation flux Φ_f (Fig. 7.20b).

Simplified equivalent circuit and phasor diagram for non-salient rotor syn-chronous generator under inductive load RL are plotted in Fig. 7.21. In this equivalent circuit $R_1 = 0$ and

$$j\mathbf{I}_a X_s \approx \mathbf{E}_f - \mathbf{V}_1 \tag{7.45}$$

Example 7.1

A three-phase, non-salient pole, Y-connected synchronous generator with syn-chronous reactance of $X_s = 14.0$ Ω delivers te active power $P_{out} = 1.68$ MW to the power system at inductive power factor. The line-to-line voltage is $V_{1L} = 11$ kV and armature current $I_a = 100.0$ A. Find the power factor $\cos\varphi$, load angle δ and EMF E_f induced in the armature winding. The armature winding resistance is negligible ($R_1 \approx 0$).

Solution

Apparent power

$$S_{out} = \sqrt{3} V_{1L} I_a = \sqrt{3} \times 11\,000 \times 100.0 = 1.905 \times 10^6 \text{ VA}$$

Power factor

$$\cos\varphi = \frac{P_{out}}{S_{out}} = \frac{1.906 \times 10^6}{1.68 \times 10^6} = 0.882 \text{ ind}$$

The angle between the current and voltage $\varphi = \arccos(0.882) = 28.14°$. Thus, $\sin\varphi = \sin(28.14°) = 0.472$.

On the basis of simplified phasor diagram for overexcited machine ($\cos\varphi$ ind, $R_1 = 0$)

$$E_f \cos(\delta + \varphi) = V_1 \cos\varphi$$

$$E_f \sin(\delta + \varphi) = X_s I_a + V_1 \sin\varphi$$

Dividing through the second equation by the first equation

$$\tan(\delta + \varphi) = \frac{X_s I_a + V_1 \sin \varphi}{V_1 \cos \varphi}$$

Phase voltage $V_1 = V_{1L}/\sqrt{3} = 11\ 000/\sqrt{3} = 6351$ V and

$$\tan(\delta + \varphi) = \frac{14.0 \times 100.0 + 6351.0 \times 0.472}{6351.0 \times 0.882} = 0.785$$

Hence

$$\delta + \varphi = \arctan(0.785) = 38.13°$$

$$\delta = 38.13° - 28.14° = 9.99°$$

EMF per phase

$$E_f = \frac{V_1 \cos \varphi}{\cos(\delta + \varphi)} = \frac{6351.0 \times 0.882}{\cos(38.13°)} = 7119 \text{ V}$$

Line-to-line EMF $E_{fL} = \sqrt{3}E_f = \sqrt{3} \times 7119.0 = 12,330.6$ V.

7.9 Phasor diagram of synchronous machine with non-salient pole rotor and saturated magnetic circuit

When the magnetic circuit is saturataed (nonlinear), magnetic fluxes and EMFs are not proportional to respective MMFs. As the magnetic circuit becomes more and more saturated, the armature leakage reactance X_1 is practically constant, but the armature reaction reactance X_a and EMF E_f decrease. Nonlinearity is caused by nonlinear magnetization curves $B = f(H)$ and $\mu_r = F(H)$ of ferromagnetic materials (Fig. 7.22).

Combining together eqns (5.54), (7.18), (7.38) and (7.39) the armature reaction reactance X_a of a non-salient pole rotor synchronous machine including magnetic saturation is [25]

$$X_a = \frac{E_a}{I_a} = 4m_1\mu_0 f \frac{(N_1 k_{w1})^2}{\pi p} \frac{\tau L_i}{k_{sat} k_c g} \qquad (7.46)$$

where k_{sat} is the saturation factor of the magnetic circuit given by eqns (5.49), (5.51), (5.100) and (5.101).

The phasor diagram of synchronous generator with non-salient pole rotor and saturated magnetic circuit is plotted in Fig. 7.23. The open circuit characteristic $E = f(F)$ determines the EMF E_f with magnetic saturation taken into account. The EMF E'_f corresponds to unsaturated magnetic circuit (air gap line).

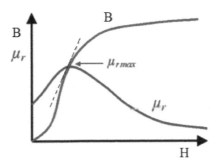

Fig. 7.22. Typical magnetization curves $B = f(H)$ and $\mu_r = F(H)$ of ferromagnetic materials. The EMF E_f is proportional to B and MMF (field current I_f) is proportional to H.

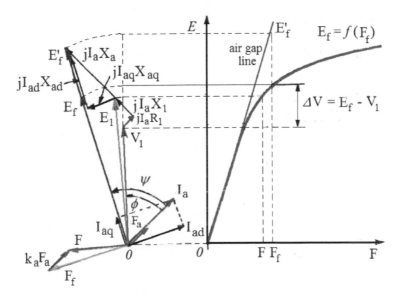

Fig. 7.23. Construction of phasor diagram of synchronous generator with non-salient pole rotor and saturated magnetic circuit.

The *voltage change* is the change in voltage at the output terminals of the armature winding that takes place when the generator with open armature terminals is loaded with nominal RL load at $n = n_s = const$ and $I_f = const$, i.e.,

$$\Delta V = \frac{E_f - V_{1n}}{V_{1n}} \tag{7.47}$$

The voltage change and short-circuit current ratio are given in Table 7.1.

Table 7.1. The voltage change and steady-state short-circuit current ratio at no-load excitation for synchronous generators.

Type of generator	Voltage change, % $\cos\phi = 1$	$\cos\phi = 0.8$	Steady-state short-circuit current ratio
Three-phase:			
High-speed	$8\ldots15$	$18\ldots31$	$2.0\ldots1.4$
Low speed	$9\ldots13$	$18\ldots25$	$2.5\ldots2.0$
Turboalternators	$16\ldots25$	$30\ldots48$	$1.2\ldots0.7$
Single-phase generators for railway substations	$17\ldots20$	$35\ldots40$	$1.2\ldots0.8$

7.10 Steady-state characteristics of synchronous turboalternator

The most important steady-state characteristics of synchronous generators are the open-circuit (no-load) characteristic $E_f = f(I_f)$ $(Z_L \to \infty)$ and short-circuit characteristic $I_a = f(I_f)$ $(Z_L = 0,\ V_1 = 0)$ plotted in Fig. 7.17.

On the basis of the no-load characteristic (Fig. 7.17a) and short-circuit characteristic (Fig. 7.17b) the so-called *short-circuit triangle* can be built (Fig. 7.24). This is a right triangle with sides (catheti) AB= $I_a X_1$ and BC= $k_a F_a$, where $I_a X_1$ is the voltage drop across the armature winding leakage reactance, and $k_a F_a$ is the d-axis armature reaction MMF reduced to the field excitation winding. The distance AC is hypotenuse of the short-circuit triangle.

Using the open-circuit characteristic $E_f = f(I_f)$ and short-circuit triangle, the load characteristics $V_1 = f(I_f)$ at $n = n_s = const$ and $I_a = const$ for different power factors $\cos\varphi$ can be plotted (Fig. 7.25). The load characteristics for inductive loads (lagging power factor) are below the no-load characteristic $E_f = f(I_f)$ and load characteristics for capacitive loads (leading power factor) are above the no-load characteristic $E_f = f(I_f)$.

The external characteristics $V_1 = f(I_a)$ at $n = n_s = const$, $I_f = const$ and $\cos\varphi = const$ are plotted in Fig. 7.26a. The regulation characteristics $I_f = f(I_a)$ at $n = n_s = const$ and $V_1 = const$ of unsaturated synchronous generator are plotted in Fig. 7.26b.

7.11 Losses and efficiency

In general, the power losses in synchronous machines can be classified as *basic losses* and *additional losses* (Fig. 7.27). The basic losses consist of main copper losses in the armature and field excitation windings, stator core losses, i.e., hysteresis and eddy-current losses, and rotational (mechanical) losses, i.e., friction in bearings, windage and ventilation losses. If a synchronous machine is equipped with slip rings, also the slip ring–brush losses must be included

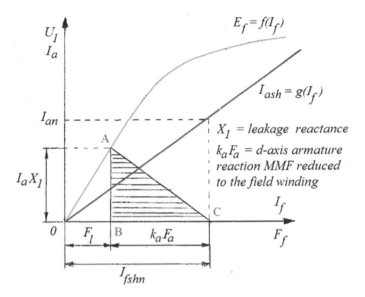

Fig. 7.24. Construction of the short-circuit triangle.

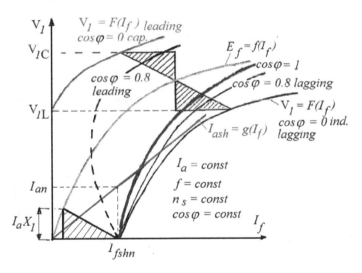

Fig. 7.25. Load characteristics $V_1 = f(I_f)$ at $n = n_s = const$ and $I_a = const$ and different power factors $\cos \phi$.

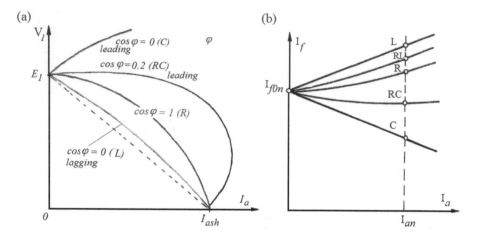

Fig. 7.26. Characteristics of synchronous generator: (a) external characteristics $V_1 = f(I_a)$ at $n = n_s = const$, $I_f = const$ and $\cos\phi = const$; (b) regulation characteristics $I_f = f(I_a)$ at $n = n_s = const$ and $V_1 = const$.

into friction losses. The additional losses consist of losses in the armature and field windings due to leakage fluxes, losses in the stator and rotor core due to leakage fluxes, higher MMF harmonics and tooth-ripple harmonics, and losses in metallic parts where the leakage flux penetrates (shields, coil-retaining rings, clamping rings, bandages, etc.)

Table 7.2. Losses and efficiency of a 25-MVA, 3000-rpm turboalternator at rated load [25].

Losses	Value, kW
Losses due to rotor air friction	102
Other windage losses	100
Bearing friction losses	70
Core losses	85
Stator winding losses	60
Rotor winding losses	90
Additional short-circuit losses	74
Additional no-load losses	35
Total losses	615
Efficiency at full load	97.6%

To minimize the losses in windings made of rectangular conductors due to eddy-currents induced by leakage fluxes, the so-called *transpositions* are used (Fig. 7.28). Using transpositions, the leakage fluxes are linked with every conductor along the core in the same way.

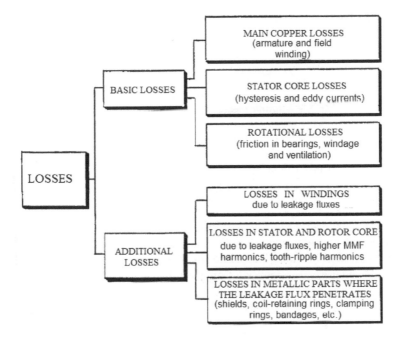

Fig. 7.27. Power losses in synchronous machine.

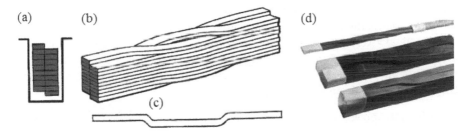

Fig. 7.28. Transpositions of stator windings: (a) cross-section of a slot; (b) a bunch of conductors with transpositions; (c) single conductor; (d) practical implementation.

The efficiency of a synchronous generator is expressed by the following formula:

$$\eta = \frac{P_{out}}{P_{in}} = \frac{P_{out} + \sum \Delta P - \sum \Delta P}{P_{out} + \sum \Delta P} = 1 - \frac{\sum \Delta P}{P_{out} + \sum \Delta P} \qquad (7.48)$$

where the input mechanical power

$$P_{in} = P_{out} + \sum \Delta P \qquad (7.49)$$

Table 7.3. Losses and efficiency of a 20-MVA, 187-rpm hydrogenerator at rated load [25].

Losses	Value, kW
Rotational losses	125.7
Core losses including additional losses	141.2
Stator winding losses	102
Rotor winding losses	101.0
Additional short-circuit losses	58.1
Total losses	528.0
Efficiency at full load	96.8%

and $\sum \Delta P$ is the sum of all power losses.

The losses and efficiency of a 25 MVA, 3000 rpm turboalternator at rated load are specified in Table 7.2, while the losses and efficiency of 20 MVA, 187 rpm hydrogenerator at rated load are specified in Table 7.3.

7.12 Exciters

The main role of the *field excitation system* is to deliver the direct current (DC) to the field winding of synchronous machine in order to excite the required magnetic field in the air gap at given operating conditions.

The field excitation system should secure the synchronous operation of a synchronous generator at any condition, i.e., at steady-state conditions, at transients and in emergency situations. The field excitation system must meet the following requirements:

- Adequate power must be delivered to the field winding of synchronous machine;
- Reliability of operation;
- Fast-acting (fast-response) field regulator;
- Possibility of field forcing;
- Possibility of fast magnetic field extinction.

The field excitation system at steady-state operation should provide continuous and correct operation of a synchronous generator in power system both (1) at rated (nominal) load and (2) under different than rated conditions, but considered as normal operating conditions.

The field excitation system consists of the source of excitation power, i.e., *exciter*, rectifier and field extinction system.

The regulation of the field current is done through the variation of control voltage at the input terminals of the excitation power source or through direct interaction in the path *excitation source–armature winding*.

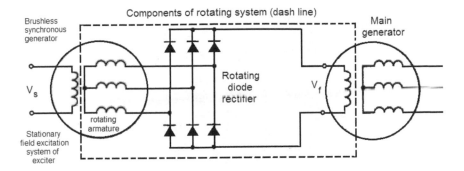

Fig. 7.29. Power circuit of brushless exciter.

Brush exciters have been used in old designs of synchronous machines. A rotor of a synchronous machine was equipped with two slip rings connected to the field excitation winding. The DC current generated by a DC brush (commutator) generator was delivered to the slip rings via two brushes. The DC brush generator was mounted on the shaft of a synchronous machine. Sometimes, an external rectifier was used instead of a DC generator.

Nowadays, *brushless exciters* are used almost exclusively for all synchronous machines. Brushless exciters show the following advantages:

- No brushes and slip rings;
- No friction losses slip rings brushes;
- Increased power of the field excitation system.

A brushless exciter consists of a reversed synchronous generator and rotating rectifier. Reversed construction means that the three-phase armature is on the rotor and the field excitation system on the stator. The power circuit diagram of brushless exciter and its connection to the main synchronous generator is shown in Fig. 7.29. A 3D image of rotating elements is shown in Fig. 7.30. By removing the brushes and slip rings the excitation system is more compact, which reduces the length of the synchronous machine. Terminal leads of the field excitation system of the main synchronous machine are drawn through the hollow shaft.

Excitation systems with brushless exciters show the following advantages:

- High reliability of DC power delivery to main generator;
- Limited maintenance;
- Very good dynamics.

Reversed synchronous exciter with controlled rectifier has become a standard excitation system for large turboalternators rated up to 2 GW with rotor water cooling system.

Fig. 7.30. Rotating elements of brushless exciter of a large-power generator: 1 – three-phase armature, 2 – semiconductor diode, 3 – shaft.

7.13 Operation of synchronous generators

7.13.1 Modes of operation of synchronous generators

Stand alone operation is such an operation, when a synchronous generator is directly connected to the load (Fig. 7.31a). Features:

- The phase angle between the voltage and current of the generator is determined by the character of the load;
- At given load current the output voltage depends on the field excitation current;
- At given load current the frequency depends on the speed of the rotor.

Operation on flexible bus bar takes place when two synchronous generators of similar nominal power operate on common bus bar (Fig. 7.31b). The main feature of such operation is that the voltage and frequency of the bus bar depend on each of the synchronous generators.

Operation on infinite bus bar takes place when a synchronous generator of smaller power operates on the bus bar connected to a certain number of synchronous generators of larger power (Fig. 7.31c). The main feature of such operation is that the generator of smaller power has no influence on the voltage and frequency of the infinite bus bar.

7.13.2 Operation on infinite bus bar

Fig. 7.21 shows a simplified equivalent circuit and phasor diagram of a non-salient pole rotor synchronous generator operating with lagging power factor (RL load), in which the armature winding resistance $R_1 = 0$. The terminal phase voltage is

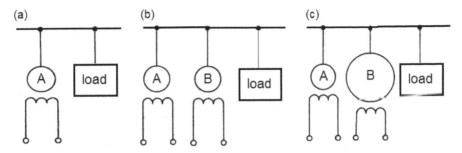

Fig. 7.31. Modes of operation of synchronous generators: (a) stand alone operation; (b) parallel operation of two generators A and B of similar nominal power on flexible bus bar ($P_A \approx P_B$); (c) operation on infinite bus bar ($P_A \ll P_B$).

$$\mathbf{V}_1 = \mathbf{E}_f - j\mathbf{I}_a X_s \tag{7.50}$$

Thus, the armature current leads the phasor $\mathbf{V}_1 - \mathbf{E}_f$ by $90°$, i.e.,

$$\mathbf{I}_a = j\frac{\mathbf{V}_1 - \mathbf{E}_f}{X_s} \tag{7.51}$$

A synchronous generator operating on an infinite bus bars shows the following properties:

(a) $V_1 = E_f$, $\delta = 0$ $\Rightarrow$ $I_a = 0$. Control of the reactive power by changing the field excitation current I_f.

(b) $E_f > V_1$, $\delta = constant$ $\Rightarrow$ Reactive power is delivered by the generator to the infinite bus bar.

(c) $E_f < V_1$, $\delta = constant$ $\Rightarrow$ The load angle δ is determined by the prime mover, e.g., turbine engine. Control of reactive power by prime mover.

7.13.3 Torque–load angle characteristics of non-salient pole rotor synchronous machine

The power at the output terminals of the armature winding

$$P_{out} = m_1 V_1 I_a \cos \phi \tag{7.52}$$

The internal electromagnetic power (power crossing the air gap) and the internal electromagnetic torque

$$P_{elm} = m_1 E_f I_a \cos \psi \tag{7.53}$$

$$T_{elm} = \frac{P_{elm}}{\Omega_s} = \frac{m_1 E_f I_a \cos \psi}{2\pi n_s} \tag{7.54}$$

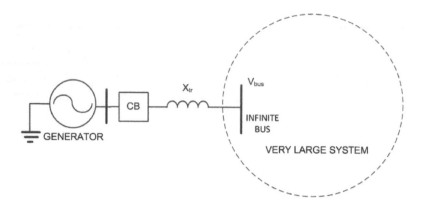

Fig. 7.32. Synchronous generator connected to an infinite system. CB – circuit braker, X_{tr} – reactance of transformer, V_{bus} – bus voltage.

where ψ is the angle between the armature current I_a and the terminal voltage V_1. According to the simplified phasor diagram (Fig. 7.21)

$$\cos \psi = \frac{AB}{I_a X_s} = \frac{V_1 \sin \delta}{I_a X_s} \tag{7.55}$$

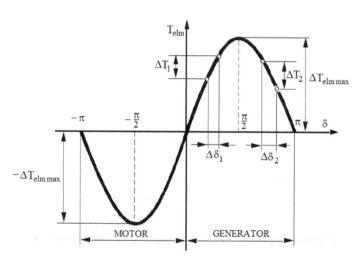

Fig. 7.33. Torque–angle characteristic $T_{elm} = f(\delta)$ of a non-salient pole rotor synchronous generator.

Thus, the electromagnetic torque as a function of the angle δ between the EMF E_f and terminal voltage V_1 is

$$T_{elm} = \frac{m_1}{2\pi n_s} \frac{V_1 E_f}{X_s} \sin \delta \qquad (7.56)$$

The angle δ is called the *load angle*, *power angle* or *torque angle*. For synchronous machine with non-salient pole rotor the electromagnetic torque T_{elm} is plotted against the load angle δ in Fig. 7.33. Conventionally, the operation of synchronous machine as a generator is for $0 \geq \delta \leq 180°$. The operation as a motor is for $0 \leq \delta \geq -180°$. The maximum electromagnetic torque T_{elmmax} of a non-salient pole rotor synchronous machine is for $\delta = |90°|$, i.e.,

$$T_{elmmax} = T_{elm}(|90°|) \qquad (7.57)$$

The unstable operation is when the electromagnetic torque T_{elm} increases with the increase in the angle δ ($\Delta\delta_1$ in Fig. 7.33). The stable operation is when the electromagnetic torque T_{elm} decreases with the increase in the angle δ ($\Delta\delta_2$ in Fig. 7.33), i.e.,

$$\frac{\Delta T_{elm}}{\Delta\delta} > 0 \qquad \text{or} \qquad \frac{dT_{elm}}{d\delta} > 0 \qquad (7.58)$$

7.13.4 Circle diagram of non-salient pole rotor synchronous machine

On the basis of simplified circuit diagram shown in Fig. 7.21 the EMF $\mathbf{E}_f$ is

$$\mathbf{E}_f = \mathbf{V}_1 + jX_s\mathbf{I}_a \qquad (7.59)$$

Dividing both sides of eqn (7.59) by jX_s

$$\frac{\mathbf{E}_f}{jX_s} = \frac{\mathbf{V}_1}{jX_s} + \mathbf{I}_a \qquad (7.60)$$

Thus, the armature current

$$\mathbf{I}_a = j\frac{\mathbf{V}_1}{X_s} - j\frac{\mathbf{E}_f}{X_s} \qquad (7.61)$$

The last equation (7.61) can be illustrated with the aid of a current circle diagram plotted in Fig. 7.34. The generating mode is for positive load angle δ and positive EMF E_{f1}. The motor mode is for negative angle $-\delta$ and negative EMF $-E_{f2}$. The straight lines $T = const$ and $P_{out} = const$ are the lines of constant torque and constant output power, respectively. The circles are locus of constant EMF $E_f = const$. The stable operation is on the right from the phasors of EMF E_f. The unstable operation is on the left from the phasors of EMF E_f.

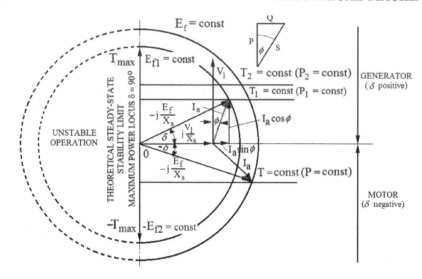

Fig. 7.34. Current circle diagram of a synchronous machine with non-salient pole rotor.

7.13.5 V-curves

The *V-curves* are the curves of the armature current versus field excitation current at constant voltage $V_1 = const$, constant torque $T = const$ and constant speed $n = n_s = const$ (Fig. 7.35). The name "V-curves" is due to the shape of these curves, similar to the letter "V". The V-curves are limited on the left-hand side with the stability limit and on the right-hand side with the maximum allowable currents of the field winding and armature winding. The minimum armature current I_a corresponds to the power factor $\cos \varphi = 1$.

7.13.6 Synchronization

Synchronization is a sequence of actions to be taken before connection of a synchronous generator to the bus bars. The objective of synchronization is to bring the generator to such a state to prevent high dangerous currents during paralleling the generator with bus bars (Fig. 7.36a).

The following conditions must be made to perform correct synchronization:

(a) The *rms* values of voltages of the generator and bus bar should be equal;
(b) The frequencies of the generator voltage and bus bar voltage should be equal;
(c) The phase sequence of the generator voltage and bus bar voltage must be the same (Fig. 7.36b);
(d) The instantaneous values of corresponding voltages of the generator and bus bar should be the same.

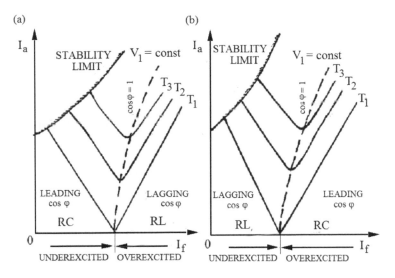

Fig. 7.35. V-curves $I_a = f(I_f)$ at $V_1 = const$, $T = const$ and $n = n_s = const$: (a) for synchronous generator; (b) for synchronous motor.

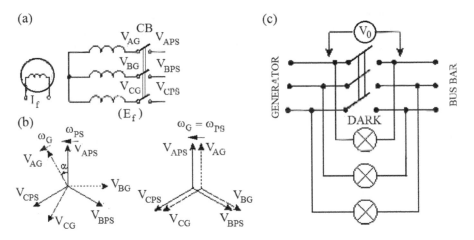

Fig. 7.36. Synchronization: (a) connection diagram; (b) generator voltage system and bus bar voltage system; (c) rotating light arrangement with one dark bulb. V_{AG}, V_{BG}, V_{CG} – three-phase generator voltage system, V_{APS}, V_{BPS}, V_{CPS} – three-phase voltages of power system, ω_G – speed of the generator voltage system, ω_{PS} – speed of the power system voltages, V_0 – zero-voltmeter, CB – circuit breaker.

In order to perform correct synchronization, the following actions should be taken:

(a) To drive the generator to the synchronous speed or very close to the synchronous speed;
(b) To excite the generator with the field current I_f to obtain the open-circuit voltage $E_f \approx V_{1n}$;
(c) To check the sequence of phases of the generator and bus bar;
(d) To close the circuit breaker when the zero-voltmeter V_0 in Fig. 7.36c shows "zero" (this is the time instant when instantaneous values of corresponding voltages of the generator and bus bar are equal).

In the case of the same phase sequences of the generator and power system an effect of rotating light appears. The speed of rotation is proportional to the difference in frequencies. In the case of different sequences all bulbs blink simultaneously.

Example 7.2

Two non-salient pole rotor synchronous generators A and B with unsaturated magnetic circuits operate in parallel. The stator windings of both generators are Wye connected. Nominal (rated) parameters of generators are as follows:

- Output powers of generators: $P_{nA} = 300$ kW, $P_{nB} = 250$ kW
- Output voltage of generators $V_{1n} = 6000$ V
- Field excitation currents: $I_{fnA} = 15$ A, $I_{fnB} = 11$ A
- d-axis synchronous reactances per unit: $x_{sdA} = 1.7$, $x_{sdB} = 1.6$
- Power factors: $\cos \varphi_{nA} = 0.8$ lagging, $\cos \varphi_{nB} = 0.8$ lagging.

It is necessary to calculate:

(a) The field excitation current of generator B to obtain $V_1 = V_{1n}$ for the following conditions: load power $P_L = 420$ kW, power factor of load $\cos \varphi_L = 0.74$ lagging, field excitation current of generator A $I_{fA} = 14$ A. Generators deliver to the utility grid the power of 210 kW each.
(b) For the same conditions as specified in (a) additional load has been connected: $P = 100$ kW, $\cos \varphi = 1$ (e.g., lighting). Keeping the same operating conditions of generator B, calculate the new value of the field excitation current of generator A, to obtain constant voltage.

Solution

(a) Calculate the field excitation current of generator B to obtain $V = V_n$

As specified, each generator delivers the same active power, i.e., $P_A = 210$ kW, $P_B = 210$ kW. Angle between the current and voltage of the load

$$\varphi_L = \arccos(0.74) = 42.3° \qquad \sin(\varphi_L) = 0.673$$

Reactive power absorbed by the load, VAr

$$Q_L = P_L \tan(\varphi_L) = 381,749.8 \text{ VAr}$$

Angle between the current and voltage of generator A under nominal operating conditions

$$\varphi_{nA} = \arccos(0.8) = 36.87° \qquad \sin(\varphi_{nA}) = 0.6$$

Angle between the current and voltage of generator B under nominal operating conditions

$$\varphi_{nB} = \arccos(0.8) = 36.87° \qquad \sin(\varphi_{nB}) = 0.6$$

EMF e_{fnA} per unit of generator A under nominal operating conditions

$$e_{fnA} = \sqrt{1 + x_{sdA}^2 + 2x_{sdA}\sin(\varphi_{nA})} = \sqrt{1 + 1.7^2 + 2 \times 1.7 \times 0.6} = 2.435$$

Per unit EMF e_{fA} of generator A under new conditions

$$e_{fA} = e_{fnA}\frac{I_{fA}}{I_{fnA}} = 2.435\frac{14.0}{15.0} = 2.273$$

Putting

$$I_{anA}\cos(\varphi_A) = \frac{P_A}{3V_{1n}I_{anA}} = \frac{P_A}{P_{nA}}\cos(\varphi_{nA})$$

into

$$e_{fA}^2 = [1 + x_{sdA}i_{aA}\sin(\varphi_A)]^2 + [x_{sdA}i_{aA}\cos(\varphi_A)]^2$$

one obtains

$$i_{aA}\sin(\varphi_A) = \frac{1}{x_{sdA}}\left[\sqrt{e_{fA}^2 - \left[x_{sdA}\frac{P_A}{P_{nA}}\cos(\varphi_{nA})\right]^2} - 1\right]$$

$$= \frac{1}{1.7}\left[\sqrt{2.273^2 - \left[1.7\frac{210,000}{300,000} \times 0.8\right]^2} - 1\right] = 0.626$$

where

$$i_{aA} = \frac{I_{aA}}{I_{anA}}$$

Reactive power of generator A

$$Q_A = 3V_1 I_{aA} \sin(\varphi_A) = 3V_1 I_{anA} i_{aA} \sin(\varphi_A) = \frac{P_{nA}}{\cos(\varphi_{nA})} i_{aA} \sin(\varphi_A)$$

$$= \frac{300,000}{0.8} \times 0.626 = 234,668 \text{ VAr}$$

Reactive power of generator B

$$Q_B = Q_L - Q_A = 381,749.8 - 234,668.0 = 147,082 \text{ VAr}$$

Apparent power of generator B

$$S_B = \sqrt{P_B^2 + Q_B^2} = \sqrt{210,000^2 + 147,082^2} = 256,384.7 \text{ VA}$$

Power factor of generator B

$$\cos(\varphi_B) = \frac{P_B}{S_B} = \frac{210,000}{256,384.7} = 0.819 \qquad \sin(\varphi_B) = 0.574 \qquad \varphi_B = 35°$$

Armature current of generator B per unit

$$i_{aB} = \frac{S_B}{S_{nB}} = \frac{P_B}{P_{nB}} \frac{\cos(\varphi_{nB})}{\cos(\varphi_B)} = \frac{210,000}{250,000} \frac{0.8}{0.819} = 0.82$$

EMF e_{fB} of generator B per unit

$$e_{fB} = \sqrt{1 + (x_{sdB} i_{aB})^2 + 2 x_{sdB} i_{aB} \sin(\varphi_B)}$$

$$= \sqrt{1 + (1.6 \times 0.82)^2 + 2 \times 1.6 \times 0.82 \times 0.574} = 2.057$$

EMF e_{fnB} per unit of generator B under nominal operating conditions

$$e_{fnB} = \sqrt{1 + 1.6^2 + 2 \times 1.6 \times 0.6} = 2.341$$

Field excitation current of generator B

$$I_{fB} = I_{fnB} \frac{e_{fB}}{e_{fnB}} = 11.0 \times \frac{2.057}{2.341} = 9.7 \text{ A}$$

(b) *Keeping the same operating conditions of generator B, calculate the new value of the field excitation current of generator A to obtain constant voltage*

Active power of loads under new conditions

$$P_L = P_L + P = 420,000 + 100,000 = 520,000 \text{ W}$$

Reactive power under new conditions remains unchanged, i.e., $Q_L = 381,749.8$ VAr. Generator B delivers the following active $P_B = 210,000$ W and reactive power $Q_B = 147,082$ VAr. Thus, the machine A must deliver $P_A = P_L - P_B = 520,000 - 210,000 = 310,000$ W, $Q_A = Q_L - Q_B = 381,749.8 - 147,082.0 = 234,667.8$ VAr and $S_A = \sqrt{P_A^2 + Q_A^2} = \sqrt{310,000^2 + 234,667.8^2} = 388,804.6$ VA. Power factor of generator A

$$\cos(\varphi_A) = \frac{P_A}{S_A} = \frac{310,000.0^2}{388,804.6^2} = 0.797 \qquad \sin(\varphi_A) = 0.604 \qquad \varphi_A = 31.12°$$

Armature current of generator A per unit

$$i_{aA} = \frac{P_A}{P_{nA}} \frac{\cos(\varphi_{nA})}{\cos(\varphi_A)} = \frac{310,000}{300,000} \frac{0.8}{0.797} = 1.037$$

EMF e_{fA} of generator A per unit

$$e_{fA} = \sqrt{1 + (x_{sdA} i_{aA})^2 + 2 x_{sdA} i_{aA} \sin(\varphi_A)}$$

$$= \sqrt{1 + (1.7 \times 1.037)^2 + 2 \times 1.7 \times 1.037 \times 0.604} = 2.497$$

EMF e_{fnA} of generator A per unit under nominal conditions

$$e_{fnA} = \sqrt{1 + x_{sdA}^2 + 2 x_{sdA} \sin(\varphi_{nA})} = \sqrt{1 + 1.7^2 + 2 \times 1.7 \times 0.6} = 2.435$$

Field current of generator A

$$I_{fA} = I_{fnA} \frac{e_{fA}}{e_{fnA}} = 15.0 \times \frac{2.497}{2.435} = 15.4A$$

7.14 Salient-pole rotor synchronous machine

7.14.1 Magnetic field in a salient-pole rotor synchronous machine

In an unsaturated non-salient rotor synchronous machine the air gap $g = const$ and thus the armature reaction d-axis reactance $X_a = X_{ad} = const$, independent of the mutual position of the rotating axis of the stator (armature) MMF with respect to the axis of the rotor.

In salient pole rotor synchronous machine two axes must be distinguished:

- Direct axis d: small air gap, large permeance (small reluctance) for the magnetic flux;
- Quadrature axis q: large air gap, small permeance (large reluctance) for the magnetic flux.

The rotor MMF F_f and rotor magnetic flux Φ_f are in the d-axis (Fig. 7.13). Neglecting the magnetic saturation, it can be assumed that the MMF F_f and magnetic flux Φ_f are proportional to the field excitation current I_f and the phasor $\mathbf{E}_f$ leads the $\mathbf{F}_f$ and Φ_f by an angle of 90°. The magnetic flux Φ_f in the air gap is practically sinusoidal.

The load current in the stator (armature) windings creates an MMF, which rotates in the same direction as the rotor, but the axis of the stator MMF can be shifted by an angle with respect to the rotor axis. The permeance of the magnetic circuit for the stator magnetic flux is dependent on the position of the stator axis with respect to the rotor axis. The stator MMF can be resolved into two components:

- F_{ad} in the d-axis
- F_{aq} in the q-axis

Change in operation mode of the machine causes changes in these components, but the permeance in each axis remains the same, if the machine is unsaturated. The components F_{ad} and F_{aq} are sinusoidal.

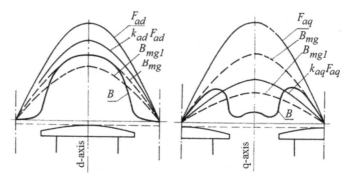

Fig. 7.37. Waveforms of the magnetic flux densities B, B_{mg}, B_{mg1}, MMFs F_{ad}, F_{aq} and equivalent MMFs $k_{ad}F_{ad}$ and $k_{aq}F_{aq}$.

The first harmonics of the curve of the magnetic flux density B_{mg} in the air gap are sinusoidal waveforms for the d and q-axis components and have amplitudes smaller than the B_{mg}, i.e.,

- The amplitude B_{mg1} in the d-axis is $(0.8 \ldots 0.9) \, B_{mg}$
- The amplitude B_{mg1} in the q-axis is $(0.2 \ldots 0.45) \, B_{mg}$

Reduction of the F_{ad} and F_{aq} components of the MMF caused by the saliency of the rotor is taken into account by multiplying F_{ad} and F_{aq} by the following coefficients:

- $k_{ad} \approx 0.8 \ldots 0.9$ in the d-axis
- $k_{aq} \approx 0.2 \ldots 0.5$ in the q-axis

i.e., equivalent MMFs $k_{ad}F_{ad}$ and $k_{aq}F_{aq}$ have been introduced. The equivalent MMFs correspond to constant air gap $g = const$ around the rotor and stator periphery, which is equal to the air gap in the central axis of the rotor poles. This is to obtain at this air gap the waveforms of the first harmonic B_{mg1} of the magnetic flux density (Fig. 7.37).

7.14.2 Form factor of the excitation field

The form factor of the field excitation results from eqn (8.1), i.e.,

$$k_f = \frac{B_{mg1}}{B_{mg}} = \frac{4}{\pi} \sin \frac{\alpha_i \pi}{2} \tag{7.62}$$

where the *pole-shoe arc–to–pole pitch* ratio $\alpha_i < 1$ is calculated according to eqn (7.13).

7.14.3 Form factors of the armature reaction

The *form factors of the armature reaction* are defined as the ratios of the *first harmonic amplitudes–to–maximum values of normal components of armature reaction magnetic flux densities* in the d-axis and q-axis, respectively, i.e.,

$$k_{fd} = \frac{B_{ad1}}{B_{ad}} \qquad k_{fq} = \frac{B_{aq1}}{B_{aq}} \tag{7.63}$$

The peak values of the first harmonics B_{ad1} and B_{aq1} of the armature magnetic flux density can be calculated as coefficients of Fourier series for $\nu = 1$, i.e.,

$$B_{ad1} = \frac{4}{\pi} \int_0^{0.5\pi} B(x) \cos x \, dx \tag{7.64}$$

$$B_{aq1} = \frac{4}{\pi} \int_0^{0.5\pi} B(x) \sin x \, dx \tag{7.65}$$

For a salient-pole motor with electromagnetic excitation and the air gap $g \approx 0$ (fringing effects neglected), the d-axis and q-axis form factors of the armature reaction are

$$k_{fd} = \frac{\alpha_i \pi + \sin \alpha_i \pi}{\pi} \qquad k_{fq} = \frac{\alpha_i \pi - \sin \alpha_i \pi}{\pi} \tag{7.66}$$

7.14.4 Reaction factor

The *reaction factors* in the d- and q-axis are defined as

$$k_{ad} = \frac{k_{fd}}{k_f} \qquad k_{aq} = \frac{k_{fd}}{k_f} \tag{7.67}$$

The form factors k_f, k_{fd} and k_{fq} of the excitation field and armature reaction and reaction factors k_{ad} and k_{aq} for salient-pole synchronous machines according to eqns (7.62), (7.63), (7.66) and (7.67) are given in Table 7.4.

Table 7.4. Factors k_f, k_{fd}, k_{fq}, k_{ad}, and k_{aq} for salient-pole synchronous machines according to eqns (7.62), (7.63), (7.66) and (7.67).

Factor	$\alpha_i = b_p/\tau$						
	0.4	0.5	0.6	$2/\pi$	0.7	0.8	1.0
k_f	0.748	0.900	1.030	1.071	1.134	1.211	1.273
k_{fd}	0.703	0.818	0.913	0.943	0.958	0.987	1.00
k_{fq}	0.097	0.182	0.287	0.391	0.442	0.613	1.00
k_{ad}	0.939	0.909	0.886	0.880	0.845	0.815	0.785
k_{aq}	0.129	0.202	0.279	0.365	0.389	0.505	0.785

7.14.5 Phasor diagram of a salient-pole rotor synchronous machine

The equivalent MMFs $k_{ad}F_{ad}$ and $k_{aq}F_{aq}$ excite their own magnetic fluxes

$$\Phi_{ad} = \frac{2}{\pi}\frac{k_{ad}F_{ad}}{R_\mu} = \frac{2}{\pi}\Lambda k_f k_{ad}F_{ad} = \frac{2}{\pi}\Lambda k_{fd}F_{ad} \tag{7.68}$$

$$\Phi_{aq} = \frac{2}{\pi}\frac{k_{aq}F_{aq}}{R_\mu} = \frac{2}{\pi}\Lambda k_f k_{aq}F_{aq} = \frac{2}{\pi}\Lambda k_{fq}F_{aq} \tag{7.69}$$

The reaction factors k_{ad} and k_{aq} are expressed by eqns (7.67). The form factor k_f of the field excitation and form factors of the armature reaction k_{fd} and k_{fq} are given by eqns (7.62) and (7.66), respectively.

The permeance for the armature reaction fluxes $\Lambda = const$ because it has been assumed that the equivalent air gap $g = const$. Neglecting the saturation of the magnetic circuit this permeance is

$$\Lambda = \mu_0\frac{\tau L_i}{g k_C} \tag{7.70}$$

where k_C is Carter's coefficient given by eqn (5.56).

Each of the magnetic fluxes (7.68) and (7.69) excites in the stator (armature) windings its own EMF of the armature reaction, i.e.,

$$E_{ad} = \pi\sqrt{2}f N_1 k_{w1}\Phi_{ad} \tag{7.71}$$

and

$$E_{aq} = \pi\sqrt{2}f N_1 k_{w1}\Phi_{aq} \tag{7.72}$$

For a synchronous machine with a non-salient pole rotor the permeances of the magnetic circuit in the d- and q-axis are the same, so that the armature reaction MMF is given by eqn (7.18). Similar equations as for MMFs in the d- and q-axis, i.e., eqns (7.16) and (7.17) can be written for the armature currents, i.e., eqns (7.19) and (7.20). These equations result from Fig. 7.14b.

Putting magnetic fluxes (7.68) and (7.69) to eqns (7.71) and (7.72), the EMFs of armature reaction in complex form are

$$\mathbf{E}_{ad} = j\pi\sqrt{2}fN_1k_{w1}\frac{2}{\pi}\Lambda k_{fd}\frac{m_1\sqrt{2}}{\pi}\frac{N_1k_{w1}}{p}I_{ad}\sin\psi$$

$$= j\frac{4}{\pi}m_1f\frac{(N_1k_{w1})^2}{p}\Lambda k_{fd}I_{ad}\sin\psi = jX_{ad}I_{ad} \tag{7.73}$$

$$\mathbf{E}_{aq} = j\pi\sqrt{2}fN_1k_{w1}\frac{2}{\pi}\Lambda k_{fq}\frac{m_1\sqrt{2}}{\pi}\frac{N_1k_{w1}}{p}I_{aq}\cos\psi$$

$$= j\frac{4}{\pi}m_1f\frac{(N_1k_{w1})^2}{p}\Lambda k_{fq}I_{aq}\cos\psi = jX_{aq}I_{aq} \tag{7.74}$$

Finally, there are the following simple relationships between the EMFs and armature reaction reactances in the d- and q-axis

$$\mathbf{E}_{ad} = jX_{ad}I_{ad} \qquad\qquad \mathbf{E}_{aq} = jX_{aq}I_{aq} \tag{7.75}$$

Putting eqn (7.70) to eqns (7.73) and (7.74), the armature reaction reactances in the d and q axis are

$$X_{ad} = 4m_1\mu_0 f\frac{(N_1k_{w1})^2}{\pi p}\frac{\tau L_i}{gk_C}k_{fd} \tag{7.76}$$

$$X_{aq} = 4m_1\mu_0 f\frac{(N_1k_{w1})^2}{\pi p}\frac{\tau L_i}{gk_C}k_{fq} \tag{7.77}$$

The influence of magnetic saturation is only in the d-axis, which can be taken into account with the aid of saturation factor k_{sat} given by eqn (5.49) and (5.100). Thus, including the saturation of magnetic circuit, the d-axis armature reaction reactance is

$$X_{ad} = 4m_1\mu_0 f\frac{(N_1k_{w1})^2}{\pi p}\frac{\tau L_i}{gk_Ck_{sat}}k_{fd} \tag{7.78}$$

For non-salient pole rotor synchronous machine, the armature reaction reactance X_a is the same in the d- and q-axis and given by eqn (7.46). Thus, the armature reaction reactances for a synchronous machine with salient pole rotor can be shortly expressed as

$$X_{ad} = k_{fd}X_a \qquad\qquad X_{aq} = k_{fd}k_{sat}X_a \tag{7.79}$$

The sum of the armature-reaction reactance X_{ad} or X_{aq} and armature leakage reactance X_1 is called *synchronous reactance*:

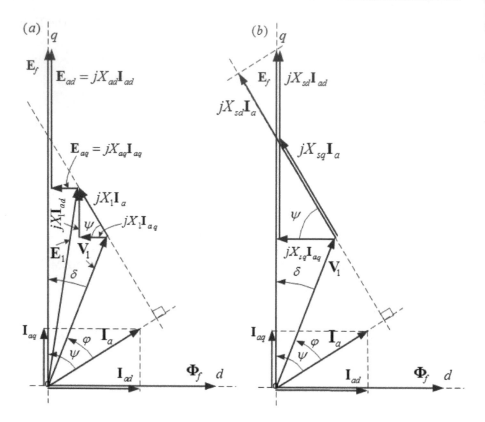

Fig. 7.38. Phasor diagrams of a salient-pole rotor synchronous generator: (a) with X_{ad}, X_{aq} and X_1 reactances; (b) with X_{sd}, X_{sq} synchronous reactances. The stator winding resistance has been neglected, i.e., $R_1 = 0$.

- d-axis synchronous reactance

$$X_{sd} = X_{ad} + X_1 \qquad (7.80)$$

- q-axis synchronous reactance

$$X_{sq} = X_{aq} + X_1 \qquad (7.81)$$

Similar to induction machines, the leakage reactance $X_1 = 2\pi f L_1$ is due to the stator (armature) leakage fluxes: slot leakage reactance, end-turn leakage reactance and differential leakage reactance. The differential leakage reactance is caused by higher space harmonics, i.e., the armature current multiplied by the differential leakage reactance gives a voltage drop due to higher space harmonics of the MMF.

The armature reaction reactances X_{ad} and X_{aq} correspond to the mutual (air gap) reactance X_m of an induction motor. Usually, $X_{sd} > X_{sq}$, except in the case of some PM synchronous machines.

Neglecting the armature winding resistance R_1, the voltage balance equation expressed with the aid of synchronous reactances X_{sd} and X_{sq} is (see Fig. 7.38b)

$$\mathbf{E}_f = \mathbf{V}_1 + jX_{sd}\mathbf{I}_{ad} + jX_{sq}\mathbf{I}_{aq} \tag{7.82}$$

The phasor diagrams of a salient-pole synchronous generator for resistive-inductive RL load and $R_1 = 0$ are plotted in Fig. 7.38. The full phasor diagram with armature winding resistance R_1 being included is shown in Fig. 8.5a (Chapter 8). In phasor diagrams plotted in Fig. 7.38 the stator winding resistance has been neglected, i.e., $R_1 = 0$.

7.14.6 Power and electromagnetic torque of a salient-pole rotor synchronous machine

Neglecting the stator winding resistance R_1 and core losses, the output power P_{out} of a generator or input power of a motor is equal to the electromagnetic power P_{elm}. On the basis of phasor diagram plotted in Fig. 7.38, the output electric power of a synchronous generator is

$$P_{out} = P_{elm} = m_1 V_1 I_a \cos \varphi = m_1 V_1 I_a \cos(\psi - \delta) \tag{7.83}$$

$$= m_1 V_1 I_a (\cos \psi \cos \delta + \sin \psi \sin \delta) = m_1 V_1 I_{ad} \sin \delta + m_1 V_1 I_{aq} \cos \delta$$

Also, Fig. 7.38 allows for finding the inductive voltage drops across the d and q axis synchronous reactances, i.e.,

$$X_{sd}I_{ad} = E_f - V_1 \cos \delta \quad \Rightarrow \quad I_{ad} = \frac{E_f - V_1 \cos \delta}{X_{sd}} \tag{7.84}$$

$$X_{sq}I_{aq} = V_1 \sin \delta \quad \Rightarrow \quad I_{aq} = \frac{V_1 \sin \delta}{X_{sq}} \tag{7.85}$$

Putting eqns (7.84) and (7.85) into eqn (7.83)

$$P_{out} = P_{elm} = m_1 V_1 \frac{V_1 \sin \delta}{X_{sq}} \cos \delta + m_1 V_1 \frac{E_f - V_1 \cos \delta}{X_{sd}} \sin \delta \tag{7.86}$$

Using trigonometric identity $\sin \delta \cos \delta = 0.5 \sin(2\delta)$

$$P_{out} = P_{elm} = m_1 \frac{V_1 E_f}{X_{sd}} \sin \delta + m_1 \frac{V_1^2}{2} \left(\frac{1}{X_{sq}} - \frac{1}{X_{sd}} \right) \sin(2\delta) \tag{7.87}$$

Neglecting R_1 and power losses, the electromagnetic torque is equal to the shaft torque, i.e., $T_{elm} = T$. The electromagnetic torque

$$T_{elm} = \frac{P_{elm}}{2\pi n_s}$$

$$= \frac{m_1}{2\pi n_s} \left[\frac{V_1 E_f}{X_{sd}} \sin\delta + \frac{V_1^2}{2} \left(\frac{1}{X_{sq}} - \frac{1}{X_{sd}} \right) \sin(2\delta) \right] = T_{elmsyn} + T_{elmrel}$$

$$(7.88)$$

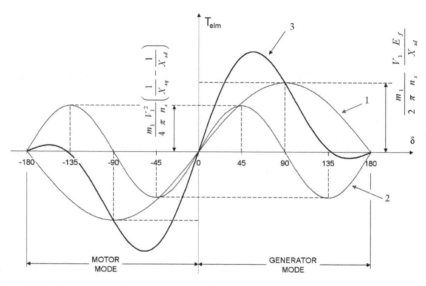

Fig. 7.39. Torque–load angle characteristics at $I_f = const$ and $n = n_s = const$ of a salient-pole rotor synchronous machine. 1 – synchronous torque, T_{elmsyn}, 2 – reluctance torque, T_{elmrel}, 3 – resultant torque, T_{elm}.

The first term in eqn (7.88)

$$T_{elmsyn} = \frac{m_1}{2\pi n_s} \frac{V_1 E_f}{X_{sd}} \sin\delta \qquad (7.89)$$

is the the same as that, which exists in a non-salient pole rotor synchronous machine. The second term

$$T_{elmsrel} = \frac{m_1}{2\pi n_s} \frac{V_1^2}{2} \left(\frac{1}{X_{sq}} - \frac{1}{X_{sd}} \right) \sin(2\delta) \qquad (7.90)$$

is a reluctance torque, which only exists in a salient-pole synchronous machine and for $X_{sd} = X_{sq}$, the reluctance torque $T_{elmrel} = 0$. The reluctance torque T_{elmrel} is independent of the EMF E_f, so it is independent of the field excitation current I_f. In order to produce the reluctance torque, a salient pole rotor synchronous machine does not need to be excited or have PMs in the rotor. The salient-pole rotor must be only connected to the three-phase power supply.

The electromagnetic torque $T_{elm}(\delta)$ and its components T_{elmsyn} and T_{drel} at $I_f = const$ and $n = n_s = const$ are plotted against the load angle δ in Fig. 7.39.

Example 7.3

A salient-pole synchronous generator has the following nominal parameters:

- Apparent power $S_n = 50$ kVA;
- Line voltage $V_{1Ln} = 380$ V;
- Frequency $f_n = 60$ Hz;
- Speed $n_n = 1800$ rpm;
- Power factor $\cos \varphi_n = 0.82$.

Laboratory tests performed on this generator have shown that:

1. While measuring synchronous reactances X_{sd} and X_{sq} with the aid of "small slip" method, the extremum currents drawn by an unexcited generator (open field excitation winding terminals) driven with speed $n = 1768$ rpm and connected to the utility grid with voltage $V_s = 115$ V and frequency $f_s = 60$ Hz are: $I_{max} = 22.2$ A, $I_{min} = 11.3$ A.
2. The field winding (wound with copper wire) resistance $R_f = 0.8$ Ω.
3. The field excitation current to obtain nominal voltage at no load and nominal speed is $I_{f0} = 10.5$ A.

Neglecting the stator (armature) resistance and assuming linear magnetization curve (straight line) of the machine, find:

(a) Synchronous reactances X_{sd} and X_{sq};
(b) Nominal field excitation current I_{fn};
(c) Requested voltage across the field winding terminals, under nominal operating conditions of the generator, if the temperature of the field excitation winding is $t = 120°C$.

Solution

Nominal phase voltage

$$V_{1n} = \frac{V_{1Ln}}{\sqrt{3}} = \frac{380}{\sqrt{3}} = 219.4 \text{ V}$$

Nominal current

$$I_{an} = \frac{S_n}{\sqrt{3}V_{1Ln}} = \frac{50\ 000}{\sqrt{3} \times 380} = 76.0 \text{ A}$$

Angle between the current and voltage under nominal operating conditions

$$\varphi_n = \arccos(\varphi_n) = \arccos(0.82) = 34.9° \quad \Rightarrow \quad \sin(\varphi_n) = \sin(34.9) = 0.572$$

(a) Synchronous reactances X_{sd} and X_{sq}

To measure synchronous reactances, an unexcited generator is connected to a grid with known voltage and frequency and driven with a speed close to synchronous speed. Neglecting the resistance of the field winding, under these operating conditions, the armature reaction magnetic flux Φ_a is proportional to $V_s/f_s = const$. The axis of armature reaction flux under subsynchronous speed of the rotor moves around the rotor periphery. Magnetic asymmetry of the rotor causes that constant armature reaction flux requires variable armature current. The armature current takes its minimum value when the axis of magnetic flux is in the d-axis. The maximum armature current corresponds to the magnetic flux in the q-axis.

- d-axis synchronous reactance

$$X_{sd} = \frac{V_s}{\sqrt{3}I_{min}} = \frac{115}{\sqrt{311.3}} = 5.876 \ \Omega$$

- q-axis synchronous reactance

$$X_{sq} = \frac{V_s}{\sqrt{3}I_{max}} = \frac{115}{\sqrt{322.2}} = 2.991 \ \Omega$$

(b) Nominal field excitation current

In a synchronous generator with linear magnetization curve at constant speed the field excitation current I_f is proportional to the EMF E_f induced in the armature winding by the field excitation flux of the rotor. To find the nominal field current I_{fn} the EMF E_{fn} must be known. The EMF E_{fn} can be found using the simplified phasor diagram of salient-pole generator at nominal operating conditions and neglecting the armature winding resistance. The following equation results from the phasor diagram (Fig. 7.38b:

- Nominal phase EMF excited by the rotor flux

$$E_{fn} = \sqrt{(V_{1n} \cos \varphi_n)^2 + (V_{1n} \sin \varphi_n + X_{sq}I_{an})^2} + (X_{sd} - X_{sq})I_{adn}$$

- d-axis armature current

$$I_{adn} = I_{an} \sin \psi_n$$

- Angle ψ_n between the armature current and the q-axis (or EMF E_{fn})

$$\sin \psi_n = \frac{V_{1n} \sin \varphi_n + X_{sq} I_{an}}{\sqrt{(V_{1n} \cos \varphi_n)^2 + (V_{1n} \sin \varphi_n + X_{sq} I_{an})^2}}$$

On the basis of the above three equations, the EMF E_{fn} has the form

$$E_{fn} = \frac{V_{1n}^2 + V_{1n}(X_{sd} + X_{sq})I_{an} \sin \varphi_n + X_{sd} X_{sq} I_{an}^2}{\sqrt{V_{1n}^2 + 2V_{1n} X_{sq} I_{an} \sin \varphi_n + X_{sq}^2 I_{an}^2}}$$

For $\varphi_n = 34.9°$

$$E_{fn} = \frac{219.4^2 + 219.4(5.876 + 2.991) \times 76.0 \times 0.572 + 5.876 \times 2.991 \times 76.0^2}{\sqrt{219.4^2 + 2 \times 219.4 \times 2.991 \times 76.0 \times 0.572 + 2.991^2 \times 76.0^2}}$$

$$= 591.2 \text{ V}$$

Nominal field excitation current

$$I_{fn} = I_{f0} \frac{E_{fn}}{V_{1n}} = 10.5 \frac{591.2}{219.4} = 28.3 \text{ A}$$

For other values of the angle φ between the voltage and current the EMF E_f and the field excitation current I_f are:

- $\varphi = 0°$, $E_f = 473.5$ V, $I_f = 22.7$ A
- $\varphi = 90°$, $E_f = 665.8$ V, $I_f = 31.9$ A
- $\varphi = -45°$, $E_f = 263.4$ V, $I_f = 12.6$ A

(c) Requested voltage across the field winding terminals, under nominal operating conditions of the generator, if the temperature of the field excitation winding is $t = 120° C$

Field winding resistance (copper) at 120°C

$$R_{f120} = R_f \frac{235 + t}{235 + t_0} = 0.8 \frac{235 + 120}{235 + 25} = 1.092 \text{ } \Omega$$

Requested voltage across the field winding terminals under nominal operating conditions

$$V_{fn} = R_{f120} I_{fn} = 1.092 \times 28.3 = 30.9 \text{ V}$$

7.15 Aircraft generators

The function of the *aircraft electrical system* is to generate, regulate and distribute electrical power throughout the aircraft. Aircraft electrical components operate on many different voltages both AC and DC. Most systems use 115 V AC (400 Hz) and 28 V DC. There are several different electric generators on large aircraft (Fig. 7.40) to be able to handle excessive loads, for redundancy, and for emergency situations, which include [12, 34]:

- Engine driven AC generators;
- Auxiliary power units (APU);
- Ram air turbines (RAT);
- External power, i.e., ground power unit (GPU).

Each of the engines on an aircraft drives one or more AC generators. The power produced by these generators is used in normal flight to supply the entire aircraft with power. The power generated by APUs is used while the aircraft is on the ground during maintenance and for engine starting. Most aircraft can use the APU while in flight as a backup power source. RATs are used in the case of an engine generator or APU failure, as an emergency power source. External power may only be used with the aircraft on the ground. A GPU (portable or stationary unit) provides AC power through an external plug on the nose of the aircraft.

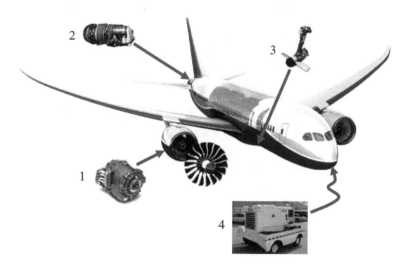

Fig. 7.40. Passenger aircraft generators: 1 – main engine starter/generator, 2 – auxiliary power unit (APU), 3 – emergency ram air turbine (RAT), 4 – ground power unit (GPU).

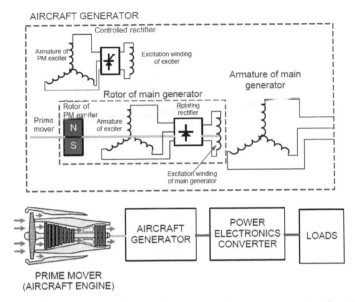

Fig. 7.41. Power circuit of wound rotor synchronous generator for aircraft.

Aircraft generators are usually wound-rotor synchronous machines with synchronous brushless exciter and PM brushless sub-exciter to provide the d.c. field excitation current to the rotor. The power circuit is shown in Fig. 7.41. PM brushless generators are rather avoided due to difficulties with shutting down the power in failure modes, e.g., interturn short circuit in the stator winding. There are also attempts of using switched reluctance (SR) generators with no windings or PMs on the rotor. A generator control unit (GCU), or voltage regulator, is used to control generator output. The generator shaft is driven by an aircraft engine with the aid of gears or directly by low spool engine shaft.

Aircraft generators are typically three-phase synchronous generators with outer stator with distributed-parameter winding and inner rotor with concentrated coil winding (Fig. 7.42). These rules do not apply to special PM synchronous generators with flux regulation and SR generators.

The stator of aircraft synchronous generators has slotted winding located in semi-closed trapezoidal or oval slots. A large number of stator slots per pole per phase and double layer chorded windings allow for reducing the contents of higher space harmonics in the air gap magnetic flux density waveforms.

The number of salient rotor poles is typically from 2 to 12. Pole faces have round semi-closed slots to accommodate the damper. The rotor core is made of the same material as the stator core, i.e., iron-cobalt thin laminations. Rotor coils are protected against centrifugal forces with the aid of metal wedges between poles which also participate in the cooling system of the rotor.

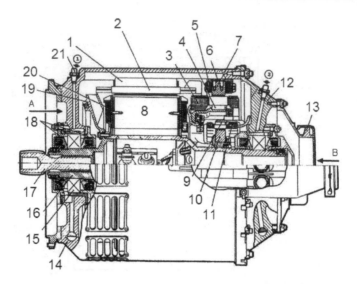

Fig. 7.42. Wound-field air-cooled synchronous generator GT40PC6 rated at 30 kVA: 1 – armature core of main generator, 2 – armature winding of main generator, 3 – armature winding of exciter, 4 – armature core of exciter, 5 – field winding of exciter, 6 – pole, 7 – field excitation system of exciter, 8 – rotor pole of main generator, 9 – armature of subexciter, 10 – PM, 11 – armature winding of subexciter, 12 – end shield, 13 – nozzle, 14 – housing, 15 – bearing, 16 – hollow shaft of rotor, 17 – shaft end, 18 – flanges, 19 – fan, 20 – field winding of main generator, 21 – point of lubrication [14].

The rotor field excitation winding is connected via rotating diode rectifier to a three-phase armature winding of a brushless exciter (Fig. 7.41). The brushless exciter has the armature winding on the rotor and stationary field excitation winding. The exciter armature system (winding and laminated stack), rectifier and excitation winding of the main generator are located on the same shaft. The DC current is to be supplied from an external DC source, usually from a small PM generator (sub-exciter) with stationary armature winding and rotating PMs. Rotating PMs are located on the shaft of the main generator.

The frequency of the armature current of a synchronous generator driven by the aircraft engine is speed dependent. Both the shaft speed and output frequency can be constant or variable. Consequently, aircraft generators are divided into the three following groups:

- Constant speed constant frequency (CSCF) generators;
- Variable speed constant frequency (VSCF) generators;
- Variable frequency (VF) generators[1].

---

[1] Sometimes called "wild frequency" (WF) generators.

A constant output frequency without an AC to AC utility converter can only be obtained if the generator is driven at a constant speed.

VSCF systems employ an AC three-phase generator and solid state converter. The solid state converter consists of (a) a rectifier which converts a variable frequency current into DC current, (b) intermediate circuit and (c) inverter which then converts the DC current into constant frequency three phase AC current.

In variable frequency (VF) systems the output frequency of an AC generator is permitted to vary with the rotational speed of the shaft. The VF is not suitable for all types of AC loads. It can be applied directly only to resistive loads, e.g., electric heaters (deicing systems).

Also, generators are turned by a differential assembly and hydraulic pumps to obtain constant speed. The purpose of the constant speed drive (CSD) is to take rotational power from the engine and, no matter the engine speed, turn the generator at a constant speed. The CSD converts variable engine speed into constant speed to run an AC generator. This is necessary because the generator output must be constant frequency, e.g., 400 Hz.

An integrated drive generator (IDG) is simply a CSD and generator combined into one unit mounted co-axially or side-by-side.

Aircraft generators use forced air or oil cooling systems. The most effective is the so-called *spray oil cooling* where end turns of stator windings are oil-sprayed. The current density of spray-oil cooled windings can exceed 28 A/mm^2. Pressurized oil can also be pumped though the channels between round conductors in slots.

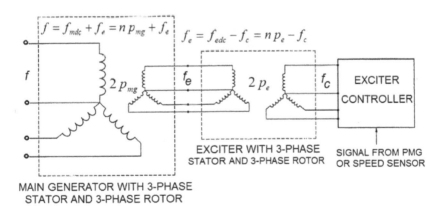

Fig. 7.43. Architecture of a doubly-fed AC generator.

In VSCF applications, the so-called doubly-fed AC generator can provide good performance. These AC main generators have three-phase stator (armature) windings and three-phase rotor armature windings (Fig. 7.43). The AC exciter has reversed construction: three-phase armature winding on the rotor, and three-phase excitation winding on the stator (Fig. 7.43). In this way, the design is totally brushless. The three-phase stationary winding of the exciter is fed from a solid state controller. Instead of a three-phase exciter, a three-phase rotary transformer can be used. The output frequency of the main generator when excited with DC current is

$$f_{mdc} = p_{mg} n_{sh} \qquad (7.91)$$

where n_{sh} is the shaft (prime mover) speed and p_{mg} is the number of pole pairs of the main generator. The output frequency of the main generator when excited with AC current

$$f = f_{mdc} + f_e \qquad (7.92)$$

The frequency of exciter (frequency of the rotor of main generator)

$$f_e = f_{edc} - f_c \qquad (7.93)$$

The output frequency of the controller f_c can be either positive or negative sequence. In the above equation (7.93) the frequency of the exciter rotor when excited with DC current is

$$f_{edc} = p_e n_{sh} \qquad (7.94)$$

where p_e is the number of pole pairs of the three-phase exciter. To keep the nominal (rated) output frequency $f = f_n = constant$ of the main generator at given shaft speed n_{sh}, the frequency f_c of the controller must be equal to

$$f_c = f_{edc} - f + f_{mdc} \qquad (7.95)$$

The output frequency of the controller f_c can be either positive (the "+" sign) or negative (the "−" sign) sequence. Neglecting the losses, the output power of the main generator is

$$P_{out} = \frac{P_{in}}{1 - |s|} \qquad (7.96)$$

where P_{in} is the required mechanical power by the main generator and s is the slip defined as

$$s = \frac{n_{sh} - f/p_{mg}}{n_{sh}} \qquad (7.97)$$

In the above equation (7.97) the speed n_{sh} is the shaft speed that appears in eqn (7.91), p_{mg} is the number of pole pairs of the main generator and f is the output frequency of the main generator, as given by eqn (7.92). For subsynchronous speed $s > 0$ and for super-synchronous speed $s < 0$. The power absorbed or delivered by the rotor of the main generator is equal to the power of the exciter P_e, i.e.,

$$P_e = |s| P_{out} \qquad (7.98)$$

Example 7.4

Calculate the output frequency of a three-phase, 4-pole doubly-fed AC generator operating with a 10-pole exciter for:

(a) Shaft speed $n_{sh} = 1200$ rpm, controller frequency $f_c = -260$ Hz (negative sequence);
(b) Shaft speed $n_{sh} = 4800$ rpm, controller frequency $f_c = +160$ Hz (positive sequence).

Solution

(a) $n_{sh} = 1200$ rpm, $f_c = -260$ Hz (negative sequence)

The output frequency of the main generator when excited with DC current acording to eqn (7.91)

$$f_{mdc} = 2\frac{1200}{60} = 40 \text{ Hz}$$

The frequency of the exciter rotor when excited with DC current according to eqn (7.94)

$$f_{edc} = 5\frac{1200}{60} = 100 \text{ Hz}$$

The frequency of exciter (frequency of the rotor of main generator) according to eqn (7.93)

$$f_e = 100 - (-260) = 360 \text{ Hz}$$

The output frequency of the main generator when excited with AC current according to eqn (7.92)

$$f = 40 + 360 = 400 \text{ Hz}$$

(b) $n_{sh} = 4800$ rpm, $f_c = +160$ Hz (negative sequence)

The output frequency of the main generator when excited with DC current according to eqn (7.91)

$$f_{mdc} = 2\frac{4800}{60} = 160 \text{ Hz}$$

The frequency of the exciter rotor when excited with DC current according to eqn (7.94)

$$f_{edc} = 5\frac{4800}{60} = 400 \text{ Hz}$$

The frequency of exciter (frequency of the rotor of main generator) according to eqn (7.93)

$$f_e = 400 - (+160) = 240 \text{ Hz}$$

The output frequency of the main generator when excited with AC current according to eqn (7.92)

$$f = 160 + 240 = 400 \text{ Hz}$$

In both cases, independent of the shaft speed n_{sh} the output frequency $f = 400$ Hz $= const$ is stablized by variable frequency f_c of the controller.

7.16 Synchronous motor

7.16.1 Fundamentals

A synchronous machine fed with utility grid and loaded with a torque operates as a *synchronous motor*. Synchronous motors are usually salient-pole rotor synchronous machines.

A significant feature of the wound-field (electromagnetically-excited) synchronous motor is the *controllability of its power factor* up to unity or leading values. To avoid the need for slip rings and brushes, almost all modern wound-field rotor synchronous motors are provided with *brushless exciters*, i.e., a small AC generator is mounted on the rotor shaft, the output of which is rectified by shaft-mounted rectifiers rotating with the rotor (Section 7.12). A cage winding is frequently mounted on salient-pole rotors that act as the so-called *damper*, damping the oscillations under transient conditions. This winding in synchronous motors is necessary to start the motor (asynchronous starting).

Synchronous motors can operate with higher efficiency than induction motors and can keep constant speed even under load variation and voltage fluctuation. They are most suitable for industrial drives such as compressors, blowers, pumps, fans, mills, crushers and motor-generator sets. Overexcited synchronous motors draw leading reactive current and can be used to compensate for a large number of induction motors that draw lagging reactive power. Synchronous motors are more expensive than their induction counterparts.

The most important characteristics of a synchronous motor are its torque–load angle characteristic $T_{elm} = f(\delta)$ (Fig. 7.39) and its armature current–field current characteristics (V-curves) $I_a = f(I_f)$ at $P_{out} = const$ or at $T = const$ (Figs 7.35 and 7.44).

Fig. 7.44 shows the V-curves for a synchronous motor and corresponding power factor curves versus the field excitation current, $\cos \varphi = f(I_f)$. The left-hand side of the graph is for the underexcited motor, i.e., when the motor

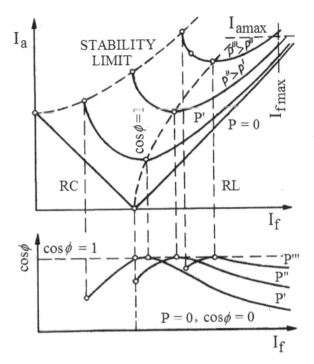

Fig. 7.44. V-curves $I_a = f(I_f)$ and power factor curves $\cos\varphi = f(I_f)$ of a synchronous motor at $P_{out} = const$ and $n = n_s = const$.

draws both active and reactive inductive power (RL load), while the right-hand side of the graph is for the overexcited motor, i.e., when the motor draws the active power and delivers the reactive inductive power, which is equivalent to drawing the capacitive power (RC). Any change in the exciting current at $P_{out} = const$ results in a change in both the armature current and the power factor. Thus, by changing the field excitation current in a synchronous motor, the power factor $\cos\varphi$ can be controlled. The overexcited synchronous motor is an RC load and in this way it compensates the reactive inductive power, which is drawn by induction motors and transformers operating without any load.

If the power $P_{out} = 0$, the synchronous motor practically draws from the utility grid only reactive power, i.e., capacitive power when the synchronous motor is overexcited and inductive power when the synchronous motor is underexcited. Because there are losses in the synchronous motor, some small active power $P_0 > 0$ is drawn from the utility grid.

The overload capacity factor (Fig. 7.39) of a synchronous motor is

$$OCF = \frac{P_{elmmax}}{P_{elm}} \qquad (7.99)$$

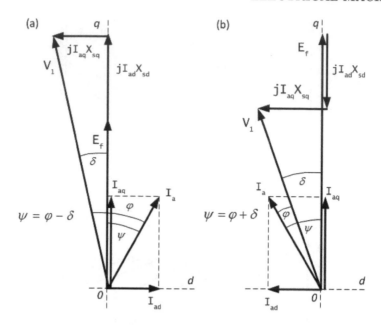

Fig. 7.45. Simplified phasor diagram for salient-pole synchronous motor ($R_1 = 0$): (a) underexcited motor; (b) overexcited motor.

The simplified phasor diagram for salient-pole synchronous motor, in which the stator winding resistance $R_1 = 0$, is given in Fig. 7.45. The full phasor diagram with $R_1 \neq 0$ is given in Fig. 8.5 (Chapter 8).

Example 7.5

A three-phase, four-pole, 500-kW, 3.3-kV (line-to-line), 50-Hz, $\cos \phi = 0.8$ (lagging), $\eta = 0.94$, Y-connected, salient-pole synchronous motor has the d-axis synchronous reactance $X_{sd} = 10.0$ Ω and the q-axis synchronous reactance $X_{sq} = 6.0$ Ω. The stator resistance R_1 can be neglected (large synchronous motor). Find the electromagnetic power, electromagnetic torque, maximum electromagnetic synchronous torque and maximum developed reluctance torque.

Solution

The number of pole pairs is $p = 2$. The synchronous speed according to eqn (7.1)

$$n_s = \frac{50}{2} = 25 \text{ rev/s} = 1500 \text{ rpm}$$

The phase voltage $V_1 = 3300/\sqrt{3} = 1905.25$ V. The armature current

$$I_a = \frac{P_{out}}{3V_1 \eta \cos\phi} = \frac{500 \times 10^3}{3 \times 1905.25 \times 0.94 \times 0.8} = 116.33 \text{ A}$$

According to Fig. 7.45 if $R_1 = 0$, the voltage drop $V_1 \sin\delta \approx I_{aq}X_{sq} = I_a X_{sq} \cos(\phi \pm \delta)$ where $\cos(\phi \pm \delta) = \cos\phi\cos\delta \mp \sin\psi\sin\delta$. Thus

$$\tan\delta \approx \frac{I_a X_{sq}\cos\phi}{V_1 \pm I_a X_{sq}\sin\phi}$$

where the "+" sign is used when $\Psi = \phi + \delta$ and the "−" sign is used when $\Psi = \phi - \delta$. For given rated conditions

$$\tan\delta \approx \frac{116.33 \times 6.0 \times 0.8}{1905.25 - 116.33 \times 6.0 \times 0.6} = 0.3756$$

where $\sin(\phi) = 0.6$, i.e., $\cos(\phi) = 0.8$. The rated power angle $\delta = 20.59^0$, $\sin(\delta) = 0.3517$ and $\cos(\delta) = 0.9361$.

The q-axis armature current results from the phasor diagram shown in Fig. 7.45 in which $R_1 = 0$, i.e.,

$$I_{aq} \approx \frac{V_1 \sin(\delta)}{X_{sq}} = \frac{1905.25 \times 0.3517}{6.0} = 111.67 \text{ A}$$

The d-axis armature current

$$I_{ad} = \sqrt{I_a^2 - I_{aq}^2} = \sqrt{116.33^2 - 111.67^2} = 32.59 \text{ A}$$

The EMF E_f excited by the rotor flux can also be found on the basis of Fig. 7.45 in which $R_1 = 0$, i.e.,

$$E_f \approx V_1 \cos\delta - I_{ad}X_{sd} = 1905.25 \times 0.9361 - 32.59 \times 10.0 = 1457.64 \text{ V}$$

The electromagnetic power P_{elm} according to eqn (7.87)

$$P_{elm} = 3[\frac{1905.25 \times 1457.64}{10.0}\sin(20.59^0)$$

$$+\frac{1905.25^2}{2}\left(\frac{1}{6.0} - \frac{1}{10.0}\right)\sin(2 \times 20.59^0)] = 531,997.7 \text{ W}$$

Since the stator winding resistance and the stator core losses have been neglected, the calculated electromagnetic power P_{elm} is too high and equal to the input power P_{in}. The electromagnetic torque according to eqn (7.88)

$$T_{elm} = \frac{531\ 997.7}{2\pi \times 25} = 3386.8 \text{ Nm}$$

The maximum synchronous torque is for $\delta = 90°$ or $\sin(\delta) = 1$, i.e.,

$$T_{elmsynmax} = \frac{3}{2\pi n_s} \frac{V_1 E_f}{X_{sd}} = \frac{3}{2\pi \times 25} \frac{1905.25 \times 1457.64}{10.0} = 5304 \text{ Nm}$$

The maximum reluctance torque is for $\delta = 45°$ or $\sin(2\delta) = 1$, i.e.,

$$T_{elmrelmax}$$

$$= \frac{3V_1^2}{4\pi n_s}\left(\frac{1}{X_{sq}} - \frac{1}{X_{sd}}\right) = \frac{3 \times 1905.25^2}{4\pi \times 25}\left(\frac{1}{6.0} - \frac{1}{10.0}\right) = 2310.91 \text{ Nm}$$

7.16.2 Starting

A synchronous motor does not produce a starting current after switching on the voltage to the stator winding. There are three *methods of starting* synchronous motors:

- Asynchronous starting;
- Starting by means of an auxiliary motor;
- Frequency-change starting.

Asynchronous starting

A synchronous motor that has a cage winding on its rotor can be started as a cage induction motor. The starting torque is produced as a result of the interaction between the stator rotating magnetic field and the rotor cage winding currents. In small synchronous motors with solid steel salient poles, the cage winding is not necessary since the eddy currents induced in the solid steel pole shoes can interact with the stator magnetic field.

At the first instant of starting, using the asynchronous starting method, the field winding of the synchronous motor should be short-circuited or else closed via a resistance, about 10 times that of the resistance of field winding itself. Otherwise, due to high inductance of the field winding, a high voltage induced at the output terminals of the field winding could damage the insulation. The stator winding is connected to an AC power mains. The electromagnetic torque is produced and the motor accelerates running similarly to an induction motor. When the speed is closed to the synchronous speed (small value of slip) the DC current should be switched on to the field winding and the synchronous motor is pulled into synchronism. Three characteristic torques can be distinguished (Fig. 7.46):

- The starting electromagnetic torque T_{elmst} developed by the motor at standstill, i.e., slip $s = 1$;
- The pull-in torque T_{pin} developed by the synchronous motor operating as an induction motor at speed $n = 0.95 n_s$ or slip $s = 0.05$;

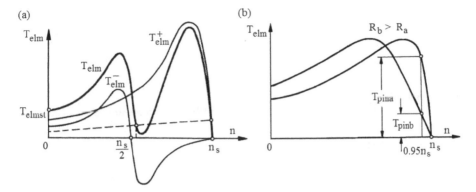

Fig. 7.46. Asynchronous starting characteristics of synchronous motor: (a) asynchronous torque T_{elm}^{+} due to forward-traveling magnetic field, asynchronous torque T_{elm}^{-} due to backward-traveling magnetic field, and resultant asynchronous torque T_{elm}; (b) influence of cage winding resistance $R_a < R_b$ on the starting electromagnetic torque and pull-in torque $T_{pina} > T_{pinb}$.

- The maximum or breakdown torque T_{elmmax} at $s = s_{cr}$, $V_1 = V_{1n}$, $I_f = I_{fn}$ above which the motor falls out of step.

The forward-traveling magnetic field produced by the stator winding moves in the same direction as the rotor. Its speed relative to the rotor is sn_s. The backward-traveling magnetic field moves in opposite direction to the rotor with the speed $n_s(1-s) + sn_s$ or $n_s(1-s) - sn_s$, respectively. Thus,

$$n_s(1-s) + sn_s = n_s \qquad\qquad n_s(1-s) - sn_s = (1-2s)n_s \qquad (7.100)$$

The electromagnetic torque created by the forward-traveling field T_{elm}^{+} is the same as that in an induction motor and is added to the torque of the cage winding. The electromagnetic torque created by the backward-traveling field T_{elm}^{-} depends on the slip s and is added to or subtracted from the resultant torque of the cage winding and the torque created by the backward-traveling field. At $s = 0.5$ this torque $T_{elm}^{-} = 0$ (Fig. 7.45a).

The starting cage winding should have an appropriate resistance (Fig. 7.45b). If the resistance R_a is small, the pull-in torque T_{pina} is high at subsynchronous speed close to the synchronous speed. If the resistance R_b is large, the pull-in torque is small, which gives greater difference between the synchronous and subsynchronous speed. This makes it more difficult to pull-out the rotor into synchronism (Fig. 7.45b).

Starting by means of an auxiliary motor

The synchronous motor has an auxiliary starting motor on its shaft, capable of bringing it up to the synchronous speed at which synchronizing with the power circuit is possible. This auxiliary motor is commonly an induction motor. At start-up, the unexcited synchronous motor is accelerated almost to synchronous speed using the induction motor. At first, the synchronous motor is connected to the mains with an additional resistance across its rotor field excitation winding. The value of this resistance is about 10 times that of the resistance of the field winding itself. If the latter were left open when the motor was started, such a high voltage would be induced across its terminals (owing to the great number of turns in the winding) that there would be a real danger of a breakdown of the insulation. Once the speed is close to the synchronous speed, the additional field winding resistance is removed and a DC excitation field current is supplied. The synchronous motor is then pulled into synchronism.

The disadvantage of this method is that it is impossible to start the motor under load. Larger auxiliary motors and very costly installation would be involved, and this is generally completely impractical.

Frequency-change starting

In this method, the frequency of the voltage applied to the motor is smoothly changed from zero up to the rated value. The motor runs synchronously during the entire starting period. Solid-state converters are commonly used to achieve this (Fig. 7.47).

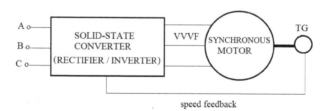

Fig. 7.47. Frequency-change starting.

7.16.3 Comparison of synchronous motors with induction motors

Table 7.4 gives a comparison of the speed, power factor $\cos \varphi$, air gap, torque-voltage characteristic, and cost of wound-field rotor synchronous motors and induction motors. The large air gap in synchronous motors, which actually makes them much more reliable than induction motors, is also a practical requirement, which minimizes the effect of the armature reaction, minimizes the synchronous reactance and improves stability.

Table 7.5. Comparison of wound-field rotor synchronous motors and induction motors

Specifications	Wound-field rotor synchronous motor	Induction motor
Speed	Constant, independent of the load	As the load increases, the speed decreases slightly
$\cos \varphi$	Adjustable $\cos \varphi$ Operation at $\cos \phi = 1$ is possible	Impossible to change $\cos \varphi$ (except for inverter-fed motors) $\cos \varphi \approx 0.8...0.9$ at rated load $\cos \varphi = 0.1$ at no load
Air gap	Large, from a few millimeters up to centimeters	Small, from less than 1 mm to maximum 3 mm
Torque-voltage characteristic	Torque directly proportional to the input voltage Better starting performance than that of an induction motor	Torque directly proportional to the square of the input voltage
Cost	Expensive machine	Cost-effective machine

7.17 Synchronous reluctance motors

The *synchronous reluctance motor* runs with an unexcited rotor that has unequal reluctance in its *d*- and *q*-axis. Since the rotor has no excitation system, the EMF in eqns (7.82), (7.84), (7.86), (7.87), (7.88) and (7.89) is zero ($E_f = 0$).

The input current I_a of a reluctance motor is higher than that of ordinary synchronous and induction motors since the EMF $E_f = 0$. This correspondingly affects the efficiency because of the resultant high power losses in the stator winding.

The electromagnetic torque (or reluctance torque) developed by the motor is obtained from eqn (7.88), once again by setting $E_f = 0$. The T_{elmrel} given by eqn (7.90) exists as a result of a change in the energy of the magnetic field in the air gap due to any mismatch between the field and rotor axes. Variation of the reluctance in the motor air gap is afforded by a proper selection of the rotor shape and material (see Figs 7.48 and 7.49). The electromagnetic torque of a reluctance motor can be increased by magnifying the difference in the synchronous reactances in the *d*- and *q*-axis, so as to make the ratio X_{sd}/X_{sq} as high as possible. However, this in turn involves a heavier magnetizing current resulting in a further increase in the input current due to the high reluctance of the magnetic circuit in the *q*-axis. A reluctance motor, like a PM synchronous motor, has good heat exchange properties, since most of its losses are produced in the stator.

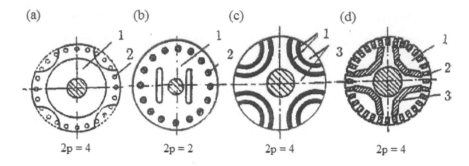

Fig. 7.48. Rotors for synchronous reluctance motors: (a) salient-pole rotor, (b) cylindrical rotor with internal channels, (c) and (d) cylindrical composite rotor (with flux barriers): 1 — ferromagnetic core, 2 — cage winding, 3 — aluminum.

Fig. 7.49. Rotor with four magnetic flux barriers for synchronous reluctance motors. Photo courtesy *Easitech*, Hangzhou, China, www.easi-tech.com.

The reluctance torque, T_{elmrel}, corresponding to the fundamental harmonic of the AC component of magnetic reluctance can be plotted against δ on the basis of eqn (7.90) (disregarding the stator winding resistance). The

(a)

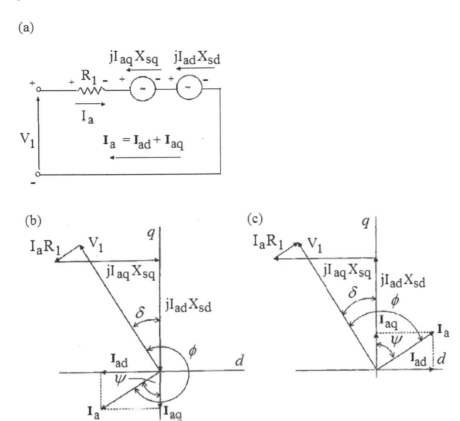

Fig. 7.50. Equivalent circuit and phasor diagrams for an AC reluctance motor: (a) equivalent circuit per phase, (b) phasor diagram for generator arrow system, (c) phasor diagram for consumer arrow system.

torque T_{elmrel} varies as $\sin(2\delta)$ in accordance with line 2 in Fig. 7.39. The steady-state condition of the motor is attained at the particular angle δ at which $T_{elmrel} = T$, where T is the steady-state external torque on the motor shaft.

If the rotor and the stator rotating field revolve at different speeds, the angle δ becomes a periodic function of time and the average value of the reluctance torque is zero. That is why, during the starting period, reluctance motors run as induction machines. In the course of starting reluctance motors, an AC component of magnetic flux appears because of the variation in the reluctance. This induces additional EMF in the stator winding.

The salient-pole rotor of Fig. 7.48a has a laminated core with a starting cage winding. If the laminated core is replaced by a solid steel core, the cage winding is removed because the eddy currents in the solid pole shoes produce an asynchronous starting torque. In practical small synchronous reluctance motors, the rotor *pole-arc–to–pole-pitch ratio* $\alpha = b_p/\tau$ lies in the region of 0.5 to 0.6, and the *maximum–to–minimum air gap ratio* is somewhere between 10 and 12. The internal channels of the cylindrical rotor shown in Fig. 7.48b ensure a variation of the reluctance along the motor periphery. As for the rotor illustrated in Figs 7.48c, 7.48d and 7.49, this variation is achieved by the use of two materials with different magnetic properties. Axially laminated rotors can also be used [33].

The equivalent circuit for synchronous reluctance motors is shown in Fig. 7.50a, while the phasor diagrams are shown in Figs 7.50b and 7.50c. An AC reluctance motor behaves as an inductive load, drawing from mains the reactive power necessary for the magnetization of its magnetic circuit (it is an underexcited motor).

The efficiency and power factor of typical reluctance motors are lower than those of cage induction motors. The low power factor is explained by the fact that the magnetic flux of a reluctance motor is entirely due to the stator magnetizing current. Heavy magnetizing current is the result of high reluctance in the q-axis of the magnetic circuit. The greater the inequality of magnetic reluctances and of inductive reactances X_{sd} and X_{sq}, the higher is the reluctance torque, T_{elmrel}, and the lower are the motor's power factor and efficiency.

7.18 Written pole motors

Written pole motors were developed in the early 1980s, [18, 30]. The stator of a *written pole machine* has a three-phase or single-phase winding distributed in slots. In addition, there is a concentrated-parameter winding consisting of a single coil placed around an exciter pole (Fig. 7.51). The rotor is furnished with a high resistance cage winding to produce asynchronous torque. The rotor active surface is also coated with a magnetic material layer, which can be magnetized by the stator exciter pole during operation. Any desired magnetic pole pattern can be "written" on the magnetic layer to provide a hysteresis torque [30]. In this way the starting parameters can be set independently of the operating parameters. The frequency of the exciter winding current is equal to the line frequency (50 or 60 Hz). The machine can be fabricated with either inner rotor (Fig. 7.51a) or outer rotor (Fig. 7.51b) [12].

The written pole motor can develop large synchronous torque below synchronous speed (Fig. 7.52). The hysteresis torque and asynchronous torque start the motor before the exciter pole winding is energized. At subsynchronous speed the exciter winding is energized and writes magnetic poles on the rotor surface. The number of written poles on the rotor surface is inversely proportional to the speed. The rotor is pulled into synchronism when the number of rotor poles is the same as the number of poles of the stator rotating field. By choosing a proper ratio of the exciter current to the stator rotating magnetic

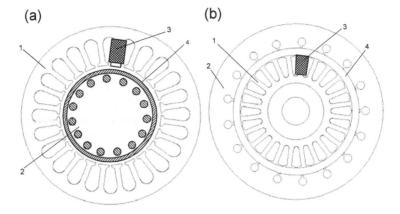

Fig. 7.51. Construction of a written pole machine: (a) with inner rotor; (b) with outer rotor. 1 – stator, 2 – rotor, 3 – stator exciter pole, 4 – rotor ferromagnetic layer [12].

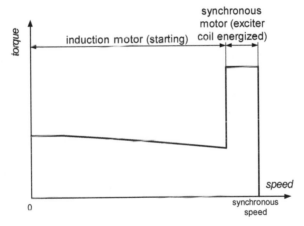

Fig. 7.52. Torque–speed characteristic of a written pole motor [12].

field, the rotor can be accelerated with maximum synchronous torque. When the rotor reaches the synchronous speed, the exciter current is switched off and the motor operates as a PM synchronous motor (Fig. 7.52). The rotor cage winding designed for high slip (about 20%) reduces the starting current to less than one third of that of a conventional induction motor [30]. At low starting current a written pole motor can accelerate to synchronous speed even with very large moment of inertia.

The written pole motor shows the following advantages [12]:

- High efficiency;
- Very low start-up current;

- Ability to start and synchronize loads with very large inertia;
- Ability to operate with about 20% variation in input voltage;
- Riding through power outages of up to 15 s;
- Restarting after power outages longer than 15 s.

Written pole motors are now available in sizes from 7.5 to 75 kW and are used in remote rural areas with single phase reticulation systems. These motors have low starting current across the line, unity power factor, high power factor during start, smooth (constant torque) start, instant restart after momentary power interruptions and can start with high inertia loads.

Certain rural applications require high output power motors capable of operating on single-phase power supplies, e.g., oil fields, gas fields, sewage lift stations, crop irrigation, grain drying and handling, lake aeration and roller mills. Written pole motors provide a feasible solution with both economic and environmental benefits.

Written pole motors are also used to provide power quality and ride through from utility power sags, surges and interrupts for remotely located weather radars and various aviation applications.

A written pole motor can be combined with a three-phase written pole generator. Such a *written pole motor-generator set* has a shared, external, high-inertia rotor and can deliver the highest quality power. Higher harmonics cannot be transferred electromagnetically from the utility grid (motor side) to generator side. The motor–generator sets provide solutions for many three-phase power quality problems, e.g., data reduction, test analysis and laboratories. In industrial manufacturing written pole motor-generator sets are used for process automation computers, distributed information technology (IT) networks, security systems, programmable logic controller (PLCs), telephone systems and data centers. Typical applications in the health care industry are critical care monitoring, intensive care monitoring, nuclear medicine, heart catheter labs, operating rooms and medical records.

Summary

A *synchronous machine* operates at a constant speed in absolute synchronism with the line frequency. The relationship between the synchronous speed n_s, frequency f and number of pole pairs p is given by eqn (7.1).

There are two types of rotors of wound-field synchronous machines: *non-salient pole rotor* (Fig. 7.2a), also called cylindrical rotor and *salient-pole rotor* (Fig. 7.2b). Salient-pole shoes are equipped with a cage winding, the so-called *damper* or *amortisseur* placed near the air gap.

Synchronous machines can be classified as *turboalternators* (turbogenerators) driven by steam turbines at speeds from 1500 to 3600 rpm, *hydroalternators* driven by hydro-power turbines at speeds from 60 to 1000 rpm, *combustion engine driven generators, gas turbine driven generators, microtur-bines,* i.e., generators integrated with gas turbines, *wind turbine generators, synchronous motors* and *synchronous condensers.*

The *rotor outer diameter* of turboalternators is limited to 1.33 m and its length is 2.5 to 6.0 times greater than the outer diameter. Hydrogenerators have large diameter rotors up to 15 m and the diameter-to-length ratio of the rotor is 0.15 to 0.2.

The *stator* of a typical synchronous machine is the same as that of an induction machine, i.e., a three-phase winding is distributed in slots. The EMF induced in the stator winding per phase is expressed by equation (7.11), the same as for induction machines.

Under operation, the stator (armature) produces its own magnetic flux, the so-called *armature reaction magnetic flux.*

The *base* or *nominal impedance* of a synchronous machine is calculated as the ratio of nominal phase voltage to nominal armature current, as given by eqn (7.23).

In synchronous generators the *output terminal voltage* V_1 can be controlled by the DC field excitation current I_f.

The *position of the armature current* I_a in the q–d coordinate systems (Fig. 7.15) determines the operation of synchronous machine: 1st quadrant – overexcited generator, 2nd quadrant – underexcited generator, 3rd quadrant – underexcited motor, 4th quadrant – overexcited motor.

The *fundamental laboratory tests* on synchronous machines are the *no-load test* and *short-circuit test*. The steady-state *open-circuit characteristic* is the EMF E_f curve versus the field current I_f at constant synchronous speed n_s and load impedance $Z_L \to \infty$. The *short-circuit characteristic* is the armature current I_a curve versus the field current I_f at constant synchronous speed n_s and shorted armature terminals ($Z_L = 0$). The *short-circuit ratio* is defined as the ratio of the field current I_{f0} corresponding to rated phase voltage V_{1n} at no load, to the nominal field current I_{fsh} corresponding to rated armature current I_{an} at shorted armature terminals, as given by eqns (7.25) and (7.26).

The *synchronous reactance of a non-salient pole synchronous machine* is the sum of the armature (stator) leakage reactance X_1 and the armature reaction reactance X_a (7.32).

The *equivalent circuit* of a synchronous machine with non-salient pole rotor is shown in Fig. 7.19. *Phasor diagrams* of non-salient pole unsaturated synchronous generators for RL and RC loads are plotted in Fig. 7.20. In a simplified equivalent circuit and phasor diagram (Fig. 7.21) the stator (armature) winding resistance $R_1 = 0$.

The *armature reaction reactance* (7.45) is inversely proportional to the equivalent air gap gk_Ck_{sat}, where k_C is *Carter's coefficient* that includes the slotted surface of the stator and rotor core and k_{sat} is the *saturation factor* of the magnetic circuit. The higher the magnetic saturation, the lower the armature reaction reactance, also called the *mutual reactance*. The *phasor diagram* of synchronous generator with non-salient pole rotor and magnetic saturation being included (Fig. 7.23) can be constructed on the basis of the open circuit characteristic $E_f = f(I_f)$ at constant synchronous speed n_s and load impedance $Z_L \to \infty$.

The *voltage change* (7.47) is the change in voltage at the output terminals of the armature winding that takes place when the generator with open armature terminals is loaded with nominal RL load at constant synchronous speed n_s and constant field current I_f.

The *short circuit triangle* is a right-angle triangle with one leg equal to the voltage drop $I_a X_1$ across the armature winding leakage reactance and second leg $k_a F_a$ equal to the d-axis armature reaction MMF reduced to the field excitation winding. This trangle can be constructed on the basis of the open circuit characteristic $E_f = f(I_f)$ and short-circuit characteristic $I_a = f(I_f)$, as shown in Fig. 7.24.

The *load characteristics* $V_1 = f(I_f)$ at constant synchronous speed n_s and constant armature current I_a for various power factors $\cos \varphi$ can be found on the basis of the open circuit characteristic $E_f = f(I_f)$ and short circuit triangle, as shown in Fig. 7.25.

The *external characteristics* (Fig. 7.26a) of a synchronous generator are the curves of phase voltage V_1 plotted against the armature current I_a at constant synchronous speed n_s and constant field current I_f for various power factors $\cos \varphi$. At $I_a = 0$ the voltage $V_1 = E_f$ and at the short circuit armature current $I_a = I_{ash}$ the voltage $V_1 = 0$.

The *regulation characteristics* (Fig. 7.26b) express the variation of the field current I_f with the armature current I_a at constant synchronous speed n_s and constant voltage V_1 for different loads: resistive R, pure inductive L, pure capacitive C, resistive-inductive RL and resisitive-capacitive RC.

In general, the *power losses* in synchronous machines can classified as *basic losses* and *additional losses*. The basic losses consist of *main copper losses* in the armature and field excitation windings, *stator core losses* and *rotational (mechanical) losses*, i.e., friction losses in bearings, windage and ventilation losses. The additional losses consist of *losses in the armature and field winding due to leakage fluxes, losses in the stator and rotor core due to leakage fluxes, higher MMF harmonics and tooth-ripple harmonics*, and *losses in metallic parts where the leakage flux penetrates*. The diagram in Fig. 7.27 explains the details. The *efficiency* of large synchronous generators is high and can exceed 98%.

The main role of the *field excitation system* is to deliver the DC current to the field winding of a synchronous machine in order to excite the required magnetic field in the air gap at given operating conditions. Nowadays almost exclusively brushless exciters are used. A *brushless exciter* consists of a *reversed synchronous generator* (three-phase armature winding is on the rotor and the field excitation system is on the stator) and *rotating diode rectifier* (Fig. 7.29).

Stand alone operation is such an operation when a synchronous generator is directly connected to the load (Fig. 7.31a). Operation on *flexible bus bar* takes place when two synchronous generators of similar nominal power operate on a common bus bar (Fig. 7.31b). Operation on an *infinite bus bar* takes place when a synchronous generator of smaller power operates on the bus bar connected to a certain number of synchronous generators of larger power (Fig. 7.31c). Dependent on the value of the EMF E_f, terminal voltage V_1 and load angle δ, in the case of a synchronous generator operating on an infinitive bus bar the reactive power can be controlled by the field excitation current I_f, the reactive power can be delivered to the infinitive bus or the reactive power can be controlled by the prime mover.

The *electromagnetic power* of a nonsalient-pole synchronous machine is proportional to the *power crossing the air gap* (7.53). In terms of the load angle δ, the *electromagnetic torque* is proportional to the $\sin \delta$ (7.56). The electromagnetic torque is also directly proportional to the EMF E_f, terminal voltage V_1 and inversely proportional to the synchronous reactance X_s (7.56).

The *current circle diagram* is the locus of the armature current at variable load angle δ, constant torque T and constant EMF E_f (Fig. 7.34). The current circle diagram determines the generator mode $(+)$, motor mode $(-)$ and stability limit.

The *V-curves* are the curves of the armature current I_a versus field excitation current I_f at constant voltage V_1, constant torque T and constant synchronous speed n_s (Fig. 7.35). The minimum armature current I_a corresponds to the power factor $\cos \varphi = 1$.

Synchronization is a sequence of actions to be taken before connection of a synchronous generator to the power system bus bars. The following conditions must be made to perform correct synchronization:

(a) The *rms* values of voltages of the generator and bus bar should be equal;
(b) The frequencies of the generator voltage and bus bar voltage should be equal;
(c) The phase sequence of the generator voltage and bus bar voltage must be the same (Fig. 7.36b);
(d) The instantaneous values of corresponding voltages of the generator and bus bar should be the same.

In a salient-pole rotor synchronous machine *two axes* must be distinguished:

- Direct axis d: small air gap, large permeance for the magnetic flux;
- Quadrature axis q: large air gap, small permeance for the magnetic flux.

The *stator MMF* is resolved into two components: F_{ad} in the d-axis and F_{aq} in the q-axis. Reduction of the F_{ad} and F_{aq} components of the MMF caused by the saliency of the rotor is taken into account by multiplying F_{ad} and F_{aq} by the following coefficients: $k_{ad} = 0.8 \ldots 0.9$ in the d-axis and $k_{aq} = 0.2 \ldots 0.5$ in the q-axis i.e., equivalent MMFs $k_{ad}F_{ad}$ and $k_{aq}F_{aq}$ have been introduced. The equivalent MMFs $k_{ad}F_{ad}$ and $k_{aq}F_{aq}$ excite their own magnetic fluxes Φ_{ad} (7.68) and Φ_{aq} (7.69). Each of the magnetic fluxes (7.68) and (7.69) excites in the stator (armature) windings own EMF of the armature reaction, i.e., E_{ad} (7.71) and E_{aq} (7.72). The d-axis armature reaction reactance X_{ad} (7.76) is different than the q-axis armature reaction reactance X_{aq} (7.77). Only the d-axis armature reaction reactance X_{ad} (7.78) depends on the saturation of the magnetic circuit. The synchronous reactance in the d-axis is $X_{sd} = X_1 + X_{ad}$ and the synchronous reactance in the q-axis is $X_{sq} = X_1 + X_{aq}$, where X_1 is the stator (armature) leakage reactance. The phasor diagrams for a salient-pole synchronous generator for inductive load are plotted in Fig. 7.38.

On the basis of the *phasor diagram* shown in Fig. 7.38b, the *output power* given by eqn (7.87) has two components: proportional to $V_1 E_f \sin \delta$ and proportional to $V_1^2 \sin(2\delta)$. This output power is derived on assumption that the stator winding losses ($R_1 = 0$), core losses, additional and mechanical losses are neglected so that the ouput power P_{out} is equal to the *electromagnetic power* P_{elm}. Thus, the *electromagnetic torque* $T_{elm} = P_{elm}/(2\pi n_s) = P_{out}/(2\pi n_s)$. Similar to the electromagnetic power, the electromagnetic torque (7.88) has a component dependent on the EMF E_f (field excitation flux) proportional to $V_1 E_f \sin \delta$ and component independent of the E_f proportional to $V_1^2 \sin(2\delta)$. The first component is the *synchronous torque* and the second component is the *reluctance torque*. The reluctance torque is produced without any field excitation flux of the rotor. It is enough to create a difference in the d-axis and q-axis synchronous reactances ($X_{sd} \neq X_{sq}$). The synchronous torque, reluctance torque and resultant electromagnetic torque of a salient-pole synchronous machine as a function of the load angle δ is plotted in Fig. 7.39.

Aircraft generators driven by *turbine engines* are salient-pole synchronous generators with electromagnetic excitation equipped with synchronous brushless exciter, rotating diode rectifier and PM brushless sub-exciter. The *three-machine set* architecture is sketched in Fig. 7.41. To minimize the mass and volume envelope, the aircraft generators operate at frequency of 400 Hz or higher, have magnetic circuit made of *cobalt alloy* laminations and are intensively *cooled by oil*. PM generators cannot be used due to the impossibility of reduction of the field excitation to zero in failure modes, e.g., inter turn short circuit. Both the shaft speed and output frequency of a generator can be constant or variable. Consequently, generators can be divided into the three following groups: (a) constant speed constant frequency (CSCF) generators; (b) variable speed constant frequency (VSCF) generators; (c) variable frequency (VF) generators sometimes called "wild frequency" (WF) generators. In VSCF applications, the so-called *doubly-fed AC generators* (Fig. 7.43) can provide good performance.

Synchronous motors are usually salient-pole rotor synchronous machines. A significant feature of the wound-field rotor synchronous motor is the controllability of its power factor up to unity or leading values. To avoid the need for slip rings and brushes, almost all modern wound-field rotor synchronous motors are provided with *brushless excitation*, i.e., a small synchronous generator (exciter) is mounted on the rotor shaft, the output of which is rectified by shaft-mounted diode rectifiers rotating with the rotor. A cage winding is frequently mounted on salient-pole rotors that is necessary to start the motor (asynchronous starting). The V-curves $I_a = f(I_f)$ and power factor curves $\cos \varphi = f(I_f)$ at constant output power P_{out} and constant synchronous speed n_s are plotted in Fig. 7.44. A synchronous motor does not produce a starting torque after switching on the voltage to the stator winding. There are three methods of *starting synchronous motors*:

- Asynchronous starting using the rotor cage winding;
- Starting by means of an auxiliary motor;
- Frequency-change starting with the aid of solid-state converter.

In *comparison with induction motor*, the synchronous motor has constant speed independent of the load, adjustable power factor $\cos \varphi$, larger air gap, torque directly proportional to the voltage (not to voltage square as an induction motor). On the other hand, it is a much more expensive machine requesting more attention during operation than an equivalent induction motor.

The *synchronous reluctance motor* runs with an unexcited rotor that has unequal reluctance in its *d*- and *q*-axis. Since the rotor has no excitation system, the EMF E_f in eqns (7.82), (7.84), (7.86), (7.87), (7.88) and (7.89) is zero ($E_f = 0$). The input stator (armature) current I_a of a reluctance motor is higher than that of ordinary synchronous and induction motors since the EMF $E_f = 0$. This correspondingly affects the efficiency because of high power losses in the stator winding. The *electromagnetic torque* of a synchronous reluctance motor is produced due to the difference in the *d*-axis and *q*-axis synchronous reactances (inductances) and is expressed by eqn (7.90). Typical rotors for synchronous reluctance motor are sketched in Fig. 7.48. Nowadays, almost exclusively, the rotors with magnetic flux barriers are used (Fig. 7.49). Equivalent circuit and phasor diagram for a synchronous reluctance motor are plotted in Fig. 7.50.

The stator of a *written pole motor* has a three-phase or single-phase winding distributed in slots. In addition, there is a concentrated-parameter winding consisting of a single coil placed around an *exciter pole* (Fig. 7.51). The rotor is furnished with a high resistance cage winding to produce asynchronous torque. The rotor active surface is also coated with a magnetic material layer, which can be magnetized by the stator exciter pole during operation. Any desired magnetic *pole pattern* can be "written" on the magnetic layer to provide a hysteresis torque. In this way the starting parameters can be set independently of the operating parameters. The hysteresis torque and asynchronous torque start the motor before the exciter pole winding is energized. At subsynchronous speed the exciter winding is energized and writes magnetic poles on the rotor surface. The number of written poles on the rotor surface is inversely proportional to the speed. The rotor is pulled into synchronism when the number of rotor poles is the same as the number of poles of the stator rotating field. When the rotor reaches the synchronous speed, the exciter current is switched off and the motor operates as a PM synchronous motor (Fig. 7.52). The rotor cage winding designed for high slip (about 20%) reduces the starting current to less than one third of that of a conventional induction motor. The written pole motor shows the following advantages: high efficiency, very low start-up current, ability to start and synchronize loads with very large inertia, ability to operate with about 20% variation in input voltage, riding through power outages of up to 15 s, restarting after power outages longer than 15 s. Written pole motors are now available in sizes from 7.5 to 75 kW and are used in remote rural areas with single phase reticulation systems.

Problems

1. A three-phase, 50-Hz, non-salient pole synchronous generator has a synchronous inductance $L_s = 25.47$ mH. Determine the ratio V_1/E_f when the machine has the following load per phase:

   (a) $L = 17.0$ mH pure inductive load;
   (b) $C = 300$ mF pure capacitive load.

   Assume that the armature winding resistance $R_1 = 0$.

   Answers: (a) 0.4; (b) 4.067.

2. A three-phase, 3000-rpm, 50-Hz, 13.8 kV (line-to-line), Y-connected non-salient pole synchronous generator has a synchronous reactance 17 Ω. The generator is operating at rated voltage and speed with the EMF $E_f = 11.7$ kV and the load angle $\delta = 18°$. Calculate:

   (a) Stator (armature) current;
   (b) Power factor;
   (c) Output power;
   (d) Maximum electromagnetic power;
   (e) Armature current to obtain the maximum electromagnetic power.

   Assume that the armature winding resistance $R_1 = 0$.

   Answers: (a) $I_a = 282.46$ A; (b) $\cos \varphi = 0.753$; (c) $P_{out} = 5.083$ MW; (d) $P_{elm} = 16.450$ MW (at $\delta = 90°$); (e) $I_a = 832.66$ A.

3. A three-phase, 500-MVA, 24-kV (line-to-line), 50-Hz, Y-connected, non-salient pole synchronous generator is operating at nominal voltage, nominal frequency, and nominal apparent power with a power factor of 0.85 lagging. The EMF per phase is $E_f = 19$ kV. Find:

   (a) The load angle δ;
   (b) The synchronous reactance X_s.

   The stator (armature) winding resistance can be assumed $R_1 = 0$.

   Answers: (a) $\delta = 19.9°$; (b) $X_s = 0.633$ Ω.

4. A three-phase, Y-connected, non-salient pole synchronous generator delivers active power $P_{out} = 250$ MW to the power system at inductive power factor $\cos\varphi = 0.85$ and load angle $\delta = 10°$. The line-to-line voltage is $V_{1L} = 30$ kW. Calculate:

(a) Armature current I_a;
(b) Synchronous reactance X_s;
(c) EMF E_f induced in the armature winding.

The stator (armature) winding resistance can be assumed $R_1 = 0$.

Answers: (a) $I_a = 5660.3$ A; (b) $X_s = 0.713$ Ω; (c) $E_f = 1975$ V.

5. Two non-salient pole rotor synchronous generators A and B with unsaturated magnetic circuits operate in parallel. The stator windings of both generators are Y-connected. Nominal (rated) parameters of generators are as follows:

- Output powers of generators: $P_{nA} = 500$ kW, $P_{nB} = 350$ kW
- Output voltage of generators $V_{1n} = 6300$ V
- Field excitation currents: $I_{fnA} = 21.0$ A, $I_{fnB} = 16.0$ A
- d-axis synchronous reactances per unit: $x_{sA} = 1.5$, $x_{sB} = 1.6$
- Power factors: $\cos\varphi_{nA} = 0.85$ lagging, $\cos\varphi_{nB} = 0.85$ lagging.

It is necessary to calculate:

(a) The field excitation current of generator B, to obtain $V = V_n$ for the following conditions: load power $P_L = 720$ kW, power factor of load $\cos\varphi_L = 0.8$ lagging, field excitation current of generator A $I_{fA} = 20$ A. Generators deliver to the utility grid the power of 360 kW each.
(b) For the same conditions as specified in (a) additional load has been connected: $P = 130$ kW, $\cos\varphi = 1$ (e.g., lighting). Keeping the same operating conditions of generator B, calculate the new value of the field excitation current of generator A to obtain constant voltage.

Answer: (a) $I_{fB} = 15.7$ A; (b) $I_{fA} = 21.6$ A.

6. A three-phase, 15.0-kV, Y-connected, 50-Hz, 250-MVA salient-pole synchronous generator has the d-axis synchronous reactance $X_{sd} = 1.05$ Ω and q-axis synchronous reactance $X_{sq} = 0.52$ Ω. The generator is supplying nominal apparent power at power factor $\cos\varphi = 0.88$ lagging. Find the EMF E_f per phase and load angle δ. Assume that the armature winding resistance $R_1 = 0$.

Answers: $E_f = 15.8$ kV, $\delta = 21.75°$.

7. A three-phase, Y-connected, salient-pole synchronous generator with the following nominal parameters:

- Apparent power $S_n = 25$ MVA
- Voltage line-to-line $V_{1Ln} = 12.5$ kV
- Armature current $I_{an} = 1105$ A

- Frequency $f_n = 50$ Hz
- Power factor $\cos \varphi = 0.86$ lagging
- Field excitation current $I_{fn} = 475$ A
- d-axis synchronous reactance $X_{sd} = 4.2$ Ω
- q-axis synchronous reactance $X_{sq} = 2.4$ Ω

is driven with nominal speed, excited with field current $I_f = 425$ A and operates as a stand alone generator. The terminal voltage is $V_{1L} = 10.5$ kV and reactive inductive power delivered to the load is $Q = 10.6$ MVAr. Neglecting the stator winding resistance R_1, find:

(a) Load angle δ of the generator;
(b) Active power P_{out} delivered to the load;
(c) Load current I_a and load power factor $\cos \varphi$.

Hint: Simplified phasor diagram for salient-pole synchronous generator and EMF E_{fn} equation derived in Example 7.3 is necessary.

Answers: (a) $E_{fn} = 10.452$ kV, $\delta = 17.48°$; (b) $P_{out} = 17.8$ MW; (c) $I_a = 1139.2$ A, $\cos \varphi = 0.859$.

8. A three-phase, six-pole, 50 Hz, 380 V (line-to-line), Y-connected synchronous motor has a synchronous reactance $X_{sd} = X_{sq} = 1.2$ Ω and negligible stator resistance $R_1 = 0$. The motor draws 62.5 kVA at $\cos \phi = 0.8$ (leading). Calculate: (a) the armature current $\mathbf{I}_a$, (b) the stator EMF E_f (line-to-neutral) excited by the rotor magnetic flux, (c) the power angle δ and (d) the maximum electromagnetic torque T_{elmmax}.

Answer: (a) $\mathbf{I}_a = 94.96 \exp(36.87°)$ A, (b) $E_f = 301.87$ V, (c) $\delta = 17.5°$, (d) $T_{elmmax} = 1581.13$ Nm.

9. A three-phase, four-pole, 460 V (line-to-line), $\eta = 88\%$, Y-connected, salient pole synchronous motor has the d-axis synchronous reactance $X_{sd} = 9.0$ Ω and the q-axis synchronous reactance $X_{sq} = 5.0$ Ω. The phase EMF $E_f = 205$ V at load angle $\delta = 32°$. Find the armature current, input power, output power, electromagnetic power, electromagnetic torque and its synchronous and reluctance components. The stator winding resistance R_1 can be neglected.

Answer: $I_{ad} = -2.25$ A, $I_{aq} = 27.1$ A, $I_a = 27.16$ A, $P_{in} = 17.34$ kW, $P_{out} = 15.26$ kW, $P_{elm} = P_{in}$ $(R_1 = 0)$, $T_{elm} = 91.98$ Nm, $T_{elmsyn} = 51.02$ Nm, $T_{elmrel} = 40.96$ Nm.

10. A three-phase, Y-conenected, salient-pole, synchronous motor rated at $P_n = 10$ MW, $V_{1L} = 6.6$ kV (line-to-line) , $f_n = 60$ Hz, $n_n = 1800$ rpm is loaded with the nominal power at power factor $\cos \varphi = 0.8$ leading. Synchronous reactances are $X_{sd} = 10.0$ Ω and $X_{sq} = 5.5$ Ω. Calculate:

(a) Apparent power S_n at the stator terminals;

(b) Armature current I_a;

(c) EMF E_f per phase induced in the stator (armature) winding;

(d) Electromagnetic power P_{elm} and electromagnetic torque T_{elm}.

The stator winding resistance R_1 can be neglected.

Answer: (a) $S_n = 12.5$ MVA; (b) $I_a = 1093.5$ A; (c) $E_f = 13.46$ kV; (d) $P_{elm} = 10$ MW, $T_{elm} = 53.05$ kNm, $T_{elmsyn} = 44.42$ kNm, $T_{elmrel} = 8.63$ kNm.

11. A three-phase, 4-pole, 380-V (line-to-line), 50-Hz, 5.5-kW, Y-connected synchronous reluctance motor has the stator winding resistance per phase $R_1 = 0.49$ Ω, d-axis synchronous reactance $X_{sd} = 15.5$ Ω and q-axis synchronous reactance $X_{sq} = 5.7$ Ω. The rotational losses $\Delta P_{rot} = 110$ W, the stray losses $\Delta P_{str} = 0.06 P_{out}$ and the stator core losses can be neglected. For the power angle $\delta = 21°$ find the armature current, output power P_{out}, shaft torque T, efficiency η and power factor $\cos \varphi$.

Answer: $I_a = 19.6$ A, $P_{out} = 5160.4$ W, $T_{sh} = 32.85$ Nm, $\eta = 0.84$, $\cos \varphi = 0.476$.

PERMANENT MAGNET BRUSHLESS MOTORS

Availability of high energy density rare earth SmCo in the 1970s and NdFeB since 1983 was a breakthrough in PM machine technology and their performance. Today, the energy density of NdFeB PMs exceeds 400 kJ/m^3. Rare earth PMs improve the output power-to-mass ratio, efficiency, dynamic performance, and reliability [13].

8.1 Permanent magnet motor drives

A permanent magnet brushless motor (PMBM) has the magnets mounted on the rotor and the armature winding mounted on the stator. In a DC brush motor the power losses occur mainly in the internal rotor with the armature winding which limits the heat transfer through the air gap to the stator and consequently the armature winding current density. In PM brushless motors, all power losses are practically dissipated in the stator where heat can be easily transferred through the ribbed frame or, in larger machines, liquid cooling systems, e.g., water or oil jackets can be used. Comparison of PM DC brush and brushless motors is given in Table 8.1.

PMBM drives fall into the two principal classes of *sinusoidally excited* and *square wave* (trapezoidally excited) motors. Sinusoidally excited motors are fed with three-phase sinusoidal waveforms (Fig. 8.1a) and operate on the principle of a rotating magnetic field. The speed of the rotor is equal to the synchronous speed n_s of the stator magnetic rotating field, given by eqns (5.80), (6.2) and (7.1). They are simply called *sinewave motors* or *PM synchronous motors*. All phase windings conduct current at a time.

Square wave motors are also fed with three-phase waveforms shifted by 120° one from another, but these waveshapes are rectangular or trapezoidal (Fig. 8.1b). Such a shape is produced when the armature current (MMF) is precisely synchronized with the rotor instantaneous position and frequency (speed). The most direct and popular method of providing the required rotor position information is to use an absolute angular position sensor mounted

Table 8.1. Comparison of PM DC brush (commutator) and brushless motors.

	Brush DC motor	Brushless DC motor
Commutator (inverter)	Mechanical commutator	Electronic commutator
Maintenance	Commutator and brushes need periodical maintenance	Minimal maintenance
Reliability	Low	High
Moment of inertia	High	Can be minimized
Power density	Medium	High
Heat dissipation	Poor (rotor armature winding)	Good (stator armature winding)
Speed control	Simple (armature rheostat or chopper)	Solid state converter required

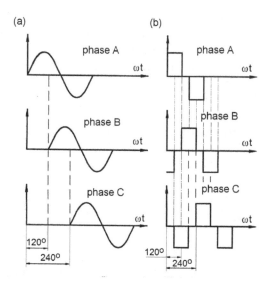

Fig. 8.1. Basic armature waveforms for three phase PMBMs: (a) sinusoidally excited, (b) square wave.

on the rotor shaft. Usually, only two phase windings out of three conduct current simultaneously. Such a control scheme or *electronic commutation* is functionally equivalent to the mechanical commutation in DC brush motors. This explains why motors with square wave excitation are called *DC brushless motors*.

The PMBM motor shows more advantages than its induction or synchronous reluctance counterparts in motor sizes up to $10 - 15$ kW. The largest commercially available motors are rated at least at 750 kW (1000 hp). There have also been successful attempts to build rare-earth PMBMs rated above

1 MW in Germany and 36.5 MW PMBM by *DRS Technologies*, Parsippany, NJ, USA.

The armature winding of PMBMs is usually distributed in slots. When cogging (detent) torque needs to be eliminated, slotless windings are used. In comparison with slotted windings, the slotless windings provide higher efficiency at high speeds, lower torque ripple and lower acoustic noise. On the other hand, slotted motors provide higher torque density, higher efficiency in lower speed range, lower armature current and use less PM material.

An auxiliary DC field winding located in the rotor or magnetic flux diverters with additional DC winding located in the stator can help to increase the speed range over constant power region or control the output voltage of variable speed generators. These machines are called PMBMs with hybrid field excitation system.

8.2 Permanent magnet synchronous motors

Recent developments in rare-earth PM materials and in power electronics have opened new prospects for the design, construction and applications of PM motors. Servo drives with PMBMs fed from solid-state converters are becoming commercially available on an increasing scale.

The application of rare-earth PMs to synchronous motors not only eliminates the need for an exciter, but also offers much lower motor volume, limited maintenance, high efficiency over a broad range of loads, very good dynamic performance and reduced noise.

PM synchronous motors are usually built with one of the following rotor configurations:

(a) Surface-magnet rotor (Fig. 8.2a);
(b) Spoke-type magnet rotor (Fig. 8.2b);
(c) Interior-magnet rotor with flat PMs (Fig. 8.2c);
(d) Inset-magnet rotor (Fig. 8.2d);
(e) Rotor with double-layer interior-magnets (Fig. 8.2e);
(f) Rotor with buried magnets asymmetrically distributed (Fig. 8.2f).

The *surface magnet rotor* can have magnets magnetized radially (as in Fig. 8.2a) or circumferentially. An external non-ferromagnetic cylinder (sleeve) is sometimes used. It protects the PMs against damage due to centrifugal forces, the demagnetizing action of the armature reaction provides an asynchronous starting torque and acts as a damper.

The *spoke-type magnet rotor* has circumferentially magnetized PMs (Fig. 8.2b). An asynchronous starting torque can be produced with the aid either of a cage winding (if the core of the motor is laminated) or of salient poles (in the case of a motor with a solid steel core). If the shaft is ferromagnetic, however, a large amount of useless magnetic flux will be directed through it.

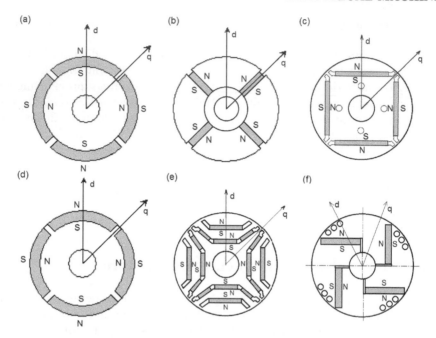

Fig. 8.2. Rotor configurations for PM synchronous motors: (a) surface-magnet rotor; (b) spoke-type magnet rotor; (c) interior-magnet rotor with flat PMs; (d) inset-magnet rotor; (e) rotor with double-layer interior-magnets; (f) rotor with buried magnets asymmetrically distributed.

To increase the linkage flux, therefore, a spoke-type magnet rotor should always be equipped with a nonferromagnetic shaft or nonferromagnetic bushing between the ferromagnetic shaft and rotor core.

The *interior-magnet rotor* of Fig. 8.2c has radially magnetized and alternately poled magnets. Because the magnet pole area is smaller than the pole area at the rotor surface, the no-load air gap flux density is less than the flux density in the magnet.

Comparison of PM synchronous motors with surface, spoke-type and interior magnets symetrically distributed is given in Table 8.2.

The *inset-type PM rotor* shown in Fig. 8.2d is very similar to the surface-magnet rotor. The magnets are placed in slots made in the rotor core.

The rotor with double-layer interior-magnets (Fig. 8.2e) magnetized radially can create strong magnetic flux density in the air gap, when rare-earth PMs are used. Very often, rare-earth PMs are replaced with cost-effective ferrite magnets, usually distributed in three or more layers, which can provide a cost-effective alternative solution. If no PMs are in axial slot, the machine will operate as a synchronous reluctance machine with flux barriers in the rotor core. Single-layer V-shaped or double-layer IV shaped magnet, especially in high count pole rotors are also common. Sometimes, the machine with PMs

Table 8.2. Comparison of PM synchronous motors with surface, spoke-type and interior magnets symetrically distributed.

Surface magnets	Spoke-type magnets	Interior magnets
Air gap magnetic flux density smaller than B_r	Air gap magnetic flux density can be greater than B_r (with more than four poles)	Air gap magnetic flux density smaller than B_r
Simple motor construction	Relatively complicated motor construction. A nonferromagnetic shaft is common.	More complex construction than surface PM rotor
Small armature reaction flux	Higher armature reaction flux	Higher armature reaction flux
PMs not protected against armature fields	PMs protected against armature fields	PMs protected against armature fields
Eddy-current losses in PMs when their conductivity is greater than zero	No eddy-current losses in PMs	Negligible eddy-current losses in PMS

distributed in one or more layers of the rotor is called a *synchronous-reluctance machine*.

An alternative construction involves a rotor with *asymmetrically distributed buried magnets* and cage winding (Fig. 8.2f) according to German Patent 1173178 assigned to Siemens, also called *Siemosyn*. The magnets are magnetized radially. Owing to the cage winding, the motor is self-starting and can be directly plugged-in to 50 or 60 Hz utility grid, without any solid state converter (line start PM synchronous motor).

8.3 Air gap magnetic flux density

Fig. 8.3 shows the magnetic flux density distribution in a surface-type PM synchronous machine.

The first harmonic of the air gap magnetic flux density is

$$b_{mg1} = \frac{2}{\pi} \int_{-0.5\alpha_i\pi}^{0.5\alpha_i\pi} B_{mg} \cos\alpha \, d\alpha = \frac{4}{\pi} B_{mg} \sin\frac{\alpha_i\pi}{2} \qquad (8.1)$$

where the amplitude of the first harmonic of the magnetic flux density

$$B_{mg1} = \frac{4}{\pi} B_{mg} \qquad (8.2)$$

Neglecting the saturation of the magnetic circuit, the magnetic flux density $B_{mg} = \mu_0 F_{exc}/(g'k_C)$ under the pole shoe can be found on the basis of the

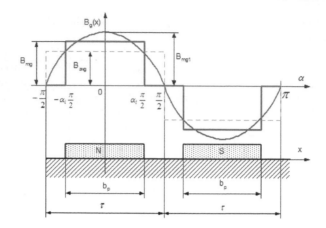

Fig. 8.3. Distribution of the normal component of the magnetic flux density in the air gap of a surface-type PM brushless machine (flat model). B_{mg} – magnitude of the idelized rectangular waveform of the normal component of magnetic flux density excited by surface PMs, B_{mg1} – amplitude of the first harmonic, B_{av} – magnitude of normal component averaged over one pole pitch τ.

excitation MMF F_{exc}, equivalent air gap g' which includes the PM height h_M and Carter's coefficient k_C. Carter's coefficient is defined by eqn (5.56). For $\alpha_i = 1$ the fundamental harmonic component B_{mg1} is $4/\pi$ times the B_{mg} peak flat-topped value – eqn (8.2).

The coefficient α_i is defined as the ratio of the *average–to–maximum value* of the normal component of the air gap magnetic flux density, i.e.,

$$\alpha_i = \frac{B_{avg}}{B_{mg}} \tag{8.3}$$

If the magnetic field distribution in the air gap is sinusoidal, $\alpha_i = 2/\pi$. For zero magnetic voltage drop in the ferromagnetic core and uniform air gap

$$\alpha_i = \frac{b_p}{\tau} \tag{8.4}$$

The coefficient α_i is also called the *pole-shoe arc b_p–to–pole pitch τ ratio*.

8.3.1 Electromotive force EMF (voltage induced)

The no-load *rms* EMF in one phase of the stator winding (EMF) excited by the DC magnetic flux Φ_f of the rotor is given by the same equation (7.11) as for wound-field synchronous machines, i.e.,

$$E_f = \pi\sqrt{2}fN_1k_{w1}\Phi_f \tag{8.5}$$

where N_1 is the number of the stator turns per phase, k_{w1} is the stator winding factor (eqn 5.23), and the fundamental harmonic Φ_{f1} of the excitation magnetic flux density Φ_f without armature reaction is expressed by the same equation as eqn (6.29), i.e.,

$$\Phi_{f1} = L_i \int_0^\tau B_{mg1} \sin\left(\frac{\pi}{\tau}x\right) dx = \frac{2}{\pi}\tau L_i B_{mg1} \tag{8.6}$$

The EMF E_{ad} induced by the d-axis armature reaction flux Φ_{ad} and the EMF E_{aq} induced by the q-axis flux Φ_{aq} are given by eqns (7.71) and (7.72), respectively,

The first harmonics of armature reaction magnetic fluxes are expressed by eqns (7.68) and (7.69), which can be also written as

$$\Phi_{ad} = \frac{2}{\pi}B_{mad1}\tau L_i \tag{8.7}$$

$$\Phi_{aq} = \frac{2}{\pi}B_{maq1}\tau L_i \tag{8.8}$$

where B_{mad1} and B_{maq1} are the peak values of the first harmonic of armature reaction magnetic flux density in the d- and q-axis, respectively.

As shown in Fig 8.2 the *direct* or d-axis is the center axis of the magnetic pole while the *quadrature* or q-axis is the axis parallel (90° electrical) to the d-axis. The EMFs E_f, E_{ad}, E_{aq}, and magnetic fluxes Φ_f, Φ_{ad}, and Φ_{aq} are used in construction of phasor diagrams and equivalent circuits. The EMF E_i per phase with the armature reaction taken into account is

$$E_i = \pi\sqrt{2}fN_1k_{w1}\Phi_g \tag{8.9}$$

where Φ_g is the air gap magnetic flux under load (excitation flux Φ_f reduced by the armature reaction flux). At no-load (very small armature current) $\Phi_g \approx \Phi_f$. Including the saturation of the magnetic circuit

$$E_i = 4\sigma_f fN_1k_{w1}\Phi_g \tag{8.10}$$

The form factor σ_f depends on the magnetic saturation of armature teeth, i.e., the sum of the air gap MVD and the teeth MVD divided by the air gap MVD.

8.3.2 Armature line current density and current density

The peak value of the stator (armature) line current density (A/m) or *specific electric loading* is defined as the number of conductors in all phases $2m_1N_1$ times the peak armature current $\sqrt{2}I_a$ divided by the armature circumference πD_{1in}, i.e.,

$$A_m = \frac{2m_1\sqrt{2}N_1I_a}{\pi D_{1in}} = \frac{m_1\sqrt{2}N_1I_a}{p\tau} = \frac{m_1\sqrt{2}N_1J_as_a}{p\tau} \qquad (8.11)$$

where J_a is the current density (A/m^2) in the stator (armature) conductors and s_a is the cross-section of armature conductors including parallel wires. For air cooling systems $J_a \leq 7.5$ A/mm^2 (sometimes up to 10 A/mm^2) and for liquid cooling systems $10 \leq J_a \leq 28$ A/mm^2. The top value is for very intensive oil spray cooling systems.

8.3.3 Electromagnetic power

For an m_1-phase salient pole synchronous motor with negligible stator winding resistance $R_1 = 0$, the electromagnetic power is expressed by eqn (7.86).

8.3.4 Synchronous reactance

For a salient pole synchronous motor the d-axis and q-axis synchronous reactances are expressed by eqns (7.80) and (7.81), i.e.,

$$X_{sd} = X_1 + X_{ad} \qquad\qquad X_{sq} = X_1 + X_{aq} \qquad (8.12)$$

where $X_1 = 2\pi f L_1$ is the stator leakage reactance, X_{ad} is the d-axis armature reaction reactance, also called d-axis mutual reactance, and X_{aq} is the q-axis armature reaction reactance, also called q-axis mutual reactance. The reactance X_{ad} is sensitive to the saturation of the magnetic circuit while the influence of the magnetic saturation on the reactance X_{aq} is minimal and depends on the rotor construction. In salient-pole wound-field synchronous machines X_{aq} is practically independent of the magnetic saturation. Usually, $X_{sd} > X_{sq}$ except for some PM synchronous machines, e.g., according to Figs 8.2c) and 8.2f).

The leakage reactance X_1 consists of the slot, end-connection, differential and tooth–top leakage reactances.

8.3.5 Electromagnetic (developed) torque

The electromagnetic torque developed by the synchronous motor is determined by the electromagnetic power P_{elm} and angular synchronous speed $\Omega_s = 2\pi n_s$, which is equal to the mechanical angular speed of the rotor. Similar to wound-field synchronous machines, the electromagnetic torque is expressed by eqn (7.88). This equation neglects the stator winding resitance R_1.

In a salient pole-synchronous motor the electromagnetic torque has two components (Fig. 7.39): the fundamental (synchronous) torque $T_{elsmsyn}$ given by eqn (7.89) and the reluctance torque T_{elmrel} given by eqn (7.90). The synchronous torque $T_{elsmsyn}$ is the function of both the input voltage V_1

and the EMF E_f. The reluctance torque T_{elmrel} depends only on the voltage V_1 and also exists in an unexcited machine ($E_f = 0$) provided that $X_{sd} \neq X_{sq}$. For salient-pole wound-field synchronous machines $X_{sd} > X_{sq}$. For some PM synchronous motors as, for example, according to Figs 8.2c and 8.2f, $X_{sd} < X_{sq}$. The proportion between X_{sd} and X_{sq} strongly affects the shape of curves 2 and 3 in Fig. 7.39. For non-salient pole rotor synchronous machines $X_{sd} - X_{sq}$, and

$$T_{elm} = T_{elmsyn} = \frac{m_1}{2\pi n_s} \frac{V_1 E_f}{X_{sd}} \sin \delta \tag{8.13}$$

8.3.6 Equivalent field MMF

Assuming $g = 0$, the equivalent d-axis field MMF, which produces the same fundamental wave flux as the armature-reaction MMF, is

$$F_{excd} = k_{ad} F_{ad} = \frac{m_1 \sqrt{2}}{\pi} \frac{N_1 k_{w1}}{p} k_{ad} I_a \sin \Psi \tag{8.14}$$

where I_a is the armature current and Ψ is the angle between the resultant armature MMF F_a and its q-axis component $F_{aq} = F_a \cos \Psi$. Similarly, the equivalent q-axis MMF is

$$F_{excq} = k_{aq} F_{aq} = \frac{m_1 \sqrt{2}}{\pi} \frac{N_1 k_{w1}}{p} k_{aq} I_a \cos \Psi \tag{8.15}$$

8.3.7 Armature reaction reactance

The d-axis armature reaction reactance (7.78) with the magnetic saturation being included is

$$X_{ad} = k_{fd} X_a = 4 m_1 \mu_0 f \frac{(N_1 k_{w1})^2}{\pi p} \frac{\tau L_i}{g'} k_{fd} \tag{8.16}$$

where μ_0 is the magnetic permeability of free space, L_i is the effective length of the stator core and the inductive reactance of the armature of a non-salient-pole (cylindrical rotor) synchronous machine X_a is given by eqn (7.46). Similarly, for the q-axis

$$X_{aq} \approx k_{fq} X_a = 4 m_1 \mu_0 f \frac{(N_1 k_{w1})^2}{\pi p} \frac{\tau L_i}{g'_q} k_{fq} \tag{8.17}$$

See also eqns (7.76) and (7.77). For most PM configurations the equivalent air gap g' in eqn (8.16) should be replaced by

$$g' = g k_C k_{sat} + \frac{h_M}{\mu_{rrec}} \tag{8.18}$$

and g_q' in eqn (8.17) by

$$g_q' = g_q k_C k_{satq} \tag{8.19}$$

where g_q is the mechanical clearance in the q-axis, k_C is the Carter's coefficient for the air gap according to eqn (5.56) and $k_{sat} \geq 1$ is the saturation factor of the magnetic circuit. For salient pole rotors with electromagnetic excitation the saturation factor $k_{satq} \approx 1$, since the q-axis armature reaction fluxes, closing through the large air spaces between the poles, depend only slightly on the saturation. The saturation factor is defined by eqns (5.51) and (5.100).

8.4 Phasor diagram

When drawing phasor diagrams of synchronous machines, two arrow systems are used:

(a) Generator arrow system, i.e.,

$$\mathbf{E}_f = \mathbf{V}_1 + \mathbf{I}_a R_1 + j\mathbf{I}_{ad} X_{sd} + j\mathbf{I}_{aq} X_{sq}$$

$$= \mathbf{V}_1 + \mathbf{I}_{ad}(R_1 + jX_{sd}) + \mathbf{I}_{aq}(R_1 + jX_{sq}) \tag{8.20}$$

(b) Consumer (motor) arrow system, i.e.,

$$\mathbf{V}_1 = \mathbf{E}_f + \mathbf{I}_a R_1 + j\mathbf{I}_{ad} X_{sd} + j\mathbf{I}_{aq} X_{sq}$$

$$= \mathbf{E}_f + \mathbf{I}_{ad}(R_1 + jX_{sd}) + \mathbf{I}_{aq}(R_1 + jX_{sq}) \tag{8.21}$$

where

$$\mathbf{I}_a = \mathbf{I}_{ad} + \mathbf{I}_{aq} \tag{8.22}$$

and

$$I_{ad} = I_a \sin \Psi \qquad\qquad I_{aq} = I_a \cos \Psi \tag{8.23}$$

When the current arrows are in the opposite direction the phasors $\mathbf{I}_a$, $\mathbf{I}_{ad}$, and $\mathbf{I}_{aq}$, are reversed by $180°$. The same applies to the voltage drops. The location of the armature current $\mathbf{I}_a$ with respect to the d- and q-axis for generator and motor mode is shown in Fig. 7.15.

Phasor diagrams for synchronous generators are constructed using the generator arrow system. An overexcited generator (Fig. 8.5a) delivers an inductive current and a corresponding reactive power to the line. The same system can be used for motors; however, the consumer arrow system is more convenient. To draw the phasor diagram for underexcited motor shown in Fig. 8.5a the d-q coordinate system shown in Fig. 7.15 has been rotated $180°$ to obtain the

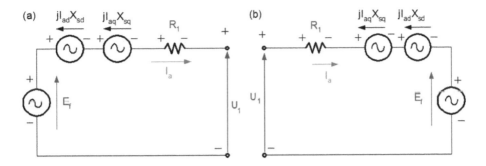

Fig. 8.4. Equivalent circuit of PM synchronous machine for: (a) generator arrow system; (b) consumer (motor) arrow system.

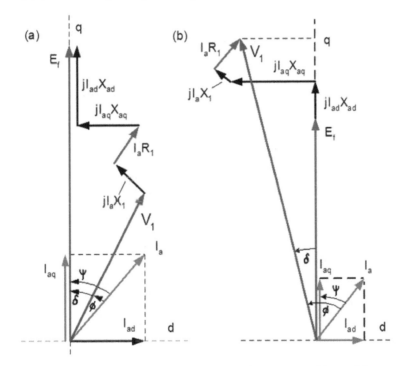

Fig. 8.5. Phasor diagrams of salient-pole synchronous machines: (a) overexcited generator (generator arrow system), (b) underexcited motor (consumer arrow system).

3rd quadrant in the position of the 1st quadrant. Fig. 8.5b shows the phasor diagram using the consumer arrow system for a load current $\mathbf{I}_a$ lagging the vector $\mathbf{V}_1$ by the angle ϕ. At this angle the motor is, conversely, underexcited and induces with respect to the input voltage $\mathbf{V}_1$ a capacitive current component $I_a \sin \Psi$. An overexcited motor, consequently, draws a leading current from the circuit and delivers reactive power to it.

In the phasor diagrams shown in Fig. 8.5 the stator core losses have been neglected. This assumption is justified only for power frequency synchronous motors with unsaturated armature cores.

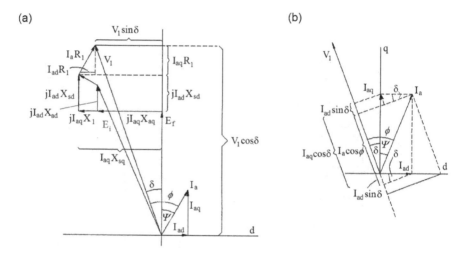

Fig. 8.6. Phasor diagrams for finding: (a) axis currents I_{ad} and I_{aq}; (b) input power P_{in} as a function of I_{ad}, I_{aq}, and δ.

Fig. 8.6 shows the phasor diagram of an underexcited synchronous motor with some necessary details for finding the *rms* axis currents I_{ad} and I_{aq}. The input voltage V_1 projections on the d- and q-axis are

$$V_1 \sin \delta = I_{aq} X_{sq} - I_{ad} R_1$$

$$V_1 \cos \delta = E_f + I_{ad} X_{sd} + I_{aq} R_1 \tag{8.24}$$

For an overexcited motor

$$V_1 \sin \delta = I_{aq} X_{sq} + I_{ad} R_1$$

$$V_1 \cos \delta = E_f - I_{ad} X_{sd} + I_{aq} R_1 \tag{8.25}$$

The currents of an underexcited motor

$$I_{ad} = \frac{V_1 (X_{sq} \cos \delta - R_1 \sin \delta) - E_f X_{sq}}{X_{sd} X_{sq} + R_1^2} \tag{8.26}$$

$$I_{aq} = \frac{V_1(R_1 \cos\delta + X_{sd}\sin\delta) - E_f R_1}{X_{sd}X_{sq} + R_1^2} \tag{8.27}$$

are obtained by solving the set of eqns (8.24). Similarly, the currents of an overexcited motor are found by solving the set of eqns (8.25). The *d*-axis current of an overexcited motor is

$$I_{ad} = \frac{V_1(R_1 \sin\delta - X_{sq}\cos\delta) + E_f X_{sq}}{X_{sd}X_{sq} + R_1^2} \tag{8.28}$$

and the *q*-axis current is expressed by eqn (8.27). The *rms* armature current of an underexcited motor as a function of V_1, E_f, X_{sd}, X_{sq}, δ, and R_1 is

$$I_a = \sqrt{I_{ad}^2 + I_{aq}^2} = \frac{V_1}{X_{sd}X_{sq} + R_1^2}$$

$$\times \sqrt{\left[(X_{sq}\cos\delta - R_1\sin\delta) - \frac{E_f X_{sq}}{V_1}\right]^2 + \left[(R_1\cos\delta + X_{sd}\sin\delta) - \frac{E_f R_1}{V_1}\right]^2} \tag{8.29}$$

The angle between the phasor $\mathbf{I}_a$ and *q*-axis is $\psi = \phi \mp \delta$ where the "$-$" sign is for an underexcited motor and the "$+$" sign is for an overexcited motor.

The phasor diagram (Fig. 8.6b) can also be used to find the input power, i.e.,

$$P_{in} = m_1 V_1 I_a \cos\phi = m_1 V_1(I_{aq}\cos\delta - I_{ad}\sin\delta) \tag{8.30}$$

Putting $V_1\sin\delta$ and $V_1\cos\delta$ according to eqns (8.24) into eqn (8.30)

$$P_{in} = m_1[I_{aq}E_f + I_{ad}I_{aq}X_{sd} + I_{aq}^2 R_1 - I_{ad}I_{aq}X_{sq} + I_{ad}^2 R_1]$$

$$= m_1[I_{aq}E_f + R_1 I_a^2 + I_{ad}I_{aq}(X_{sd} - X_{sq})]$$

Because the stator core loss has been neglected, the electromagnetic power is the motor input power minus the stator winding loss $\Delta P_{1w} = m_1 I_a^2 R_1 = m_1(I_{ad}^2 + I_{aq}^2)R_1$. Thus

$$P_{elm} = P_{in} - \Delta P_{1w} = m_1[I_{aq}E_f + I_{ad}I_{aq}(X_{sd} - X_{sq})]$$

$$= \frac{m_1[V_1(R_1\cos\delta + X_{sd}\sin\delta) - E_f R_1)]}{(X_{sd}X_{sq} + R_1^2)^2} \tag{8.31}$$

$$\times [V_1(X_{sq}\cos\delta - R_1\sin\delta)(X_{sd} - X_{sq}) + E_f(X_{sd}X_{sq} + R_1^2) - E_f X_{sq}(X_{sd} - X_{sq})]$$

The electromagnetic torque developed by a salient-pole synchronous motor is

$$T_{elm} = \frac{P_{elm}}{2\pi n_s} = \frac{m_1}{2\pi n_s} \frac{1}{(X_{sd}X_{sq} + R_1^2)^2}$$

$$\times \{V_1 E_f(R_1 \cos\delta + X_{sd}\sin\delta)[(X_{sd}X_{sq} + R_1^2) - X_{sq}(X_{sd} - X_{sq})]$$

$$-V_1 E_f R_1(X_{sq}\cos\delta - R_1\sin\delta)(X_{sd} - X_{sq})$$

$$+V_1^2(R_1\cos\delta + X_{sd}\sin\delta)(X_{sq}\cos\delta - R_1\sin\delta)(X_{sd} - X_{sq})$$

$$-E_f^2 R_1[(X_{sd}X_{sq} + R_1^2) - X_{sq}(X_{sd} - X_{sq})]\} \tag{8.32}$$

The last term is the constant component of the electromagnetic torque independent of the load angle δ. Putting $R_1 = 0$, eqn (8.32) becomes the same as eqn (7.88). Small synchronous motors have a rather high stator winding resistance R_1 that is comparable with X_{sd} and X_{sq}. That is why eqn (8.32) is recommended for calculating the performance of small motors.

8.5 Characteristics

The load characteristics, i.e., the armature current I_a, shaft torque T_{sh}, input power P_{in}, power factor $\cos\phi$, and efficiency η, are plotted against the relative output power (output power–to–rated output power) in Fig. 8.7.

   For $I_{ad} = 0$ (Fig. 8.8) the angle $\Psi = 0$ (between the armature current $I_a = I_{aq}$ and EMF E_f). Therefore, the angle φ between the current and voltage is equal to the load angle δ between the voltage V_1 and EMF E_f, i.e.,

$$\cos\varphi = \frac{E_f + I_a R_1}{V_1} \tag{8.33}$$

and

$$V_1^2 = (E_f + I_a R_1)^2 + (I_a X_{sq})^2 \approx E_f^2 + I_a^2 X_{sq}^2 \tag{8.34}$$

Thus

$$\cos\varphi \approx \sqrt{1 - \left(\frac{I_a X_{sq}}{V_1}\right)^2} + \frac{I_a R_1}{V_1} \tag{8.35}$$

At constant voltage V_1 and frequency (speed) the power factor $\cos\phi$ decreases with the load torque (proportional to the armature current I_a). The power factor can be kept constant by increasing the voltage in proportion to the current increase, i.e., keeping $I_a X_{sq}/V_1 = const$.

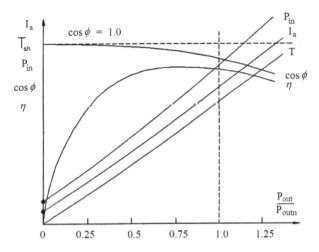

Fig. 8.7. Performance characteristics of a PM synchronous motor: armature current I_a, shaft torque T_{sh}, input power P_{in}, power factor $\cos\varphi$, and efficiency η plotted against P_{out}/P_{outn}, where P_{outn} is the rated ouput power.

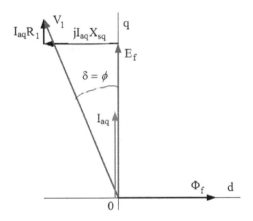

Fig. 8.8. Phasor diagram of a PM synchronous motor at $I_{ad} = 0$.

Example 8.1

A three-phase, 4-pole, 50-Hz, 1.5-kW, 380-V (line-to-line), Y-connected synchronous motor with a interior PM rotor has the stator winding resistance per phase $R_1 = 4.98\ \Omega$, d-axis synchronous reactance $X_{sd} = 18.5\ \Omega$ and q-axis synchronous reactance $X_{sq} = 40.5\ \Omega$ (in a many interior magnet motors $X_{sd} < X_{sq}$). The number of turns per phase $N_1 = 240$, the winding factor $k_{w1} = 0.96$, the effective length of the stator core $L_i = 0.103$ m and the inner stator core diameter $D = 0.0825$ m. At the power angle $\delta = 45^0$, the air gap magnetic flux density $B_{mg} = 0.685$ T, the rotational losses $\Delta P_{rot} = 40$ W, the stray losses $\Delta P_{str} = 0.05 P_{out}$ and the stator core losses can be neglected. Find the armature current, electromagnetic torque T_{elm}, output power P_{out}, efficiency η and power factor $\cos \varphi$.

Solution

The number of pole pairs is $p = 2$. The synchronous speed according to eqn (7.1)

$$n_s = \frac{50}{2} = 25 \text{ rev/s} = 1500 \text{ rpm}$$

The phase voltage $V_1 = 380/\sqrt{3} = 219.4$ V. The pole pitch

$$\tau = \frac{\pi D}{2p} = \frac{\pi \times 0.0825}{4} = 0.0648 \text{ m}$$

The magnetic flux (fundamental space harmonic) produced by the rotor

$$\Phi_f = \frac{2}{\pi} \tau L_i B_{mg} = \frac{2}{\pi} 0.0648 \times 0.103 \times 0.685 = 0.00291 \text{ Wb}$$

The EMF per phase induced in the stator winding by the rotor flux Φ_f according to eqn (8.5)

$$E_f = \pi \sqrt{2} \times 50 \times 240 \times 0.96 \times 0.00291 = 148.94 \text{ V}$$

For the power angle $\delta = 45^0$ or $\delta = 45 \times \pi/180 = 0.7854$ rad, $\cos \delta = \sin \delta \approx 0.707$. The components of the armature current according to eqns (8.26) and (8.27)

$$I_{ad} = \frac{219.4(40.5 \times 0.707 - 4.98 \times 0.707) - 148.94 \times 40.5}{18.5 \times 40.5 + 4.98^2} = -0.68 \text{ A}$$

$$I_{aq} = \frac{219.4(4.98 \times 0.707 + 18.5 \times 0.707) - 148.94 \times 4.98}{18.5 \times 40.5 + 4.98^2} = 3.75 \text{ A}$$

This is an overexcited motor. The rms armature current

$$I_a = \sqrt{I_{ad}^2 + I_{aq}^2} = \sqrt{(-0.68)^2 + 3.75^2} = 3.81 \text{ A}$$

The input apparent power

$$S_{in} = 3V_1 I_a = 3 \times 219.4 \times 3.81 = 2507.7 \text{ W}$$

On the basis of the phasor diagram (Fig. 8.6b) the following relationship can be obtained:

$$I_a \cos \varphi = I_{aq} \cos \delta - I_{ad} \sin \delta$$

$$= 3.75 \times 0.707 - (-0.68) \times 0.707 = 3.13 \text{ A}$$

The input active power

$$P_{in} = 3V_1 I_a \cos \phi = 3 \times 219.4 \times 3.13 = 2060.2 \text{ W}$$

The stator winding losses

$$\Delta P_{1w} = 3I_a^2 R_1 = 3 \times 3.81^2 \times 4.98 = 216.9 \text{ W}$$

Neglecting the stator core losses, the electromagnetic power is

$$P_g \approx P_{in} - \Delta P_{1w} = 2060.2 - 216.9 = 1843.3 \text{ W}$$

The electromagnetic torque

$$T_{elm} = \frac{P_{elm}}{2\pi n_s} = \frac{1843.3}{2\pi \times 25} = 11.73 \text{ Nm}$$

The output power

$$P_{out} = P_{elm} - \Delta P_{rot} - \Delta P_{str} = P_{elm} - \Delta P_{rot} - 0.05\Delta P_{out}$$

Thus

$$P_{out} = \frac{1}{1.05}(P_g - \Delta P_{rot}) = \frac{1}{1.05}(1843.3 - 40.0) = 1717.4 \text{ W}$$

The shaft torque

$$T_{sh} = \frac{P_{out}}{2\pi n_s} = \frac{1717.4}{2\pi \times 25} = 10.93 \text{ Nm}$$

The efficiency

$$\eta = \frac{1717.4}{2060.2} = 0.834$$

The power factor

$$\cos \phi = \frac{P_{in}}{S_{in}} = \frac{2060.2}{2507.7} = 0.821$$

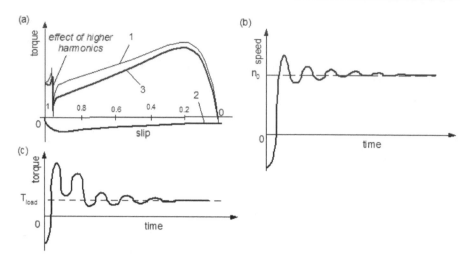

Fig. 8.9. Characteristics of a line start PM brushless motor: (a) steady-state torque–slip characteristic; (b) speed–time characteristic; (c) torque–time characteristic. 1 – asynchronous torque, 2 – braking torque produced by PMs, 3 – resultant torque, n_0 – steady-state speed, T_{load} – load torque.

8.6 Starting

8.6.1 Asynchronous starting

A synchronous motor is not self-starting. To produce an *asynchronous starting torque*, its rotor must be furnished with a cage winding or mild steel pole shoes. The starting torque is produced as a result of the interaction between the stator rotating magnetic field and the rotor currents induced in the cage winding or mild steel pole shoes [19].

PM synchronous motors that can produce asynchronous starting torque are commonly called *line start PM synchronous motors*. These motors can operate without solid state converters. After starting, the rotor is pulled into synchronism and rotates with the speed imposed by the line input frequency. The efficiency of line start PM motors is higher than that of equivalent induction motors and the power factor can be equal to unity.

The rotor bars in line start PM motors are unskewed because PMs are embedded axially in the rotor core. In comparison with induction motors, line start PM motors produce much higher content of higher space harmonics in the air gap magnetic flux density distribution, current and electromagnetic torque. Further, the line start PM synchronous motor has a major drawback during the starting period as the magnets generate a brake torque which decreases the starting torque and reduces the ability of the rotor to synchronize a load. Starting characteristics of a line start PM brushless motor are plotted in Fig. 8.9.

8.6.2 Starting by means of an auxiliary motor

Auxiliary induction motors are frequently used for starting large wound-field synchronous motors with electromagnetic excitation (Section 7.16.2). Auxiliary induction motors are very rarely used for starting PM synchronous motors.

8.6.3 Frequency-change starting

Frequency-change starting is a common method of starting synchronous motors with electromagnetic (wound-field) and PM field excitation. The frequency of the voltage applied to the motor is smoothly changed from the value close to zero to the rated value. The motor runs synchronously during the entire starting period being fed from a variable voltage variable frequency (VVVF) solid state inverter (Fig. 7.46).

8.7 Permanent magnet DC brushless motors

In DC brush motors 90% of maintenance relates to commutator and brushes. The *electromechanical commutator and brushes* are the least reliable parts of DC brush machines. The electromechanical commutator can be replaced with an *electronic commmutator* as shown in Fig. 8.10. The PM excitation system of a DC brush machine is moved from the stator to the rotor and the rotor armature winding is moved to the stator. The stator armature winding, usually three-phase, is connected to a solid state converter. In the case of a DC brushless motor the solid state converter operates as an inverter and in the case of a DC brushless generator the solid state converter operates as a rectifier.

The electric and magnetic circuits of PM synchronous motors and PM DC brushless motors are similar, i.e., polyphase armature windings are located in the stator and moving magnet rotor serves as the field excitation system. PM synchronous motors are fed with three-phase sinusoidal voltage waveforms and operate on the principle of magnetic rotating field. For *constant voltage-to-frequency* control technique no rotor position sensors are required. PM DC brushless motors operate from a DC voltage source and use direct feedback of the rotor angular position, so that the input armature current can be switched, among the motor phases, in exact synchronism with the rotor motion. This concept is known as *electronic commutation*. The solid state inverter and position sensors are equivalent to the mechanical commutator in DC brush motors. Variable DC bus voltage can be obtained using

- Variable voltage transformer and diode rectifier;
- Controlled rectifier, e.g., IGBT[1] bridge.

---

[1] Insulated gate bipolar transistor.

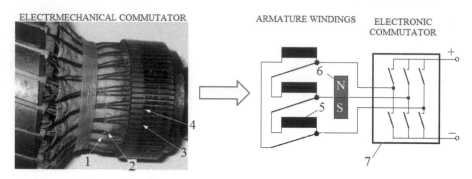

Fig. 8.10. Replacement of an electromechanical commutator with electronic commutator. 1 – mica insulation, 2 – copper conductors of rotor (armature) winding, 3 – commutator segment, 4 – mica insulation between segments, 5 – armature winding placed on the stator, 6 – high energy PM placed on the rotor, 7 – solid state converter.

In the second case the DC bus voltage is a function of the firing angle of the controlled rectifier.

8.7.1 Electromagnetic torque

The electromagnetic torque developed by a synchronous motor is usually expressed as a function of the angle ψ between the q-axis (EMF E_f axis) and the armature current I_a, i.e.,

$$T_{elm} = c_T \Phi_f I_a \cos \psi \tag{8.36}$$

where c_T is the *torque constant*. For a PM motor

$$T_{elm} = k_T I_a \cos \psi \tag{8.37}$$

where $k_T = c_T \Phi_f$ is a new torque constant. The magnetic flux $\Phi_f = const$ if the armature reaction is negligible.

The maximum torque is when $\cos \psi = 1$ or $\psi = 0°$. It means that the armature current $I_a = I_{ad}$ is in phase with the EMF E_f.

For the DC brushless motor the torque equation is similar to eqn (4.18) for a DC brush (commutator) motor, i.e.,

$$T_{elm} = c_{Tdc} \Phi_f I_a = k_{Tdc} I_a \tag{8.38}$$

where c_{Tdc} and $k_{Tdc} = c_{Tdc} \Phi_f$ are torque constants.

Assuming that the flux linkage in the stator winding produced by the PM rotor is $\Psi_f = M_{12} i_2$, the *electromagnetic torque* is

$$T_{elm}(i_a, \theta) = i_a \frac{d\Psi_f}{d\theta} \tag{8.39}$$

where $i_1 = i_a$ is the stator current, i_2 is the current in a fictitious rotor winding which is assumed to have a mutual inductance M_{12} with the stator winding and θ is the rotor angular position.

8.7.2 Linear and rotational speed of brushless motors

The *linear speed* in m/s is the full angle of rotation or 2τ divided by the period of full rotation $T = 1/(pn)$ i.e.,

$$v = \frac{2\tau}{T} = 2\tau pn = \pi Dn \tag{8.40}$$

where τ is the pole pitch, p is the number of pole pairs, D is the rotor outer diameter and n is the rotational speed in rev/s. Surface linear speed of the rotor cannot exceed the permissible value for a given rotor construction.

8.7.3 Commutation of PM brushless motors

Unipolar-driven PM brushless motor

The *unipolar* or *half-wave* operation of PMBMs is explained in Fig. 8.11. The three phase winding is Y-connected and a neutral point is available. The DC voltage V_{dc} is switched across phase to neutral terminals with the aid of one solid state switch per phase. Each phase terminal receives positive voltage and the neutral wire is of negative polarity. For the current sequence $i_{aA}, i_{aB}, i_{aC}, \ldots$ the MMF phasors F_A, F_B, F_C rotate counterclockwise. If the switching sequence is reversed, i.e., i_{aA}, i_{aC}, i_{aB}, the direction of rotation of MMF phasors will be clockwise. For a linear magnetic circuit the phase magnetic fluxes are proportional to MMFs. The EMF, current and torque waveforms are shown in Fig. 8.12. This type of operation (commutation) is called *half-wave operation* (commutation) because conduction occurs only during the positive half of the EMF waveform. The shape of the EMF waveforms depends on the design of PMs and stator windings. Phase EMF waveforms shown in Fig. 8.12 have sinusoidal shapes. In practice, both the shapes of the EMF and current differ from those shown in Fig. 8.12. The DC voltage V_{dc} is higher than the peak phase EMF so that the current flows from the phase terminal to the neutral wire during the 120° conduction period. Neglecting the winding inductance and assuming zero switching time, the armature instantaneous current is

$$i_a = \frac{V_{dc} - e_f}{R_1} \tag{8.41}$$

The electromagnetic power per phase at a given time instant is $p_{elm} = i_a e_f$ and the electromagnetic torque is

$$T_{elm} = \frac{1}{2\pi n}(e_{fA}i_{aA} + e_{fB}i_{aB} + e_{fC}i_{aC}) \tag{8.42}$$

Commuting at the zero crossing of the EMF waveform should always be avoided as the torque is zero no matter how much current is injected into the phase winding.

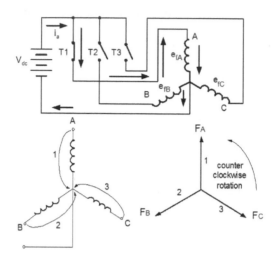

Fig. 8.11. Switching sequence and MMF phasors for three-phase unipolar-driven Y-connected DC PMBM.

For unipolar operation the torque ripple is high, not acceptable in some PMBM applications.

Bipolar-driven PM brushless motor, two phases on

The term "bipolar" indicates the capability of providing the motor phase current of either positive or negative polarity. A PM DC brushless motor is driven by a three-phase inverter bridge and all six solid state switches are used. In Fig. 8.13 the DC voltage V_{dc} is switched between phase terminals and for the Y connection two windings belonging to different phases are series connected during each conduction period. The current is

$$i_a = \frac{V_{dc} - e_{fL}}{2R_1} \tag{8.43}$$

where e_{fL} is line-to-line EMF. The current sequence is $i_{aAB}, i_{aAC}, i_{aBC}, i_{aBA},$ $i_{aCA}, i_{aCB}, \ldots$ For this current sequence, the MMFs $F_{AB}, F_{AC}, F_{BC}, F_{BA},$ $F_{CA}, F_{CB}, \ldots$ rotate counterclockwise (Fig. 8.13). This operation is called *bipolar* or *full-wave operation* because conduction occurs for both the positive and negative half of the EMF waveform. For sinusoidal EMF waveforms the currents can be regulated in such a way as to obtain approximately square

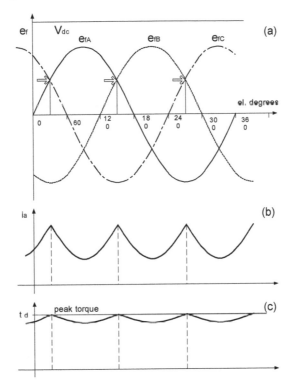

Fig. 8.12. Ideal three-phase unipolar operation of a Y-connected DC PMBM: (a) sinusoidal EMF waveforms; (b) current waveforms; (c) electromagnetic torque waveforms. Switching points are marked with arrows.

waves. The electromagnetic power and torque are always positive because negative EMF times negative current gives a positive product. Each conduction period (one step) for line currents is 60°. Therefore, this is a *six-step commutation* with only two phases on (120° current conduction) at any time leaving the remaining phase floating. As a result, the torque ripple is substantially reduced.

At non-zero speed, the maximum *torque–to–current* ratio is achieved at the peak of EMF waveforms. The current is in phase with the EMF. The commutation timing is determined by the rotor position sensors or estimated on the basis of the motor parameters, e.g., EMF.

The average torque can be maximized and torque ripple can be minimized if the EMF waveform has a trapezoidal shape (Fig. 8.14). For *trapezoidal operation* the peak line-to-line EMF occurs during the whole conduction period, i.e., 60° for line current as given in Figs 8.14 and 8.15. The EMF, i.e., $e_{fAC} = e_{fA} - e_{fC} = -e_{fCA} = e_{fC} - e_{fA}$ and the current, e.g., $i_{aAC} = -i_{aCA}$. The trapezoidal shape of the line–to–line EMFs is obtained by proper shaping and magnetizing the PMs and proper designing of the stator winding. The-

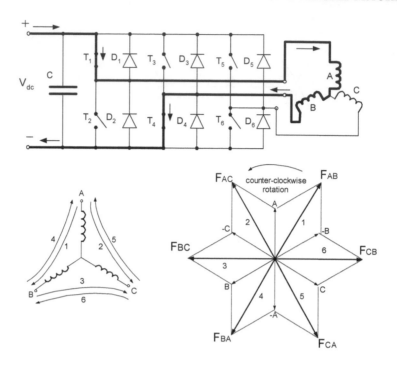

Fig. 8.13. Switching sequence and MMF phasors for six-step commutation of a Y-connected DC PMBM. Commutation sequence is AB, AC, BC, BA, CA, CB, etc.

oretically, the flat top EMF waveforms at DC voltage $V_{dc} = const$ produce square current waveforms and a constant torque independent of the rotor position (Fig. 8.15). Owing to the armature reaction and other parasitic effects, the EMF waveform is never ideally flat. However, the torque ripple below 10% can be achieved. Torque ripple can further be reduced by applying more than three phases.

Bipolar-driven PM brushless motor, three phases on

In six-step mode operation only one upper and one lower solid state switch are turned on at a time (120° conduction). With more than two switches on at a time, a 180° current conduction can be achieved, as shown in Fig. 8.16. If the full current flows, say, through one upper leg, two lower legs conduct half of the current. All three phases always conduct the current.

8.7.4 EMF

Similar to DC brush motors – eqns (4.9) and (4.63), the EMF of a PMBM can simply be expressed as a function of the rotor speed n, i.e.,

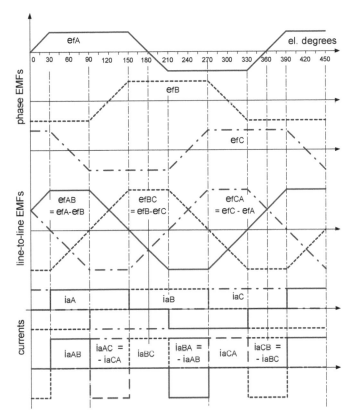

Fig. 8.14. Phase and line-to-line trapezoidal EMF and square current waveforms of a bipolar-driven PMBM with 120° current conduction.

- Phase–to–neutral EMF (e.g., a half-wave operation)

$$E_f = c_E \Phi_f n = k_E n \qquad (8.44)$$

- Line–to–line EMF (e.g., a full-wave operation)

$$E_{fL-L} = c_E \Phi_f n = k_E n \qquad (8.45)$$

where c_E or $k_E = c_E \Phi_f$ is the *EMF constant* also called the *armature constant*. For PM field excitation and negligible armature reaction, $\Phi_f \approx const$.

8.7.5 Inverter AC output voltage

When a PM brushless motor is fed from a solid state converter, and the DC bus voltage is known, the fundamental harmonic of the output AC line–to–line voltage of the inverter is

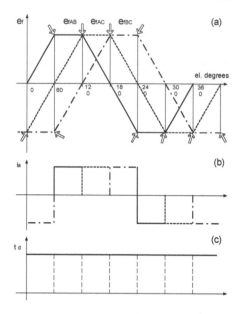

Fig. 8.15. Ideal three-phase six-step operation of a Y-connected DC PMBM: (a) trapezoidal line–to–line EMF waveforms, (b) current waveforms, (c) electromagnetic torque waveforms. Switching points are marked with arrows.

- For a six-step three-phase inverter

$$V_{1L} = \frac{\sqrt{6}}{\pi} V_{dc} \approx 0.78 V_{dc} \qquad (8.46)$$

- For a sinusoidal PWM three-phase voltage source inverter

$$V_{1L} = \sqrt{\frac{3}{2}} m_a \frac{V_{dc}}{2} \approx 0.612 m_a V_{dc} \qquad (8.47)$$

where the *amplitude modulation index*

$$m_a = \frac{V_m}{V_{mcr}} \qquad (8.48)$$

is the ratio of peak values V_m and V_{mcr} of the modulating and carrier waves, respectively. For space vector modulation $0 \leq m_a \leq 1$.

8.7.6 DC bus voltage of a controlled rectifier

The DC output voltage of a controlled rectifier is

- For a three-phase fully controlled rectifier

$$V_{dc} = \frac{3\sqrt{2} V_{1L}}{\pi} \cos \alpha \qquad (8.49)$$

(a)   (b)

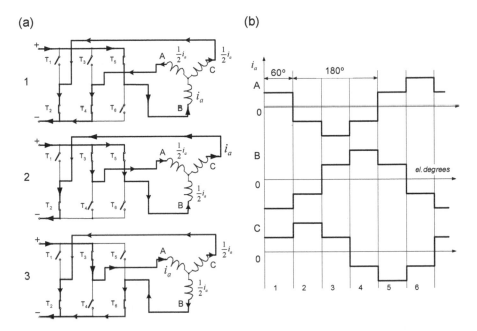

Fig. 8.16. Three-phase bipolar-driven Y-connected DC PMBM with three phases on at a time: (a) commutation, (b) current waveforms.

- For a three-phase half-controlled rectifier

$$V_{dc} = \frac{3\sqrt{2}V_{1L}}{2\pi}(1 + \cos\alpha) \qquad (8.50)$$

where V_{1L} is the line-to-line *rms* voltage and α is the firing angle.

Example 8.2

A three-phase, 4-pole, 1500-rpm PM brushless motor has the following inductances and resistance of the stator (armature) winding:

- Stator leakage inductance $L_1 = 0.853$ mH;
- d-axis armature reaction reactance $L_{ad} = 4.343$ mH;
- q-axis armature reaction reactance $L_{aq} = 3.885$ mH
- Stator winding resistance per phase $R_1 = 0.415$ Ω.

The stator three-phase winding has $N_1 = 84$ series turns per phase. The winding factor is $k_{w1} = 0.94$. The magnetic flux is $\Phi = 0.01$ Wb. The motor is fed with a PWM VSI. The DC link voltage is $V_{dc} = 500$ V and the amplitude modulation index is $m_a = 1$. Find:

(a) The armature current and the angle ψ between the current and q-axis for the load angle $\delta = 10°$;
(b) The load angle to obtain zero d-axis current $I_{ad} = 0$.

Solution

The line voltage according to eqn (8.46)

$$V_{1L} = \sqrt{\frac{3}{2}} \times 1.0 \times \frac{500}{2} = 306.2 \text{ V}$$

The phase voltage $V_1 = 306.2/\sqrt{3} = 176.8$ V and the stator current frequency is $f = (1500/60) \times 2 = 50$ Hz. The stator leakage reactance and armature reaction reactances

$$X_1 = 2\pi \times 50 \times 0.853 \times 10^{-3} = 0.268 \text{ } \Omega$$

$$X_{ad} = 2\pi \times 50 \times 4.343 \times 10^{-3} = 1.364 \text{ } \Omega$$

$$X_{aq} = 2\pi \times 50 \times 3.885 \times 10^{-3} = 1.221 \text{ } \Omega$$

Synchronous reactances

$$X_{sd} = 0.268 + 1.364 = 1.632 \text{ } \Omega \qquad\qquad X_{sq} = 0.268 + 1.221 = 1.488 \text{ } \Omega$$

EMF per phase induced by the rotor PMs according to eqn (8.5)

$$E_f = \pi\sqrt{2} \times 50 \times 84 \times 0.94 \times 0.01 = 175.4 \text{ V}$$

(a) The armature current and angle ψ between the current and q-axis for the load angle $\delta = 10°$

The d- and q-axis current are calculated on the basis of eqns (8.26) and (8.27)

$$I_{ad} = \frac{176.8 \times 1.632 \cos(10°) - 0.415 \sin(10°) - 175.4 \times 1.488}{1.632 \times 1.488 + 0.415^2} = -5.648 \text{ A}$$

$$I_{aq} = \frac{176.8 \times 0.415 \cos(10°) + 1.632 \sin(10°) - 175.4 \times 0.415}{1.632 \times 1.488 + 0.415^2} = 19.048 \text{ A}$$

The armature *rms* current

$$I_a = \sqrt{(-5.648^2 + 19.048^2} = 19.868 \text{ A}$$

Angle between the stator current and the q-axis (EMF E_f)

$$\psi = \arcsin\left(\frac{I_{ad}}{I_a}\right) = \arcsin\left(\frac{-5.648}{19.868}\right) = -16.5° = -0.288 \text{ rad}$$

(b) The load angle to obtain zero d-axis current $I_{ad} = 0$

The numerator of eqn (8.26) is equated to zero, i.e.,

$$V_1(X_{sq} \cos \delta - R_1 \sin \delta) - E_f X_{sq} = 0$$

to check if there is a real value of the load angle δ at which $I_{ad} = 0$. Putting $A = V_1 X_{sq} = 176.8 \times 1.488 = 263.13$, $B = V_1 R_1 = 176.8 \times 0.415 = 73.362$ and $C = E_f X_{sq} = 175.4 \times 1.488 = 261.088$ the following trigonometric equation is obtained:

$$A \cos \delta - B \sin \delta - C = 0 \qquad \text{or} \qquad (-B \sin \delta)^2 = (C - A \cos \delta)^2$$

After substituting $\sin^2 \delta = 1 - \cos \delta^2$, the following 2nd order linear equation is obtained:

$$(A^2 + B^2) \cos^2 \delta + 2AC \cos \delta + (C^2 - B^2) = 0$$

Putting $a = A^2 + B^2 = 74,620$, $b = -2AC = 137,400$, $c = C^2 - B^2 = 62,780$, the discriminant of the quadratic equation is

$$\Delta = b^2 - 4ac = (-137,540)^2 - 4 \times 74,620 \times 62,780 = 1.3891 \times 10^8$$

$$\sqrt{\Delta} = \sqrt{1.3891 \times 10^8} = 1.179 \times 10^4$$

Roots of the 2nd order equation

$$x_1 = \frac{-b - \sqrt{\delta}}{2a} = \frac{-137,400 - 1.179 \times 10^4}{2 \times 74,620} = 0.842$$

$$x_2 = \frac{-b + \sqrt{\delta}}{2a} = \frac{-137,400 + 1.179 \times 10^4}{2 \times 74,620} = 1.0$$

There are two solutions

$$\delta_1 = \arccos(x_1) = \arccos(0.842) = 32.68°$$

$$\delta_2 = \arccos(x_2) = \arccos(1.0) = 1.52°$$

Only $\delta_2 = 1.52°$ is correct from a physical point of view. According to eqns (8.26) and (8.27), for $\delta_2 = 1.52°$ the d-axis current $I_{ad} \approx 0$ and the q-axis current $I_{aq} = 3.155$ A. The stator *rms* current $I_a = I_{aq} = 3.155$ A.

Sometimes both roots x_1 and x_2 are complex numbers. This means that probably for motoring operation the EMF E_f is greater than the terminal phase voltage V_1.

Example 8.3

A three-phase, 4-pole, 5400-rpm PM brushless motor is fed from a six-pulse inverter with two phase on. The DC bus voltage is $V_{dc} = 540$ Vm, the line–to–line resistance $R_{1L} = 1.2$ W, the line–to–line self inductance is $L_L = 12.0$ mH. The line–to–line EMF and torque constants are $k_E = 0.07$ V/rpm and $k_T = 0.66$ Nm/A. Find the shaft torque, electromagnetic torque and losses in the stator winding.

Solution

Frequency

$$f = \frac{5400}{60} \times 2 = 180 \text{ Hz}$$

Self-inductance reactance (line-to-line)

$$X_L = 2\pi \times 180 \times 0.012 = 13.572 \ \Omega$$

Stator winding impedance (line-to-line)

$$Z_L = \sqrt{1.2^2 + 13.572^2} = 13.625 \ \Omega$$

Line-to-line EMF

$$E_{fL} = k_E \times n = 0.07 \times 5400 = 378.0 \text{ V}$$

Line-to-line voltage (8.46)

$$V_{1L} = \frac{\sqrt{6}}{\pi} V_{dc} = \frac{\sqrt{6}}{\pi} \times 540 = 389.8 \text{ V}$$

Approximate armature current

$$I_a = \frac{V_{dc} - E_{fL}}{Z_L} = \frac{540 - 378}{13.625} = 8.954 \text{ A}$$

Shaft torque

$$T = k_T I_a = 0.66 \times 8.954 = 5.91 \text{ Nm}$$

Electromagnetic power

$$P_{elm} = E_{fL} I_a = 378 \times 8.954 = 3384.8 \text{ W}$$

Electromagnetic torque

$$T_{elm} = \frac{P_{elm}}{2\pi n} = \frac{3384.8}{2\pi \times 5400/60} = 5.99 \text{ Nm}$$

Please note that for motor operation $T_{elm} > T$. Losses in the stator winding (two-phases on)

$$\Delta P_w = R_{1L} I_a^2 = 1.2 \times 8.954^2 = 96.2 \text{ W}$$

8.7.7 Rotor position sensing of DC brushless motors

Rotor position sensing in PM DC brushless motors is done by *position sensors*, i.e.,

(a) Hall elements (Fig. 8.17a);
(b) Encoders (Fig. 8.17b);
(c) Resolvers (Fig. 8.17c).

In rotary machines position sensors provide feedback signals proportional to the rotor angular position.

Fig. 8.17. Rotor position sensors: (a) Hall element; (b) encoder; (c) resolver.

The *Hall element* is a magnetic field sensor. When placed in a stationary magnetic field and fed with a DC current, it generates an output voltage

$$V_H = k_H \frac{1}{\delta} I_c B \sin \beta \qquad (8.51)$$

where k_H is Hall constant in m^3/C, δ is the semiconductor thickness, I_c is the applied current, B is the magnetic flux density and β is the angle between the vector of B and Hall element surface. The polarity depends on whether the pellet is passing a North or a South pole. Thus, it can be used as a magnetic flux detector.

In *optical encoders* a light passes through the transparent areas of a rotating disk (grating) and is sensed by a photodetector. To increase the resolution, a collimated light source is used and a mask is placed between the grating and detector. The light is allowed to pass to the detector only when the transparent sections of the grating and mask are in alignment.

In an *incremental encoder* a pulse is generated for a given increment of shaft angular position which is determined by counting the encoder output pulses from a reference. The grating has a single track. In the case of power failure an incremental encoder loses position information and must be reset to known zero point.

An *absolute encoder* is a position verification device that provides unique position information for each shaft angular location. Owing to a certain number of output channels, every shaft angular position is described by its own unique code. The number of channels increases as the required resolution increases. An absolute encoder is not a counting device like an incremental encoder and does not lose position information in the case of loss of power.

A *resolver* is a rotary electromechanical transformer that provides outputs in forms of trigonometric functions $\sin(\vartheta)$ and $\cos(\vartheta)$ of its inputs. For detecting the rotor position of brushless motors, the excitation or primary winding is mounted on the resolver rotor and the output or secondary windings are wound at right angles to each other on the stator core. As a result the output signals are sinusoidal waves in quadrature; i.e., one wave is a sinusoidal function of the angular displacement ϑ and the second wave is a cosinusoidal function of ϑ. Instead of delivering the excitation voltage to the rotor winding by brushes and slip rings, an inductive coupling system (rotary transformer) is frequently used.

There are several reasons to eliminate electromechanical position sensors:

- Cost reduction of electromechanical drives;
- Reliability improvement of the system;
- Temperature limits on Hall sensors;
- In motors rated below 1 W the power consumption by position sensors can substantially reduce the motor efficiency;
- In compact applications, e.g., computer hard disk drives, it may not be possible to accommodate position sensors.

In general, the position information of the shaft of PMBMs can be obtained using one of the following techniques:

(a) Detection of back EMF (zero crossing approach, phase-locked loop technique, EMF integration approach);

(b) Detection of the stator third harmonic voltage;

(c) Detection of the conducting interval of freewheeling diodes connected in antiparallel with the solid state switches;

(d) Sensing the inductance variation (in the d and q-axis), terminal voltages and currents.

8.7.8 Mathematical model

Assuming no rotor currents (no damper, no retaining sleeve, very high resistivity of magnets and pole faces) and the same stator phase resistances, the Kirchhoff's voltage equation for a three-phase machine can be expressed in the following matrix form (Fig. 8.18):

$$
\begin{bmatrix} v_{1A} \\ v_{1B} \\ v_{1C} \end{bmatrix} = \begin{bmatrix} R_1 & 0 & 0 \\ 0 & R_1 & 0 \\ 0 & 0 & R_1 \end{bmatrix} \begin{bmatrix} i_{aA} \\ i_{aB} \\ i_{aC} \end{bmatrix}
$$

$$+\frac{d}{dt}\begin{bmatrix} L_A & L_{BA} & L_{CA} \\ L_{BA} & L_B & L_{CB} \\ L_{CA} & L_{CB} & L_C \end{bmatrix}\begin{bmatrix} i_{aA} \\ i_{aB} \\ i_{aC} \end{bmatrix}+\begin{bmatrix} e_{fA} \\ e_{fB} \\ e_{fC} \end{bmatrix} \qquad (8.52)$$

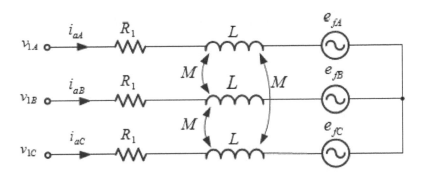

Fig. 8.18. Circuit diagram of a three-phase PM brushless motor.

For inductances independent of the rotor angular position the self-inductances $L_A = L_B = L_C = L$ and mutual inductances between phases $L_{AB} = L_{CA} = L_{CB} = M$ are equal. For no neutral wire $i_{aA} + i_{aB} + i_{aC} = 0$ and $Mi_{aA} = -Mi_{aB} - Mi_{aC}$. Hence

$$\begin{bmatrix} v_{1A} \\ v_{1B} \\ v_{1C} \end{bmatrix}-\begin{bmatrix} R_1 & 0 & 0 \\ 0 & R_1 & 0 \\ 0 & 0 & R_1 \end{bmatrix}\begin{bmatrix} i_{aA} \\ i_{aB} \\ i_{aC} \end{bmatrix}$$

$$+\begin{bmatrix} L-M & 0 & 0 \\ 0 & L-M & 0 \\ 0 & 0 & L-M \end{bmatrix}\frac{d}{dt}\begin{bmatrix} i_{aA} \\ i_{aB} \\ i_{aC} \end{bmatrix}+\begin{bmatrix} e_{fA} \\ e_{fB} \\ e_{fC} \end{bmatrix} \qquad (8.53)$$

The instantaneous electromagnetic torque is given by eqn (8.42). For a bipolar commutation and $120°$ conduction only two phases conduct at any time instant. For example, if $e_{fA} = E_f^{(tr)}$, $e_{fB} = -E_f^{(tr)}$, $e_{fC} = 0$, $i_{aA} = I_a^{(sq)}$, $i_{aB} = -I_a^{(sq)}$ and $i_{aC} = 0$, the instantaneous electromagnetic torque according to eqn (8.42) is

$$T_{elm} = \frac{2E_f^{(tr)}I_a^{(sq)}}{2\pi n} \qquad (8.54)$$

where $E_f^{(tr)}$ and $I_a^{(sq)}$ are flat topped values of trapezoidal EMF and square wave current. For constant values of EMF and currents, the torque (8.54) does not contain any pulsation.

Since $e_f = \omega\psi_f = (2\pi n/p)\psi_f$ where ψ_f is the flux linkage per phase produced by the excitation system, the instantaneous torque (8.42) becomes

$$T_{elm} = p(\psi_{fA}i_{aA} + \psi_{fB}i_{aB} + \psi_{fC}i_{aAC}) \tag{8.55}$$

For computer simulation of PM brushless motors eqns (8.42), (8.53) and (8.55) must be supplemented by the torque balance equation (1.89), i.e.,

$$J\frac{d^2\vartheta}{dt^2} + D_\vartheta\frac{d\vartheta}{dt} + K_\vartheta\vartheta = T_{elm} \pm T \tag{8.56}$$

8.7.9 Cogging torque

Cogging effect (detent effect) is defined as the interaction between the rotor magnetic flux and variable permeance of the air gap due to the stator slot geometry, i.e., slot openings. The cogging effect produces the so-called *cogging torque*.

Neglecting the armature reaction and magnetic saturation, the *cogging torque* is independent of the stator current. The fundamental frequency of the cogging torque is a function of the number of slots s_1, number of pole pairs p and input frequency f. One of the cogging frequencies (usually fundamental) can be estimated as

$$f_c = 2n_{cog}f; \qquad n_{cog} = \frac{LCM(s_1, 2p)}{2p} \quad \text{if} \quad N_{cog} = \frac{2p}{GCD(s_1, 2p)} \geq 1 \tag{8.57}$$

where $LCM(s_1, 2p)$ is the least common multiple of the number of slots s_1 and number of poles $2p$, $GCD(s_1, 2p)$ is the greatest common divisor of s_1 and $2p$ and n_{cog} is sometimes called the fundamental cogging torque index [17]. For example, for $s_1 = 36$ and $2p = 2$ the fundamental cogging torque index $n_{cog} = 18$ ($LCM = 36$, $GCD = 2$, $N_{cog} = 1$), for $s_1 = 36$ and $2p = 6$ the index $n_{cog} = 6$ ($LCM = 36$, $GCD = 6$, $N_{cog} = 1$), for $s_1 = 36$ and $2p = 8$ the index $n_{cog} = 9$ ($LCM = 72$, $GCD = 4$, $N_{cog} = 2$), for $s_1 = 36$ and $2p = 10$ the index $n_{cog} = 18$ ($LCM = 180$, $GCD = 2$, $N_{cog} = 5$), for $s_1 = 36$ and $2p = 12$ the index $n_{cog} = 3$ ($LCM = 36$, $GCD = 12$, $N_{cog} = 1$), etc. The larger the $LCM(s_1, 2p)$, the smaller the amplitude of the cogging torque.

The torque ripple can be minimized both by the proper motor design and motor control. Measures taken to minimize the cogging torque by motor design include elimination of slots, skewed slots, special shape slots and stator laminations, selection of the number of stator slots with respect to the number of poles, decentered magnets, skewed magnets, shifted magnet segments, selection of magnet width and direction-dependent magnetization of PMs. Control techniques use modulation of the stator current or EMF waveforms.

8.7.10 Concentrated-coil armature winding

The armature (stator) winding made of *concentrated non-overlapping coils* is simple to manufacture, provides short end turns and reduces winding resistance. The concentrated-coil winding is feasible when

$$\frac{N_c}{GCD(N_c, 2p)} = km_1 \tag{8.58}$$

where N_c is the total number of armature coils, GCD is the greatest common divisor of N_c and the number of poles $2p$, m_1 is the number of phases and $k - 1, 2, 3, \ldots$. Sometimes such a winding is called the *winding with fractional number of slots per pole per phase* q_1.

Coil span is equal to one tooth pitch instead of one pole pitch (like a salient pole winding). Owing to very short end turns, the winding losses are reduced. Very small mutual inductance between phases causes fault-tolerance.

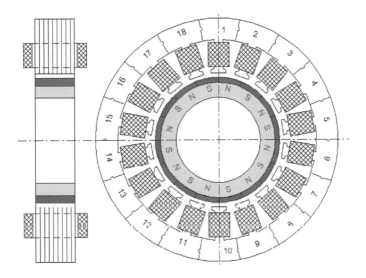

Fig. 8.19. Three-phase PM brushless motor with $2p = 16$ and $N_c = 18$.

Fig. 8.20. Three-phase in-wheel PM brushless motor with $2p = 32$ and $N_c = 36$.

For the winding shown in Fig. 8.19 with $m_1 = 3$, $2p = 16$, $N_c = 18$, eqn (8.58) yields

$$\frac{18}{GCD(18,16)} = \frac{18}{2} = k \times 3, \qquad k = 3 \qquad GCD(18,16) = 2$$

The number of slots per pole per phase

$$q_1 = \frac{18}{3 \times 16} = \frac{18}{48} = \frac{3}{8}$$

The $LCM(18,16) = 144$, so the cogging torque is low.

Fig. 8.20 shows in-wheel PM brushless motor with $m_1 = 3$, $2p = 32$ and $N_c = 36$. For this motor

$$\frac{36}{GCD(36,32)} = \frac{36}{4} = k \times 3, \qquad k = 3 \qquad GCD(36,32) = 4$$

The $LCM(36,32) = 288$, so the cogging torque is very low.

8.7.11 Electromechanical drive with PM brushless motor

On the basis of eqns (8.38), (8.44) and (8.45) the torque–speed characteristic can be expressed in the following simplified form

$$\frac{n}{n_0} = 1 - \frac{I_a}{I_{ash}} = 1 - \frac{T_{elm}}{T_{elmst}} \tag{8.59}$$

where n_0 is the no-load speed, I_{ash} is the locked rotor armature current ($n = 0$) and T_{elmst} is the stall torque ($n = 0$).

Eqn (8.59) neglects the armature reaction, rotational losses and inverter switching losses.

The torque-speed characteristics are shown in Fig. 8.21. Eqn (8.59) is very approximate and cannot be used in calculation of performance characteristisc of commercial DC PMBMs. Theoretical torque-speed characteristics (Fig. 8.21a) differ from practical characteristics (Fig. 8.21b). The *continuous torque* line is set by the maximum nominal (rated) temperature of the motor. The intermittent duty operation zone is bounded by the *peak torque* line and the maximum input voltage.

Practical electromechanical drive with a PM brushless motor and solid state inverter is shown in Fig. 8.22. The power circuit consists of a *solid state converter* (rectifier), *intermediate circuit* (capacitor for VSI) and *inverter*. The control circuit consists of *controller area network* (CAN), *microcontroller* and *gate driver*. A gate driver is a power amplifier that accepts a low-power input from a controller IC and produces a high-current drive input for the gate of a high power transistor such as IGBT or power MOSFET[2].

PMBM and solid state converter should be connected in such a way as to obtain

---

[2] Metal-oxide-semiconductor field-effect transistor.

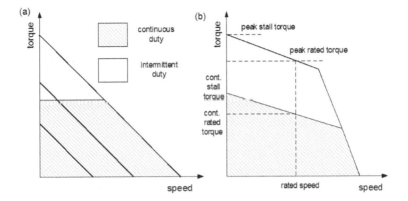

Fig. 8.21. Torque–speed characteristics of PM brushless motor: (a) theoretical; (b) practical.

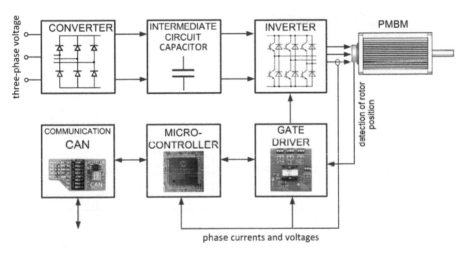

Fig. 8.22. Electromechanical drive with PM brushless motor. CAN – controller area network.

- Proper operation,
- Minimize electromagnetic interference (EMI);
- Prevent shock hazard.

8.7.12 Electric and hybrid electric vehicles

Combustion engines of automobiles are one of the major oil consumers and sources of air pollutions. Oil conservation and road traffic congestion call for new energy sources for propulsion of motor vehicles and protection of the natural environment.

An *electric vehicle* (EV) is driven solely by an electric motor fed from an on-board *rechargeable energy storage system* (RESS), e.g., battery.

A *hybrid electric vehicle* (HEV) has a conventional combustion engine (gasoline or diesel), electric motor and RESS, so that the wheels of the vehicle are driven by both a combustion engine and electric motor. All the energy wasted during braking and idling in conventional vehicles is collected, stored in RESS and utilized in HEVs. The electric motor assists in acceleration (energy saved by the RESS), which allows for a smaller and more efficient combustion engine.

In most contemporary HEVs, called "charge-sustaining," the energy for the battery charging is produced by the internal combustion engine. Some HEVs, called "plug-in" or "charge-depleting," can charge the battery from the utility grid.

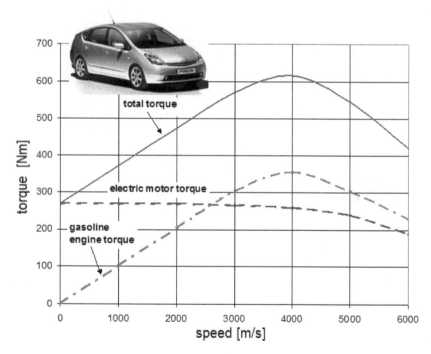

Fig. 8.23. Torque-speed characteristics of combustion engine and traction electric motor.

HEVs have many advantages over classical vehicles with gasoline or diesel engines: most important are:

(a) Smaller size of combustion engine, lower fuel consumption since part of the energy is derived from the RESS and improved efficiency (about 40% better fuel efficiency than that for conventional vehicles of similar ratings);

(b) High torque of electric motor at low speed with high torque of combustion engine in higher speed range make the torque-speed characteristic suitable for traction requirements (Fig. 8.23);

(c) Utilization of wasted energy at braking (regenerative braking), idling and low speeds;

(d) The use of electric motor reduces air pollution and acoustic noise;

(e) Wear and tear on the combustion engine components decrease, so they can work for a longer period of time;

(f) Lower maintenance costs due to reduced fuel consumption;

(g) Although the initial cost of HEVs is higher than conventional cars, their operating costs are lower over time.

EVs and HEVs use brushless electric motors, i.e., PMBMs, switched reluctance motors (SRMs) and induction motors (IMs). Simulations indicate that a 15% longer driving range is possible for an EV with PM brushless motor drive systems compared with induction types. PM brushless motor drives show the best efficiency, output power to mass, output power to volume (compactness) and overload capacity factor.

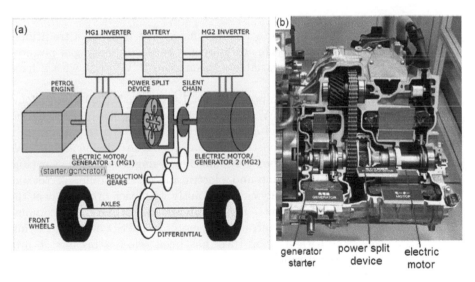

Fig. 8.24. Toyota Prius hybrid-electric drive: (a) block diagram, (b) engine cutaway.

In *series* HEVs an electric motor drives wheels, while the combustion engine drives the electric generator to produce electricity. In *parallel* HEVs the combustion engine is the main way of driving wheels and the electric motor assists only for acceleration. A *series–parallel* HEV (similar to Toyota Prius) is equipped with the so-called *power split device* (PSD), which delivers a continuously variable ratio of combustion engine–to–electric motor power to wheels. It can run in "stealth mode" on its stored electrical energy alone.

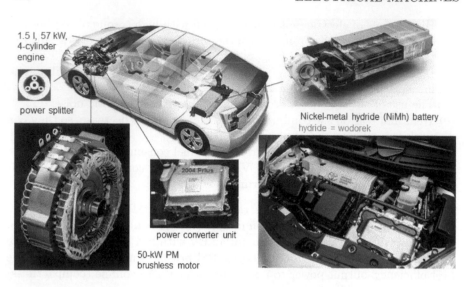

Fig. 8.25. How the Toyota Prius HEV is built.

The PSD is a planetary gear set that removes the need for a traditional stepped gearbox and transmission components in an ordinary gas powered car. It acts as a continuously variable transmission (CVT) but with a fixed gear ratio.

Toyota Prius NHW20 is equipped with a 1.5 l, 57 kW (5000 rpm), four-cylinder gasoline engine, 50 kW (1200 to 1540 rpm), 500 V (maximum) PM brushless motor and nickel-metal hydride (NiMh) battery pack as a RESS (Fig. 8.25). To simplify construction, improve transmission and achieve smoother acceleration, the gearbox is replaced by a single reduction gear (Fig. 8.24). This is because the engine and electric motor have different torque-speed characteristics, so that they can act jointly with each other to meet the driving performance requirements. Fig. 8.24b shows integration of combustion engine with generator/starter, electric motor and PSD of Toyota Prius. In PMBM, the rotor with interior PMs has been selected because it provides wider torque–speed range under the size and weight restrictions than other rotor configurations. To utilize the reluctance torque in addition to synchronous torque the q–axis permeance is maximized while keeping low d–axis permeance. A double layer PM arrangement (Fig. 8.2e) seems to be impractical in mass production due to the high cost of manufacturing, so single-layer V-shaped PMs have been used in the 8-pole rotor.

Electric motors for passenger hybrid cars are typically rated from 30 to 75 kW. Water cooling offers superior cooling performance, compactness and lightweight design over forced-air motor cooling. The water cooling permits weight reductions of 20% and size reductions of 30% as compared to forced-air designs, while the power consumption for cooling system drops by 75%. The use of a single water cooling system for the motor and solid state converter permits further size reductions.

Summary

The development of PMBMs started in the 1970s due to progress in rare-earth PM technology and progress in power electronics. PMBMs are the highest power density and highest efficiency motors with the best dynamic performance.

PMBM drives fall into the two principal classes of sinusoidally excited and square wave (trapezoidally excited) motors.

Sinusoidally excited motors are fed with three-phase sinusoidal waveforms shifted by 120° (Fig. 8.1a) and operate on the principle of a rotating magnetic field. The speed of the rotor is equal to the synchronous speed n_s of the stator magnetic rotating field, given by eqn (6.2). They are simply called *PM sinewave motors* or *PM synchronous motors*.

Square wave motors are also fed with three-phase waveforms shifted by 120° one from another, but these waveforms are rectangular or trapezoidal (Fig. 8.1b). Such a shape is produced when the armature current (MMF) is precisely synchronized with the rotor instantaneous position and frequency (speed). The most direct and popular method of providing the required rotor position information is to use an absolute angular position sensor mounted on the rotor shaft. Such electronic commutation is functionally equivalent to the mechanical commutation in DC brush motors. This explains why motors with square wave excitation are called DC brushless motors.

The *DC PMBM* is a reversed DC brush motor in which PMs are placed on the rotor and armature winding is placed on the stator (Fig. 8.10). The commutator (mechanical inverter) is replaced by a stationary solid state inverter. In the case of DC generator, the commutator (mechanical rectifier) is replaced by a stationary solid state rectifier, either controlled or uncontrolled. In PM DC brushless machine with natural air cooling system, the heat transfer and cooling conditions are much better than in a PM DC brush machine, because nearly all losses (stator winding losses and stator core losses) are dissipated in the stator and transferred via housing to the surrounding air.

Comparison of PM DC brush (commutator) and brushless motors is given in Table 8.1.

Typically, the armature winding of PMBMs is distributed in slots. When cogging (detent) torque needs to be eliminated, *slotless windings* are used.

An auxiliary DC field winding located in the rotor or magnetic flux diverters located in the stator can help to increase the speed range over constant power region or control the output voltage of variable speed generators. These machines are called PMBMs with *hybrid field excitation system*.

PM synchronous motors are usually built with one of the following rotor configurations:

(a) Surface-magnet rotor (Fig. 8.2a);
(b) Spoke-type magnet rotor (Fig. 8.2b);
(c) Interior-magnet rotor with flat PMs (Fig. 8.2c);
(d) Inset-magnet rotor (Fig. 8.2d);
(e) Rotor with double-layer interior-magnets (Fig. 8.2e);
(f) Rotor with buried magnets asymetrically distributed (Fig. 8.2f).

Comparison between PM synchronous motors with surface, spoke-type and interior magnets symmetrically distributed is given in Table 8.2.

The *open-circuit EMF* E_f induced by the rotor PM excitation system and the rotor magnetic flux Φ_f are given by the same equations as for synchronous machines with electromagnetic excitation, i.e., eqns (8.5) and (8.6), i.e.,

$$E_f = \pi\sqrt{2}fN_1k_{w1}\Phi_f$$

$$\Phi_{f1} = L_i \int_0^\tau B_{mg1} \sin\left(\frac{\pi}{\tau}x\right) dx = \frac{2}{\pi}\tau L_i B_{mg1}$$

where N_1 is the number of the stator turns per phase, f is the frequency of the magnetic flux in the stator, k_{w1} is the stator winding factor (eqn 5.23) for fundamental space harmonic, and $\alpha_i = b_p/\tau$. If the magnetic field distribution in the air gap is sinusoidal the pole-shoe arc b_p to pole pitch τ ratio is $\alpha_i = 2/\pi$.

The *armature reaction magnetic fluxes* Φ_{ad} (7.68), (8.7) and Φ_{aq} (7.69), (8.8) induce their own EMFs E_{ad} (7.71) and E_{aq} (7.72) in the d and q-axis, respectively.

The *electromagnetic power* P_{elm} and *electromagnetic torque* T_{elm} at zero stator resistance $R_1 = 0$ of PM brushless motors are given by eqn (7.83) and (7.88), respectively. Equations are exactly the same as for salient-pole synchronous machines with electromagnetic field excitation.

The synchronous reactances X_{sd}, X_{sq} in the d- and q-axis are given by eqns (7.80), (7.81) and (8.12).

The *form factors of armature reaction* k_{fd}, k_{fq} are defined as the ratios of the first harmonic amplitudes to maximum values of normal components of armature reaction magnetic flux densities in the d-axis and q-axis, respectively, eqns (7.66).

The *reaction factors* k_{ad}, k_{aq} in the d- and q-axis are defined as the ratios of form factors of armature reaction-to-the form factor of the excitation field k_f — eqns (7.67).

The values of the form factor of the excitation field k_f, form factors of the armature reaction k_{fd}, k_{fq} and the reaction factors k_{ad}, k_{aq} are given in Table 7.4.

The d-axis and q-axis *armature reaction reactances* for PM brushless machines are expressed by eqns (7.76), (7.77), (8.16) and (8.17), i.e.,

$$X_{ad} = 4m_1\mu_0 f \frac{(N_1k_{w1})^2}{\pi p} \frac{\tau L_i}{g'} k_{fd}$$

$$X_{aq} = 4m_1\mu_0 f \frac{(N_1 k_{w1})^2}{\pi p} \frac{\tau L_i}{g_q'} k_{fq}$$

where m_1 is the number of phases, $\mu_0 = 0.4\pi \times 10^{-6}$ H/m is the magnetic permeability of free space, f is the stator current frequency, N_1 is the number of series turns per phase, k_{w1} is the winding factor for fundamental space harmonic, p is the number of pole pairs, τ is the pole pitch, L_i is the effective length of the stator stack, k_{fd} and k_{fq} are the reaction factors. For surface PM configurations the *equivalent air gaps* are

$$g' = gk_C k_{sat} + \frac{h_M}{\mu_0} \qquad\qquad g_q' \approx gk_C$$

In the above equations for the equivalent air gap: g is the mechanical clearance, k_C is Carter's coefficient, k_{sat} is the saturation factor of the magnetic circuit, h_M is the height of PM per pole. If the rotor is equipped with non-ferromagnetic retaining sleeve, the thickness of the sleeve should be added to the mechanical clearance g.

Equivalent circuits for a PM synchronous machine are shown in Fig. 8.4, while *phasor diagrams* in Fig. 8.5. *Armature currents* derived on the basis of phasor diagram for underexcited PM synchronous motor with the stator winding resistance R_1 taken into account are given by eqn (8.26), (8.27) and (8.29), i.e.,

$$I_{ad} = \frac{V_1(X_{sq}\cos\delta - R_1\sin\delta) - E_f X_{sq}}{X_{sd}X_{sq} + R_1^2}$$

$$I_{aq} = \frac{V_1(R_1\cos\delta + X_{sd}\sin\delta) - E_f R_1}{X_{sd}X_{sq} + R_1^2}$$

$$I_a = \sqrt{I_{ad}^2 + I_{aq}^2}$$

The *electromagnetic torque* of a PM synchronous motor with the stator winding resistance R_1 taken into account is expressed by eqn (8.31).

The *load characteristics*, i.e., the armature current I_a, shaft torque T_{sh}, input power P_{in}, power factor $\cos\varphi$ and efficiency η are plotted against the relative output power P_{out}/P_{outn} (output power to nominal output power) in Fig. 8.7.

The d-axis current I_{ad} is the *magnetizing current*. The q-axis current I_{aq} is the *torque producing current*. For $I_{ad} = 0$ the angle $\psi = 0$ and the total armature current $I_a = I_{aq}$ is the torque producing current. The angle ψ is the angle between the armature current $I_a = I_{aq}$ and the EMF E_f. Therefore, the angle φ between the current I_{aq} and voltage V_1 is equal to the load angle δ between the voltage V_1 and EMF E_f, and the power factor is (8.33)

$$\cos\varphi = \frac{E_f + I_a R_1}{V_1}$$

A PM synchronous motor is not self-starting. The following *methods are used for starting* PM synchronous motors:

- Asynchronous starting with the aid of additional cage winding in the rotor;
- Starting by means of auxiliary motor;
- Frequency-change starting using a solid state converter, e.g., VVVF inverter.

There are three *modes of commutation* of DC PMBM

- *Unipolar commutation* with phase sequence A, B, C, A, B,... (Figs 8.11 and 8.12) where the current is always conducted by only one phase (the neutral point of the stator winding must be available);
- *Bipolar six-step commutation* with phase sequence AB, AC, BC, BA, CA, CB, ... (Figs 8.13, 8.14 and 8.15) where the current is always conducted by two phases (120° current conduction, one step equivalent to 60°);
- *Bipolar commutation with three phases on* and conduction interval 180° (Fig. 8.16).

The *EMF induced in a phase winding* of a DC PMBM neglecting saturation and armature reaction can be simply expressed as a function of speed n (8.44)

$$E_f = k_E n$$

where k_E is the EMF constant provided by manufacturers of PM brushless machines. Similar to EMF, the *electromagnetic torque* can be expressed as a function of armature current I_a (8.38)

$$T_{elm} = k_T I_a$$

where k_T is the torque constant provided by manufacturers of PM brushless machines.

The *inverter output voltage* is given by eqns (8.46) to (8.48) and the *DC bus voltage of a rectifier* is given by eqns (8.49) and (8.50).

Rotor position sensing in DC PMBM is done by *position sensors*, i.e.,

(a) Hall elements (Fig. 8.17a);
(b) Encoders (Fig. 8.17b);
(c) Resolvers (Fig. 8.17c).

The *Hall element* is a magnetic field sensor.

In *optical encoders* a light passes through the transparent areas of a grating and is sensed by a photodetector.

In an *incremental encoder* a pulse is generated for a given increment of shaft angular position which is determined by counting the encoder output pulses from a reference. The rotating disk (grating) has a single track. In the case of power failure an incremental encoder loses position information and must be reset to known zero point.

An *absolute encoder* is a position verification device with multichannel output that provides unique position information for each shaft angular location. An absolute encoder is not a counting device like an incremental encoder and does not lose position information in the case of loss of power.

A *resolver* is a rotary electromechanical transformer that provides outputs in forms of trigonometric functions $\sin(\vartheta)$ and $\cos(\vartheta)$ of its inputs. As a result the output signals are sinusoidal waves in quadrature; i.e., one wave is a sinusoidal function of the angular displacement ϑ and the second wave is a cosinusoidal function of ϑ. Instead of delivering the excitation voltage to the rotor winding by brushes and slip rings, an inductive coupling system (rotary transformer) is frequently used.

Sensorless methods of detection of the shaft position of PM brushless motors usually use one of the following techniques:

(a) Detection of back EMF (zero crossing approach, phase-locked loop technique, EMF integration approach);
(b) Detection of the stator third harmonic voltage;
(c) Detection of the conducting interval of freewheeling diodes connected in antiparallel with the solid state switches;
(d) Sensing the inductance variation (in the d and q axis), terminal voltages and currents.

The *mathematical model* of a PM brushless motor (electrical circuit diagram) is shown in Fig. 8.18 and is expressed by eqn (8.53), i.e.,

$$
\begin{bmatrix} v_{1A} \\ v_{1B} \\ v_{1C} \end{bmatrix} = \begin{bmatrix} R_1 & 0 & 0 \\ 0 & R_1 & 0 \\ 0 & 0 & R_1 \end{bmatrix} \begin{bmatrix} i_{aA} \\ i_{aB} \\ i_{aC} \end{bmatrix}
$$

$$
+ \begin{bmatrix} L-M & 0 & 0 \\ 0 & L-M & 0 \\ 0 & 0 & L-M \end{bmatrix} \frac{d}{dt} \begin{bmatrix} i_{aA} \\ i_{aB} \\ i_{aC} \end{bmatrix} + \begin{bmatrix} e_{fA} \\ e_{fB} \\ e_{fC} \end{bmatrix}
$$

where v_{1A}, v_{1B}, v_{1C} are instantaneous values of input voltages, i_{aA}, i_{aB}, i_{aC} are the instantaneous values of phase currents, e_{fA}, e_{fB}, e_{fC} are instantaneous values of EMFs, R_1 is the stator winding resistance per phase, L is the stator winding self-inductance, and M is the mutual inductance between phases. A symmetrical stator winding has been assumed. For computer simulation of

PMBMs this equation must be supplemented by the torque balance equation (1.88), i.e.,

$$J\frac{d^2\vartheta}{dt^2} + D_\vartheta\frac{d\vartheta}{dt} + K_\vartheta\vartheta = T_{elm} \pm T$$

where J is the moment of inertia, D_ϑ is the torsional damping, K_ϑ is the spring constant, T_{elm} is the electromagnetic torque and T is the external torque.

Cogging torque is produced due to interaction of the PM rotor magnetic flux and variable permeance of the air gap due to the stator slot geometry, i.e., slot openings. The larger the least common multiple of the number of slots s_1 and number of poles $2p$, i.e., $LCM(s_1, 2p)$, the smaller the amplitude of the cogging torque.

In armature winding made of *concentrated non-overlapping coils* the coil span is equal to one tooth pitch. The concentrated coil winding is feasible, when (8.58)

$$\frac{N_c}{GCD(N_c, 2p)} = km_1$$

where N_c is the total number of armature coils, $2p$ is the number of poles, m_1 is the number of phases, $k = 1, 2, 3, \ldots$ and GCD is the greatest common divisor of N_c and $2p$. Sometimes such a winding is called the winding with *fractional number of slots per pole per phase* q_1. Owing to very short end turns, the winding losses are reduced. Very small mutual inductance between phases causes fault-tolerance.

Simplified torque-speed characteristics of PMBMs are plotted in Fig. 8.21.

Electromechanical drive system with PMBM and solid state converter is shown in Fig. 8.22. PM brushless motor and solid state converter should be connected in such a way as to obtain

- Proper operation,
- Minimize electromagnetic interference (EMI),
- Prevent shock hazard.

A typical *hybrid electric vehicle* (HEV) uses a PM brushless motor in addition to combustion engine. HEVs have many advantages over classical vehicles with gasoline or diesel engines. The most important are:

(a) Smaller size of combustion engine, lower fuel consumption since part of the energy is derived from the rechargeable energy storage system RESS and improved efficiency (about 40% better fuel efficiency than that for conventional vehicles of similar ratings);
(b) High torque of electric motor at low speed with high torque of combustion engine in higher speed range make the torque speed characteristic suitable for traction requirements (Fig. 8.23);
(c) Utilization of wasted energy at braking (regenerative braking), idling and low speeds;

(d) The use of electric motor reduces air pollution and acoustic noise;

(e) Wear and tear on the combustion engine components decrease, so they can work for a longer period of time;

(f) Lower maintenance costs due to reduced fuel consumption;

(g) Although the initial cost of HEVs is higher than conventional cars, their operating costs are lower over time.

Problems

1. In a three-phase 6-pole, 60-Hz PM synchronous machine, the stator inner diameter is $D = 0.09$ m, the effective stack length $L_i = 0.14$ m, width of pole face $b_p = 0.019$ m, $q_1 = 2$, air gap magnetic flux density $B_{mg} = 0.72$ T. The stator winding is wound of full pitch coils. Find the first harmonic of the magnetic flux density in the air gap b_{mg1} – eqn (8.1), amplitude of magnetic flux densities B_{mg1}, B_{mg5}, B_{mg7} for the 1st, 5th and 7th space harmonics, magnetic flux Φ_1 for fundamental harmonic and phase EMF e_f per one turn.

   Answers: $b_{mg1} = 0.875$ T. $B_{mg1} = 0.917$ T, $B_{mg5} = 0.183$ T, $B_{mg7} = 0.131$, $\Phi_{f1} = 0.00193$ Wb, $e_f = 0.496$ V.

2. A three-phase, 6-pole, 8000-rpm PM brushless motor has the following inductances and resistance of the stator (armature) winding:

   - Stator leakage inductance $L_1 = 0.25 \times 10^{-3}$ H;
   - d-axis armature reaction reactance $L_{ad} = 2.28 \times 10^{-3}$ H;
   - q-axis armature reaction reactance $L_{aq} = 2.19 \times 10^{-3}$ H;
   - Stator winding resitance per phase $R_1 = 0.78$ Omega.

   The stator winding has $N_1 = 24$ series turns per phase. The winding factor is $k_{w1} = 0.94$. The magnetic flux is $\Phi = 0.0026$ Wb. The motor is fed with a six-step three-phase inverter. The DC link voltage is $V_{dc} = 256$ V.

   (a) Find the armature current for the load angle $\delta = 10°$ electrical;

   (b) Find the load angle to obtain zero d-axis current $I_{ad} = 0$.

   Answer: (a) $I_{ad} = 1.038$ A, $I_{aq} = 3.395$ A, $I_a = 3.55$ A; (b) At $\delta = 18.94°$ $I_{ad} = 0$ and $I_{aq} = 6.1$ A.

3. A three-phase, 8-pole, 6000-rpm PMBM is fed from a six-pulse inverter with two phases on. The DC bus voltage is $V_{dc} = 600$ V, the line-to-line winding resistance $R_{1L} = 0.9$ Ω, the line–to–line self inductance is $L_{1L} = 5.0$ mH. The line–to–line EMF and torque constants are $k_E = 0.072$ V/rpm and $k_T = 0.83$ Nm/A. Find the shaft torque, electromagnetic torque and losses in the stator winding.

   Answer: $T_{sh} = 11.07$ Nm, $T_{elm} = 11.44$ Nm, $\Delta P_w = 160$ W.

4. A three-phase, 4-pole, 1800-rpm PMBM has the following inductances and resistance of the stator (armature) winding:
   - Stator leakage inductance $L_1 = 0.925 \times 10^{-3}$ H;
   - d-axis armature reaction reactance $L_{ad} = 5.825 \times 10^{-3}$ H;
   - q-axis armature reaction reactance $L_{aq} = 5.685 \times 10^{-3}$ H;
   - Stator winding resitance per phase $R_1 = 0.78$ Omega.

   The stator winding has $N_1 = 72$ series turns per phase. The winding factor is $k_{w1} = 0.94$. The magnetic flux is $\Phi = 0.0126$ Wb. The motor is fed with a six-step three-phase inverter. The DC link voltage is $V_{dc} = 515$ V.

   (a) Find the armature current and electromagnetic torque for the load angle $\delta = 20°$ electrical including the stator resistance R_1;
   (b) Find the electromagnetic torque for the same load angle without armature winding resistance R_1 taken into account.

   Answer: (a) $I_{ad} = -9.65$ A, $I_{aq} = 29.86$ A, $I_a = 31.38$ A, $T_{elm} = 107.8$ Nm; (b) $T_{elm} = 116.3$ Nm.

5. A three-phase, six-pole, 160-kW, 346-V (line-to-line), 180-Hz, Y-connected synchronous motor with a inset-type PM rotor has the stator winding resistance per phase $R_1 = 0.011$ Ω, d-axis synchronous reactance $X_{sd} = 0.335$ Ω and q-axis synchronous reactance $X_{sq} = 0.377$ Ω ($X_{sd} < X_{sq}$). The power angle $\delta = 26°$, the air gap magnetic flux density $B_{mg} = 0.65$ T, the number of turns per phase $N_1 = 24$, the winding factor $k_{w1} = 0.925$, the effective length of the stator core $L_i = 0.19$ m, the inner stator core diameter $D = 0.33$ m, the rotational losses $\Delta P_{rot} = 1200$ W, the stray losses $\Delta P_{str} = 0.05 P_{out}$ and the stator core losses can be neglected. Find the armature current, output power P_{out}, electromagnetic torque T_{elm}, shaft torque T_{sh}, efficiency η and power factor $\cos\phi$.

   Answer: $I_{ad} = -191.44$ A, $I_{aq} = 226.7$ A, $I_a = 296.72$ A, $P_{out} = 160\,281.5$ W, $T_{elm} = 449.6$ Nm, $T_{sh} = 425.2$ Nm, $\eta = 0.93$, $\cos\phi = 0.9695$.

6. In a three-phase, 8-pole, 50-Hz PMBM the number of stator slots per pole per phase is $q_1 = 1.5$.

   (a) Find the fundamental cogging torque index n_{cog}, fundamental cogging frequency f_c at nominal speed and evaluate the cogging torque;
   (b) If the stator is redesigned and the number of stator slots per pole per phase increases to $q_1 = 2.0$, how will the cogging torque and frequency of cogging torque be affected?

   Answer: (a) $n_{cog} = 9$, $n_n = 750$ rpm, $f_c = 900$ Hz, $LCM = 72$, so the cogging torque is neither high nor low; (b) $n_{cog} = 6$, $n_n = 750$ rpm, $f_c = 600$ Hz, $LCM = 48$, so the cogging torque will increase.

7. In a PMBM the DC link voltage is $V_{dc} = 42$ V, the EMF constant $k_E = 0.021$ V/rpm, the torque constant $k_T = 0.5$ Nm/A and the stator winding resistance $R = 1.0$ Ω. Find the locked rotor armature current I_{ash}, starting electromagnetic torque T_{elmst}, nominal speed n_n and nominal electromagnetic torque T_{elmn} at nominal armature current $I_{an} = 10.5$ A. Sketch simplified speed-current and speed-torque characteristics as in Fig. 8.21a.

Answer: $I_{ash} = 42.0$ A, $T_{elmst} = 21.0$ Nm, $n_n = 1500$ rpm, $T_{elmn} = 5.25$ Nm.

Symbols and Abbreviations

A	line current density
a	number of parallel current paths of the armature winding of AC motors; number of pairs of parallel current paths of the armature winding of DC brush (commutator) motors
B	magnetic flux density
b	instantaneous value of the magnetic flux density
b_p	pole shoe width
C	number of commutator segments; capacitance
C_c	cost of ferromagnetic core
C_0	cost of all other components independent of the shape of the machine
c_E	armature constant (EMF constant)
c_T	torque constant
D	diameter
E	EMF, rms value
E_f	EMF per phase induced by the rotor of a synchronous machine
E_i	internal EMF per phase
e	instantaneous EMF
F	force; MMF
F_{exc}	MMF of the rotor excitation system
F_a	armature reaction MMF
f	frequency
f_c	frequency of cogging torque
$GCD(N_c, 2p)$	greatest common divisor of N_c and $2p$
g	air gap (mechanical clearance)
g'	equivalent air gap
H	magnetic field intensity
h	height
h_M	height of the PM
I	electric current
I_a	armature DC or rms current
I_{ash}	armature current at zero speed ("short circuit" current)

i	instantaneous value of current or stepping motor current
J	moment of inertia
J_a	current density in the armature winding
K	lumped stiffness
k	coefficient, general symbol
k_{1R}	skin effect coefficient for armature conductors
k_C	Carter's coefficient
k_{ad}	reaction factor in d-axis; coefficient of additional losses in armature core
k_{aq}	reaction factor in q-axis
k_{d1}	distribution factor for fundamental space harmonic $\nu = 1$
k_E	EMF constant $k_E = c_E \Phi_f$
k_f	form factor of the field excitation $k_f = B_{mg1}/B_{mg}$
k_i	stacking factor of laminations
k_{p1}	pitch factor for fundamental space harmonic $\nu = 1$
k_{sat}	saturation factor of the magnetic circuit due to the main (linkage) magnetic flux
k_T	torque constant $k_T = c_T \Phi_f$
k_{w1}	winding factor $k_{w1} = k_{d1} k_{p1}$ for fundamental space harmonic $\nu = 1$
L	inductance; length
$LCM(s_1, 2p)$	least common multiple of s_1 and $2p$
L_i	armature stack effective length
l_M	axial length of PM
M	mutual inductance
m	number of phases; mass
m_a	amplitude modulation index
N	number of turns
N_{cog}	number of poles–to–$GCD(s_1, 2p)$ ratio
n	rotational speed in rpm
n_0	no-load speed
P	active power
P_{elm}	electromagnetic power
ΔP	active power losses
$\Delta p_{1/50}$	specific core loss in W/kg at 1T and 50 Hz
p	number of pole pairs; sound pressure
Q	reactive power
R	resistance
R_a	armature winding resistance of DC commutator motors
R_1	armature winding resistance of AC motors
R_{int}	interpole winding resistance
$R_{\mu g}$	air gap reluctance
S	apparent power; surface
s	slip; cross-section area
s_1	number of stator teeth or slots;
s_2	number of rotor teeth or slots;

T	torque
T_{elm}	electromagnetic torque
T_{elmsyn}	electromagnetic synchronous or synchronizing torque
T_{elmrel}	electromagnetic reluctance torque
T_0	constant of avarage component of the torque
T_{sh}	shaft torque (output or load torque)
T_m	mechanical time constant
t	time; slot pitch
V	electric voltage; volume
v	instantaneous value of electric voltage; linear velocity
W	energy, J
W_m	stored magnetic energy
w	energy per volume, J/m^3
w_M	width of PM
X	reactance
X_{ad}	d-axis armature reaction (mutual) reactance
X_{aq}	q-axis armature reaction (mutual) reactance
X_{sd}	d-axis synchronous reactance
X_{sq}	q-axis synchronous reactance
$\mathbf{Z}$	impedance $\mathbf{Z} = R + jX$; $\mid \mathbf{Z} \mid = Z = \sqrt{R^2 + X^2}$
α	electrical angle
α_i	effective pole arc coefficient $\alpha_i = b_p/\tau$
β	overlap angle of pole
ΔV_{br}	voltage drop across commutation brushes
δ	power (load) angle; bias error
η	efficiency
θ	rotor angular position
ϑ	temperature
Λ	permeance, H
λ	specific permeance, H/m^2
μ	magnetic permeability
μ_0	magnetic permeability of free space $\mu_0 = 0.4\pi \times 10^{-6}$ H/m
μ_r	relative magnetic permeability
ν	number of the stator νth harmonic
σ	electric conductivity, leakage factor
τ	pole pitch
Φ	magnetic flux
Φ_{ad}	d-axis armature reaction flux
Φ_{sq}	q-axis armature reaction flux
Φ_f	excitation magnetic flux
Φ_l	leakage flux
φ	power factor angle
Ψ	flux linkage $\Psi = N\Phi$; angle between $\mathbf{I}_a$ and $\mathbf{E}_f$
Ω	angular speed $\Omega = 2\pi n$
ω	angular frequency $\omega = 2\pi f$

Subscripts

a	armature
av	average
b	braking
br	brush
c	commutation
cog	cogging
Cu	copper
d	direct axis; differential
elm	electromagnetic
eq	equivalent
Fe	ferromagnetic
f	field
fr	friction
g	air gap
h	hysteresis
in	inner
k	kinetic
L	load
l	leakage
M	magnet
m	peak value (amplitude)
max	maximum
min	minimum
n	nominal
n, t	normal and tangential components
out	output, outer
q	quadrature axis
r	relative
rhe	rheostat
rot	rotational
s	synchronous
sat	saturation
sh	shaft
st	starting
str	stray, additional
syn	synchronous or synchronizing
$vent$	ventilation
w	winding
$wind$	windage
x, y, z	cartesian coordinate system
1	primary; stator; fundamental harmonic
2	secondary; rotor

Superscripts

inc	incremental
(sq)	square or trapezoidal wave

Abbreviations

AC	alternating current
APU	auxiliary power unit
AWG	American wire gauge
CAD	computer-aided design
CAN	controller area network
CD	compact disk
CSCF	constant speed constant frequency
CSD	constant speed drive
CSI	current source inverter
CVT	continuously variable transmission
DC	direct current
DSP	digital signal processor
EDF	electrodynamic force
EMF	electromotive force
EMI	electromagnetic interference
EV	electric vehicle
FEM	finite element method
GCD	greatest common divisor
GPU	ground power unit
GTO	gate turn-off (thyristor)
HEV	hybrid electric vehicle
HV	high voltage
HVIC	high voltage integrated circuit
IC	integrated circuit
IDG	integrated drive generator
IGBT	insulated-gate bipolar transistor
ISG	integrated starter-generator
IT	information technology
LCM	least common multiple
LV	low voltage
LVIC	low voltage integrated circuit
MEA	more electric aircraft
MLT	mean length of turn
MMF	magnetomotive force
MVD	magnetic voltage drop
NiMh	nickel-metal hydride
PCB	printed circuit board
PM	permanent magnet

PMBM	permanent magnet brushless motor
PSD	power split device
PWM	pulse width modulation
RAT	ram air turbine
RESS	rechargeable energy storage system
SRM	switched reluctance motor (machine)
VF	variable frequency
VSCF	variable speed constant frequency
VSI	voltage source inverter
VSD	variable-speed drive
VVVF	variable voltage variable frequency

References

1. Bajorek, Z.: **Electrical machines** (in Polish). WNT, Warsaw, 1980.
2. Bertotti, G.: *General properties of power losses in soft ferromagnetic materials,* **IEEE Trans. on Magnetics**, MAG-22, 1986, no 5, Sept. 1986.
3. Bogusz, P., Korkosz, M., Prokop, J.: *Laboratory tests on a two-phase switched reluctance motor designed for high-speed electromechanical drive* (in Polish), **Maszyny Elektryczne – Zeszyty Problemowe**, vol. 105, 2015, No 1, Katowice, Poland, 137–142.
4. Boldea, I., Tutelea, L.: **Electric machines. steady state, transients and design with Matlab®**. CRC Press, Taylor & Francis Group, Boca Raton, FL, 2010.
5. Bose, B.K.: **Modern power electronics and a.c. drives**. Prentice Hall, Upper Saddle River, NJ, 2002.
6. Dote, Y., and Kinoshita, S.: **Brushless servomotors — fundamentals and applications**. Clarendon Press, Oxford, 1990.
7. Dabrowski, M.: **Design of AC electrical machines** (in Polish). Warsaw, WNT, 1988.
8. Fitzgerald, A.E,, Kingsley, C., Umans, S.,D.: **Electric Machinery**, 6th ed., New York, McGraw-Hill, 2002.
9. Gieras, J.F.: **Linear induction motors**. Clarendon Press, Oxford, 1994.
10. Gieras, J.F., Gieras, I.A.: **Electrical energy utilization**. Adam Marszalek Publishing House, Torun, Poland, 1998.
11. Gieras, J.F., Wang, R.J., Kamper, M.J.: **Axial flux permanent magnet brushless machines**, 2nd ed. Springer, 2008.
12. Gieras, J.F.: **Advancements in electric machines**. Springer, 2009.
13. Gieras, J.F.: **Permanent Magnet Motors Technology: Design and Applications**. 3rd edition, CRC Press, Taylor & Francis Group, Boca Raton, FL, 2010.
14. Gieras, J.F.: Electric power system of Tu-154M passenger aircraft, **Przeglad Elektrotechniczny (Electrical Review)**, vol. 89, no 2a, February 2013, 300–307.
15. Glinka, T., Mizia, W., Zywiec, A., Hickiewicz, J.: **Electrical machines. Numerical problems**, in Polish, WNT, Warsaw, 1973.
16. Gray, C.B.: **Electrical machines and drive systems**. Longman Scientific & Technical, Harlow, 1989.

17. Hanselman, D.: **Brushless permanent magnet motor design**, 2nd edition. The Writers' Collective, Cranston, RI, 2003.
18. Hoffman S., Banerjee B., and Samotyj M.: *Written-pole revolution*, **IEEE Power Eng. Review**, December: 6–9, 1997.
19. Honsinger, V.B.: *Performance of polyphase permanent magnet machines*, **IEEE Trans. on Power Apparatus and Systems**, vol. PAS–99, 1980, no 4, 1510–16.
20. Kenjo, T., and Nagamori, S.: **Permanent–magnet and brushless d.c. motors**. Clarendon Press, Oxford, 1985.
21. Kenjo, T.: **Power electronics for microprocessor era**. Oxford University Press, 1990.
22. Kenjo, T.: **Electric motors and their controls**. Oxford University Press, 1991.
23. Kenjo, T.: **Stepping motors and their microprocessor controls**. Oxford University Press, 1985.
24. Kimmich, R., Doppelbauer, M., Kirtley, J.L., Peters, D.T., Cowie, J.G., Brush, E.F.: *Performance characteristics of drive motors optimized for die-cast copper cages*, Int. Conf. on Energy Efficiency in Motor Driven Systems EEMODS'05, Heidelberg, Germany, pp. 1–8.
25. Kostenko, M., Piotrovsky, L.: **Electrical machines, 2. Transformers and d.c. machines, 2. A.c. machines**. Mir Publishers, Moscow, 1974.
26. Miksiewicz, R.: **Electrical machines. Numerical problems using Mathcad®**, in Polish, Silesian University of Technology, Gliwice, 2000.
27. Miller, T.J.E.: **Brushless permanent-magnet and reluctance motor drives**. Clarendon Press, Oxford, 1989.
28. Miller, T.J.E.: **Switched reluctance motors and their control**. Magna Physics, Oxford Science Publications, Oxford, 1993.
29. Mohan, N., Undeland, T.M., and Robbins, W.P.: **Power electronics: converters, applications, and design**. John Wiley & Sons, 1989.
30. Morash R.T., Barber R.J., and Roesel J.F. *Written-pole motor: a new a.c. motor technology*, Int. Conf. on Electr. Machines in Australia, ICEMA93, Adelaide, Australia, 1993, pp. 379–384.
31. Nasar, S.A.: *DC–switched reluctahce motor*, **Proc. IEE** (London), vol. 116, no 6, June 1969, pp. 1048–1049.
32. Nasar, S.A.: **Electric machines and power systems**. McGraw-Hill, Inc., New York, 1995.
33. Nasar, S.A., Boldea, I., and Unnewehr, L.E.: **Permanent magnet, reluctance, and self-synchronous motors**. CRC Press, Taylor & Francis Group, Boca Raton, FL, 1993.
34. Ronkowski M., Michna, M., Kostro, G., Kutt, F.: **Electrical machines around us**, in Polish. Gdansk University of Technology, Gdansk, Poland, 2011.
35. Sen, P.C.: **Principles of electric machines and power electronics**. John Wiley & Sons, New York, 1989.
36. Say, M.G.: **Alternating current machines**, Pitman, 1976.
37. Steinmetz, C.P.: *On the law of hysteresis*, *AIEE Trans.*, vol. 9, 1892. pp. 3–64.
38. Turowski, J., Turowski, M.: **Engineering electrodynamics**. CRC Press, Taylor & Francis Group, Boca Raton, FL, 2014.
39. Yamamura, S.: **A.c. Motors for high-performance Applications**. Marcel Dekker Inc, New York and Basel, 1986.

Index